GAS TURBINE ENGINEERING

Gas Turbine Engineering

Applications, Cycles and Characteristics

Richard T. C. HARMAN

Senior Lecturer,
Department of Mechanical Engineering,
University of Canterbury,
Christchurch, New Zealand

A HALSTED PRESS BOOK

John Wiley & Sons
New York

First published in Great Britain 1981 by
The Macmillan Press Ltd

Published in the U.S.A.
by Halsted Press, a Division of
John Wiley & Sons, Inc., New York

Printed in Hong Kong

Library of Congress Cataloging in Publication Data

Harman, R T C
Gas turbine engineering.

Includes bibliographical references and index.
1. Gas-turbines. I. Title.
TJ778.H27 621.43′3 80-21003
ISBN 0-470-27065-9

Contents

Preface

This book has grown from a set of handout notes, written to provide background material for final-year undergraduate students. The notes served the needs for a short course which concentrated initially on the mathematics of the engine, on the assumption that the students were familiar with and could handle anything mathematical. This proved to be totally inappropriate: the students expressed a strong preference for explanations of engineering practice, hardware, experience and understanding, to fill the gap left by their courses in engineering science.

Their protestations reminded me of my own experience as a newly graduated engineer in industry, where the opportunity was taken to engage in deep technical discussions with fellow graduates. The flaws in the arguments and omissions in anyone's training (or memory) were soon revealed at that stage, but they are still encountered in later life and in technical literature.

In writing this book, therefore, I have concentrated on simple physical explanations at the expense of the mathematics, which are usually available for those who need them from a lecture course, company files or other books. These explanations cover a number of peripheral engineering fundamentals as well as the material specific to gas turbine engines. The engineering content and design appreciation are as important to me as the gas turbine engine itself, which serves as a framework around which to select the contents and as a focus for the discussion. Thus the discussion of the main components could apply equally to their separate use in other types of plant, including turbochargers, and the sections on mechanical design, SI units, compressible flow and noise are relatively general.

The gas turbine content has been severely limited by space, because it attempts to encapsulate over 80 years of experience and research. Without the mathematics, the content has been selected to some extent to complement the existing better known books. It covers the standard ground but also extends into the use of these engines in industrial and chemical processing applications. This is the field with the major scope for innovation, and in which understanding will be most required as new ways of conserving energy are found. Such understanding may be derived from aircraft engine experience, in which engine design has stabilised in a few, well-established configurations and which has explored many lesser known characteristics. This book covers both the industrial and aero fields but overlooks much of the practice in the surface transportation and marine fields, which tends to have followed the others.

The role of this book includes the linking of the various engineering disciplines always a difficult task for students. Although materials, stressing and thermodynamics are barely touched, I have attempted to show where they are relevant within the practice of engineering. Many of these topics are interdependent and inseparable, making the discussion necessarily more complicated. The extreme example of this is the design of turbine blading, which links stress, heat transfer, aerodynamics, mechanical fastening, vibration, thermal growth and corrosion with manufacture from an almost unmachinable material.

A system of internal cross-referencing has therefore been used to indicate where fuller explanations of a topic may be found, without unduly disrupting the flow of the the text. Thus, the use of brackets (section 10.3) or (section E.6) is a convenient way of referring to one of these other sections of the book. This has permitted many topics to be covered at several levels, with the more specialised material placed later in each chapter, in the later chapters or in the appendixes. Use has also been made of italics to emphasise key words, particularly where they are explained, and most are listed in the index. This also includes rules of thumb under the headings 'empirical relationships' and 'typical values', to help provide engineering feel. A guide to deeper material is given in the references: a comprehensive list is included to cover most of the topics discussed.

While much of the book is concerned with descriptions of hardware, its operation and operating characteristics, three chapters of more immediate application are included. Chapter 3 covers the method of calculation of engine cycles in sufficient detail to show how to make a preliminary assessment of engine or plant performance and requirements. This may permit a plant engineer to assess whether his plant could benefit from the installation of gas turbine equipment. Appendix D provides data on a wide range of working fluids, process gases and basic fuels to support such calculations. Chapter 15 may then help him compare the gas turbine plant with diesel or steam alternatives, assess the ancillaries required and prepare him for discussions with competing engine makers or suppliers.

A book of this type can never have too many illustrations, as the discussion of machinery is rendered quite abstract and academic unless the reader can link it to hardware which he can understand. The diagrams used show a little of the shape and function of typical hardware, but not its colour, texture, feel, size or deterioration patterns. The reader is strongly recommended to gain access to the hardware and study it, at the nearest engine manufacturing or overhaul shop, or in an airline's or other user's maintenance area. Engines of different types and makes have quite different detail design, but any exposure to hardware and to the engine installation and associated plant can only be beneficial.

Finally, I must acknowledge the considerable help I have had in amassing and improving the material for this book. It started with work and training at Rolls-Royce Ltd, where I benefited by association with many very experienced engineers: they should have written this book, had they had the time. My first notes were based on those from a course by G. K. Hensman, given in evening classes at the Derby and District College of Technology. The need for thorough and complete explanations was impressed by several years of teaching undergraduate students. Recent requests for data and assistance from Rolls-Royce Ltd personnel have met with generous responses from many people, including J. R. Cownie, D. Nicholas, M. J. T. Smith, S. I. Cowley, A. Jubb, D. McKnight, G. L. Wilde and, especially, P. C. Ruffles.

I am also grateful for the encouragement of J. R. Tyler and B. S. Page (Ruston Gas Turbines Ltd), J. Macmillan and D. E. Sharpe (G.E.C. Gas Turbines Ltd), R. Robinson (John Players Ltd), Mr. Tyler (International Combustion Ltd), Professor A. G. Smith and Dr J. S. B. Mather (University of Nottingham) and Dr I. J. Day, who gave considerable time and material assistance during my recent leave in the United Kingdom. The accommodation provided by Mr and Mrs Bamford and typing by Mrs S. Leach speeded the later stages of the work. After my return, Mr A. Perry and other staff at Air New Zealand helped with the final touches.

I am also appreciative of the support given at the University of Canterbury in terms of advice, encouragement and facilities. Particularly helpful have been Professors D. C. Stevenson and A. G. Williamson, Dr J. B. Stott and Mr I. Gilmour. Mesdames N. Jones and J. Percival did most of the typing and helped at the proof-reading stage. Ms J. Shelton and Mr T. Bird traced the illustrations.

Finally, I dedicate the book to my wife, Evelyn, and children, Emma and Ashley, for their acceptance of the hours lost to its preparation.

R. T. C. Harman

Nomenclature

The following symbols are used extensively throughout the book. A few symbols restricted specifically to a single chapter or section are not listed below. The Greek letters used are identified first. Units are given in appendix A.

Greek Letters

α	alpha
β	beta
γ	gamma
δ	delta
ϵ	epsilon
ζ	zeta
η	eta
θ	theta
λ	lambda
μ	mu
ν	nu
π	pi
ρ	rho
σ	sigma
τ	tau
ϕ	phi
ψ	psi
ω	omega
Δ	delta
Υ	upsilon
Ω	omega

Geometry

Angles

α	airflow angles } relative to
β	blade angles } axial
δ	deviation $(\alpha_2 - \beta_2)$ for compressor, $(\cos^{-1}(o/s) - \alpha_2)$ for turbine
ϵ	deflection, turning angle $(\alpha_1 - \alpha_2)$
i	incidence $(\alpha_1 - \beta_1)$
θ	camber angle $(\beta_1 - \beta_2)$
ζ	stagger angle

Dimensions

A	cross-sectional area, normal to flow
c	length of blade chord
D	diameter
h	blade height, span
k	blade tip clearance, to casing
o	throat, minimum passage area
R	radius, or radial distance
s	peripheral blade spacing
t	maximum blade thickness

Subscripts

1	entry to blade row, leading edge
2	exit from blade row, trailing edge
e	exit plane, nozzle or engine
h	hub, radius or diameter
m	mean, radius or diameter
t	tip, radius or diameter

Rotor Identification

h-p	high pressure
i-p	immediate pressure
l-p	low pressure

Heat Transfer (Turbine Blade Cooling)

A_b surface area of blade, both sides
A_c surface area inside cooling passages
ΔH heat flow, gas to blade
h_g heat transfer coeft., gas to blade
k thermal conductivity of gas stream
Nu Nusselt number ($h_g c/k$)
Re Reynolds number ($V_2 c/\nu$)
S_c total wetted periphery of cooling passages, at any radius
T_b mean temperature of blade
T_c temp. of cooling air supply
T_g effective hot gas temp.
Z cooling passage geometry factor

Performance

Parameters

C_D drag coefficient
C_L lift coefficient
C_p specific heat, constant pressure
C_v specific heat, constant volume
c velocity of sound
D degree of reaction
E internal energy per unit mass
F force or engine thrust
f frequency, noise or vibration
H enthalpy per unit mass
$\dot{m}$ mass flow
M Mach number
N rotational speed, rev/s
P power
p pressure, absolute or acoustic
Q heat input per unit mass
R characteristic gas constant
Re Reynolds number
R_{01} ram pressure ratio, at inlet
r_p pressure ratio
s entropy
T temperature, absolute
U blade velocity, tangential
V velocity

v	specific volume (per unit mass)
W	work per unit mass
$\bar{w}$	head loss in blade row
X	loss coefficient, temperature
Y	loss coefficient, pressure
γ	ratio of C_p/C_v
Δ	difference, between input and output
η	efficiency
λ	wavelength, noise
μ	absolute viscosity
ν	kinematic viscosity
ρ	density
σ	stress
τ	torque
ϕ	equivalence ratio, combustion
ϕ	flow coefficient, V_a/U
ψ	blade loading coefficient, or temperature drop coefficient
ω	rotational speed, rad/s

Subscripts

0	(as first of two subscripts) stagnation condition, T or p
0,1,2, etc.	station along flow path between major components, or successive stations though one stage
a	axial, ambient condition, or approach station (before 0).
B	burning; combustion with η
b	blade
c	compressor, with η
e	exhaust station, outlet
m	mean, or measured condition
r	reference condition
s	stage
t	turbine, with η
th	thermal, over all, with η
w	whirl, tangential condition
0A	air at combustor inlet
0B	burner or combustor outlet condition
∞	polytropic, with η

In 1902 Sebastien de Ferranti, who became famous for his work on the electric lighting of the streets of London, steam turbines and cotton machinery, made the following prediction

> 'The prime mover of the future will be an elastic fluid turbine, ultimately taking the form of one driven by the internal combustion of gas, although the latter would have to be approached through experience gained with steam as the working fluid.'

Many other engineers of the period are credited with similar foresight, and numerous patents were taken out. Some of them attempted to make engines (section 1.2) but the technology of the day did not match their concepts.

The prediction came true within 50 years: since 1945 the gas turbine has established an unassailable position as the power plant for high-speed jet aircraft flight. Gas turbine engines providing power at their output shaft compete strongly with reciprocating petrol and diesel engines in many fields and have a clear balance of advantage in some. Gas turbines are also used to advantage in a number of industrial, chemical and electricity generating processes, in some cases bringing substantial improvements over earlier types of plant (section 1.3). Their application to new land-based tasks continues to progress, demonstrating that their future remains bright. Over 50 companies are involved in their manufacture.

1

Introduction

1.1 Basic Operating Principles

Most types of engine are designed to extract as much as possible of the energy released by the combustion of a fuel, and to deliver it in usable form at a rotating output shaft.

Burning the fuel in an open fire would simply produce waste heat. The transfer of energy from the fire to the hardware is therefore achieved by the appropriate use of a gaseous *working fluid*, often air (section E.9), which is made to flow through the engine. The normal way to handle the working fluid is by the *thermodynamic cycle* of induction, compression, heating, expansion and exhaust. In a *reciprocating engine*, these processes are performed in sequence in the same closed space, formed by a piston and cylinder which operate on the working fluid one mass at a time. In contrast, the working fluid flows without interruption through a *gas turbine engine*, passing continuously from one single purpose device to the next.

The arrangement of a simple gas turbine engine is shown in figure 1.1*a*. The rotating compressor acts as a fan to drive the working fluid into the heating system. The fluid is heated by internal combustion (burning a controlled supply of fuel in the air flow) or by heat exchange from an external source (passing the compressed fluid through heated passages). The turbine acts as a windmill, being blown round by the flow to drive the compressor and an output load via rotating shafts.

The loss of *pressure* as the fluid releases its energy to the turbine is akin to the voltage drop across an electric motor, or the water-level difference required to create a useful water flow. While an incompressible fluid (for example, water) may be pressurised and depressurised, its negligible volume change means that little energy interchange is involved. In contrast, the use of a *compressible fluid* (gas) permits the absorption or release of considerable amounts of energy. For instance, in an expanding cylinder, the gas continues to push on the retreating

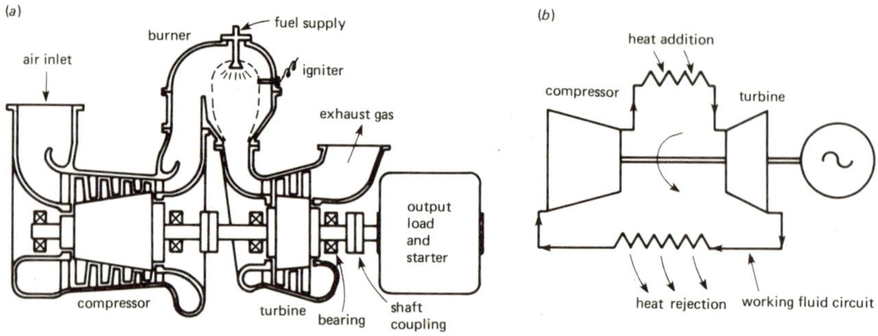

Figure 1.1 *Simple engine types: (a) open cycle; (b) closed cycle (diagrammatic)*

piston and releases energy equivalent to the mean force times the distance travelled.

Similarly, in a turbine, the energy is released as the gas presses against retreating blades while compressor blades add energy by approaching the flow. The energy content of the gas at any time is basically the kinetic energy (half the mass times velocity squared) of its molecules, which is proportional to the gas *temperature* (section B.1). The action and motion of the blades, in changing the *velocity* of the flow (section E.3), therefore also change the mean molecular velocity and energy and the gas temperature and pressure. While the pressure changes are essential to the working of the engine, its performance is assessed by the temperature changes.

The type of *pressure and temperature measurements* used is very important. Usually they are absolute, stagnation (total) values, measured by instruments which face into the approaching flow to give an indication of the energy in the fluid flow at any point (section B.2). The work done in compression or expansion is proportional to the change of stagnation temperature in the working fluid, in the form of heating during a compression process or cooling during an expansion process (equation B.10). The *temperature ratio*, between the temperatures before and after the process, is related to the *pressure ratio* across the process (equation B.12). It is the interaction between the temperature change and ratio, at different starting temperature levels, which permits the engine to generate a useful work output.

This can be demonstrated by a simple numerical example using the Kelvin scale for temperature (appendix A). For a starting temperature of 300 K (27 °C), a temperature ratio of 1.5 in compression yields a final temperature of 450 K and a change of 150 °C. Starting instead at 400 K, the same ratio would yield a change of 200 °C and a final temperature of 600 K. Note that absolute temperature levels are expressed in kelvins, while changes are in degrees Celsius. The equivalent pressure ratio would ideally be 4.13, as demonstrated in section 3.1. These numbers show that, working over the same temperature ratio, the temperature change and, therefore, the work involved in the process vary in proportion to the starting temperature level.

This conclusion can be depicted graphically in a very significant way. If the temperature changes are drawn as vertical lines ab and cd, and are separated horizontally to avoid overlap, the resultant figure 1.2*a* forms the basis for a standard type of graph used to assess engine cycles. Assuming the starting and finishing pressures to be the same for the two processes, the thin lines through ac and bd depict two of a family of lines of constant pressure, which diverge as shown. In this ideal case, expansion processes could be represented by the same diagram, simply by proceeding down the lines ba and dc. Alternatively, if ab is taken as a compression process, bd as heat addition, dc as an expansion process and ca as a heat rejection process, then the figure abdc represents the ideal cycle to which the working fluid of the engine is subjected (section 3.1).

In practice, the 600 K level at point d is too low a temperature from which to start the expansion. Figure 1.2*b* is more realistic, with line ef going from 800 to 1200 K and, clearly, much longer than line ab. Now, ab represents the work input required by the compressor. Of the expansion work capacity fe, only the

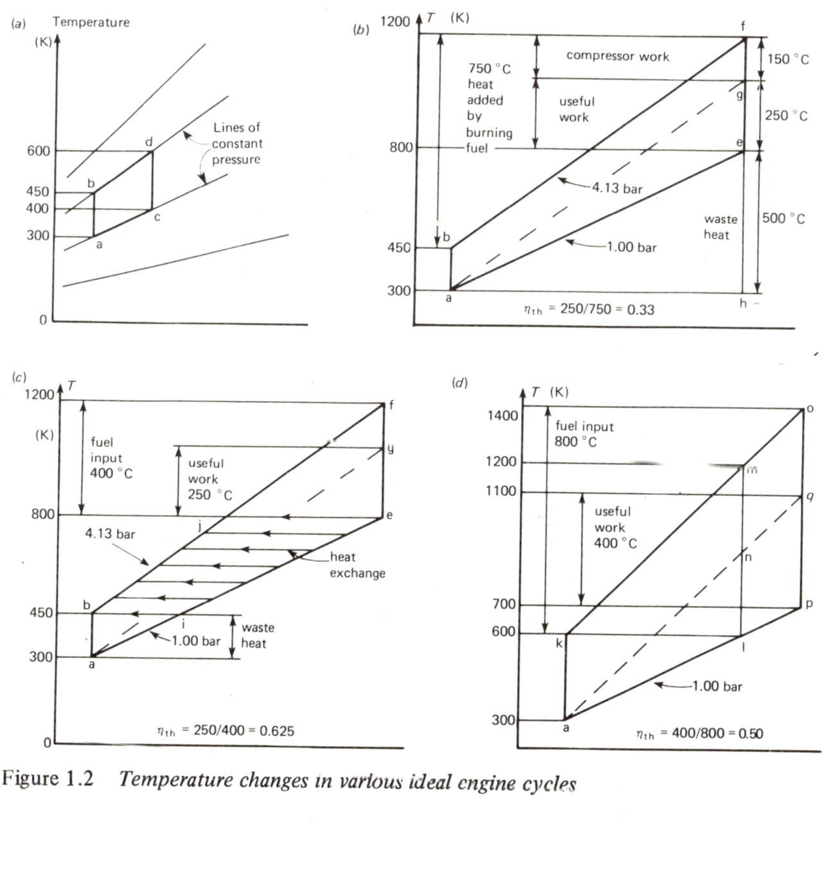

Figure 1.2 *Temperature changes in various ideal engine cycles*

fraction fg is required to drive the compressor, where ag is parallel to bf. The remaining 250 °C, line ge, is energy which can be used to perform useful external work, by further expansion through additional turbine capacity or by blowing through a nozzle to provide jet thrust (section 14.1). Thus, by expanding from a high temperature, the required temperature drop of 150 °C to drive the compressor can be obtained with a reduced temperature ratio, of only 1200/1050 = 1.143 in this case, leaving a residual exhaust gas pressure of 2.59 bar for further expansion.

Now consider line bf. Its length is proportional to the heat added, and the limit to the maximum turbine entry temperature level at f is set by the turbine technology. The ability of the engine to generate a useful output arises from its use of the energy in the input fuel flow, but not all of the fuel energy can be recovered usefully. In this example, the heat input of 1200 – 450 = 750 °C compares with the excess output of 250 °C (line ge) to represent an *efficiency* of 0.33 (250/750). If more fuel could be used, raising the maximum temperature level at the same pressure, then more useful work could be obtained at the same efficiency. This calculation of efficiency is compatible with that of equation C.6.

The line ea represents heat rejection. This could involve passing the exhaust gas through a cooler, before returning it to the compressor, and this would be a *closed cycle* (section 2.8 and figure 1.1*b*). Figure 1.1*a* shows an *open-cycle* engine, which takes air from the atmosphere and exhausts back to the atmosphere. In this case, line ea still represents heat rejection, but the path from e to a involves the whole atmosphere and very little of the gas finds its way immediately from e to a. It is fundamental to the cycle shown that the remaining 500 °C, line eh, is wasted heat because point e is at atmospheric pressure. The gas is therefore unable to expand further, so can do no more work.

An engine modification to improve the efficiency of the cycle is possible at this stage. The use of a *regenerator* or heat exchanger (section 1.3), to transfer as much as possible of the heat rejected along ea into the compressed fluid, can reduce the quantity of heat to be supplied between b and f. In figure 1.2*c*, the heat rejected along ei reappears in the compressed fluid along bj. The fuel added is then equivalent to 1200 – 800 = 400 °C which, with the same work output of 250 °C yields an over-all efficiency of 0.625. In practice, the efficiency is much lower (section 3.6.3).

An alternative modification is the use of a higher pressure ratio, which also improves the over-all efficiency. In figure 1.2*d*, the temperature ratio of 2 corresponds to a pressure ratio of 11.3. The efficiency has risen to 0.5 at the same turbine entry temperature (1200 K), because the raised compressor delivery temperature at k reduces the possible fuel input along km to 600 °C and the lowered temperature level after expansion (l) reduces the heat wastage to 300 °C. However, the useful output of 300 °C is not sufficient alone to justify the increased complexity and cost of raising the pressure ratio: it needs to be matched by a raised turbine entry temperature as shown, for example, at o. In turn, this calls for a raised level of turbine technology in terms of materials (section 8.3) and means of blade cooling (section 8.5).

The efficiency levels quoted are very high because the explanation so far has ignored many factors. *Inefficiency* of the compressor and turbine (section C.1) increases the compressor work demand while reducing the turbine work output, so drastically reduces the useful work output and efficiency. There are pressure losses in the heat addition and heat rejection processes, and the regenerator transfers less heat than the ideal case quoted. There may be variations in the fluid mass flow rate and its specific heat (energy input divided by consequent temperature rise) around the cycle. These factors can easily combine to reduce the over-all efficiency to zero, giving an unworkable engine: they are considered further in chapter 3.

1.2 A Brief History of Gas Turbine Development and Use

The use of a turbine driven by the rising flue gases above a fire dates back to Hero of Alexandria in 150 B.C., and the Chinese were operating windmills at about the same period. It was not until A.D. 1791 that John Barber patented the forerunner of the gas turbine, proposing the use of a reciprocating compressor, a combustion system and an impulse turbine. Even then, he foresaw the need to cool the turbine blades, for which he proposed water injection.

1808 saw the introduction of the first *explosion type* of gas turbine, which in later forms used valves at entry and exit from the combustion chamber to provide intermittent combustion in a closed space. The pressure thus generated blew the gas through a nozzle to drive an impulse turbine. These operated successfully but inefficiently for Karavodine and Holzwarth from 1906 onwards, and the type died out after a Brown, Boveri model designed in 1939.[1]

Developments of the continuous flow machine suffered from lack of knowledge, as different configurations were tried. Stolze in 1872 designed an engine with a seven-stage axial flow compressor, heat addition through a heat exchanger by external combustion, and a ten-stage reaction turbine. It was tested from 1900 to 1904 but did not work because of its very inefficient compressor.[1] Parsons was equally unsuccessful in 1884, when he tried to run a reaction turbine in reverse as a compressor. These failures resulted from the lack of understanding of aerodynamics prior to the advent of aircraft. As a comparison, in typical modern practice, a single-stage turbine drives about six or seven stages of axial compressor with the same mass flow.

The first successful dynamic compressor was Rateau's centrifugal type in 1905. Three assemblies of these, with a total of 25 impellers in series giving an over-all pressure ratio of 4, were made by Brown, Boveri and used in the first working gas turbine engine, built by Armengaud and Lemale in the same year. The exhaust gas heated a boiler behind the turbine to generate low-pressure steam, which was directed through turbines to cool the blades and augment the power. Low component efficiencies and flame temperature (828 K) resulted in low work output and an over-all efficiency of 0.03. By 1939, the use of *industrial* gas turbines had become well established: experience with the Velox boiler (section 2.5) led Brown, Boveri into diverging applications; a Hungarian engine (Jendrassik) with axial flow compressor and turbine used regeneration to achieve an efficiency of 0.21; and the Sun Oil Co. (U.S.A.) was using a gas turbine engine to improve a chemical process (section 2.9).

The use of gas turbine engines for *aircraft propulsion* dates from 1930, when Whittle saw that its exhaust gas conditions ideally matched the requirements for jet propulsion (section 2.1) and took out a patent.[2] His work and patent were reviewed independently before the recommendation was made to proceed.[3] His first model was built by British Thomson-Houston and ran as the Power Jets Type U in 1937, with a double-sided centrifugal compressor, a long combustion chamber which was curled round the outside of the turbine and an exhaust nozzle just behind the turbine. Problems of low compressor and turbine efficiency were matched by hardware problems and the struggle to control the combustion in a very small space. Ten reverse flow combustion chambers were introduced in 1938, the aim still being to keep the compressor and turbine as close together as possible to avoid shaft whirl problems (figure 1.3). Whittle's first flying engine was the W1, with 850 lb thrust, in 1941. It was made by Rover, whose gas turbine establishment was taken over by Rolls-Royce in 1943.

By this time there was a wide variety of configurations made by numerous British, American and German companies. A Heinkel experimental engine flew in 1939 and a General Electric version of the W1 flew in 1941. Several jet engines were operational by the end of the Second World War, but the first commercial engine did not enter service until 1953, the Rolls Royce Dart turboprop in the *Viscount*, followed by the turbojet de Havilland Ghost in the *Comet* of 1954. The subsequent growth of the use of jet engines has been visible and audible to most of the world, and has forced the growth of design and manufacturing

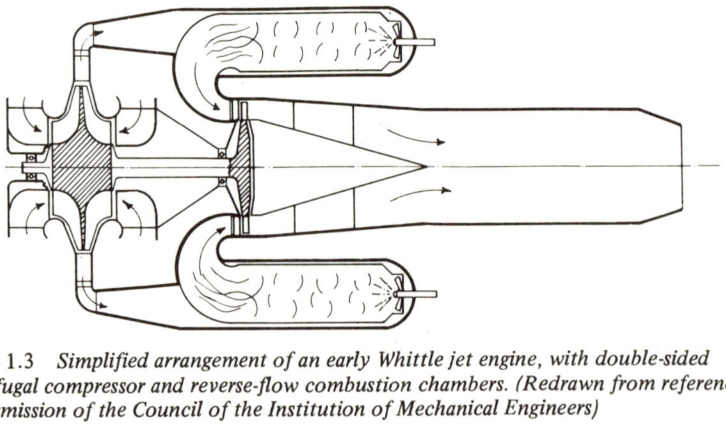

Figure 1.3 *Simplified arrangement of an early Whittle jet engine, with double-sided centrifugal compressor and reverse-flow combustion chambers. (Redrawn from reference 2 by permission of the Council of the Institution of Mechanical Engineers)*

technology.[4A] By 1970 a range of standard configurations for different tasks had become established (section 2.1), and some aircraft engines were established in industrial applications (sections 2.3 to 2.6) and in ships (section 2.2).

Gas turbines entered the *surface transportation* fields also during their early stages of development. The first railway locomotive application was in Switzerland in 1941, with a Brown, Boveri engine driving an electric generator and electric motors driving the wheels. The engine efficiency approached 0.19, using regeneration. Many other turboelectric systems and engines with other transmission systems have entered service since 1950, some involving high speed, lightweight passenger trains, but only in very small numbers of each type. Road use started with a Rover car of 1950, followed by Chrysler and other companies, but commercial use has been limited to trucks, particularly by Ford. Automotive gas turbine development has been largely independent of other types, and has forced the pace of development of regenerators (section 1.3): engines are being tested in most vehicle manufacturing countries. Hovercraft entered operation in the United Kingdom in 1963, using aircraft-derived gas turbine engines for their high power-to-weight ratio.

1.3 Component Characteristics and Capabilities

The compressors used are of the dynamic type, as distinct from the positive displacement, reciprocating and rotary types.[5] Each stage of a dynamic compressor works to increase the flow velocity, and then lets it slow down again (diffuse) to gain pressure (section 5.3). It thereby causes the working fluid to flow continuously into a region of high pressure, but this flow pattern may break down if the outlet pressure becomes excessive (section 10.1). The compressor may then enter a condition in which the flow surges backwards and forwards through the compressor (section 6.4.3) and which must be avoided by the use of adequate means of control (section 11.4).

Dynamic compressors are efficient, compact, relatively cheap and handle vast quantities of working fluid. Considering air at 1.013 bar, 288 K, the inlet flow is typically 150 kg/s (122 m^3/s or 259 000 ft^3/min) per square metre of inlet area at Mach 0.4 (section 3.3, step 7 and section B.4.3). As the intake diameters of the various types range from about 0.1 m (4 in.) to 2.44 m (8 ft), the range of mass flows varies from about 1.2 kg/s to 700 kg/s (about 1 to 570 m^3/s or 2000 to 1.2 million ft^3/min). They operate without lubrication, so the delivery is not contaminated by oil. This may be especially useful when the compressor is involved in a chemical process (section 2.9). The isentropic efficiency (section C.2) ranges typically from 0.7 to 0.9 depending on the type and quality of compressor.

The dynamic compressor is available in two forms, both represented diagrammatically as shown in figure 1.4*a*. *The centrifugal compressor* provides small to moderate air flows (up to about 50 kg/s) at a pressure ratio of up to

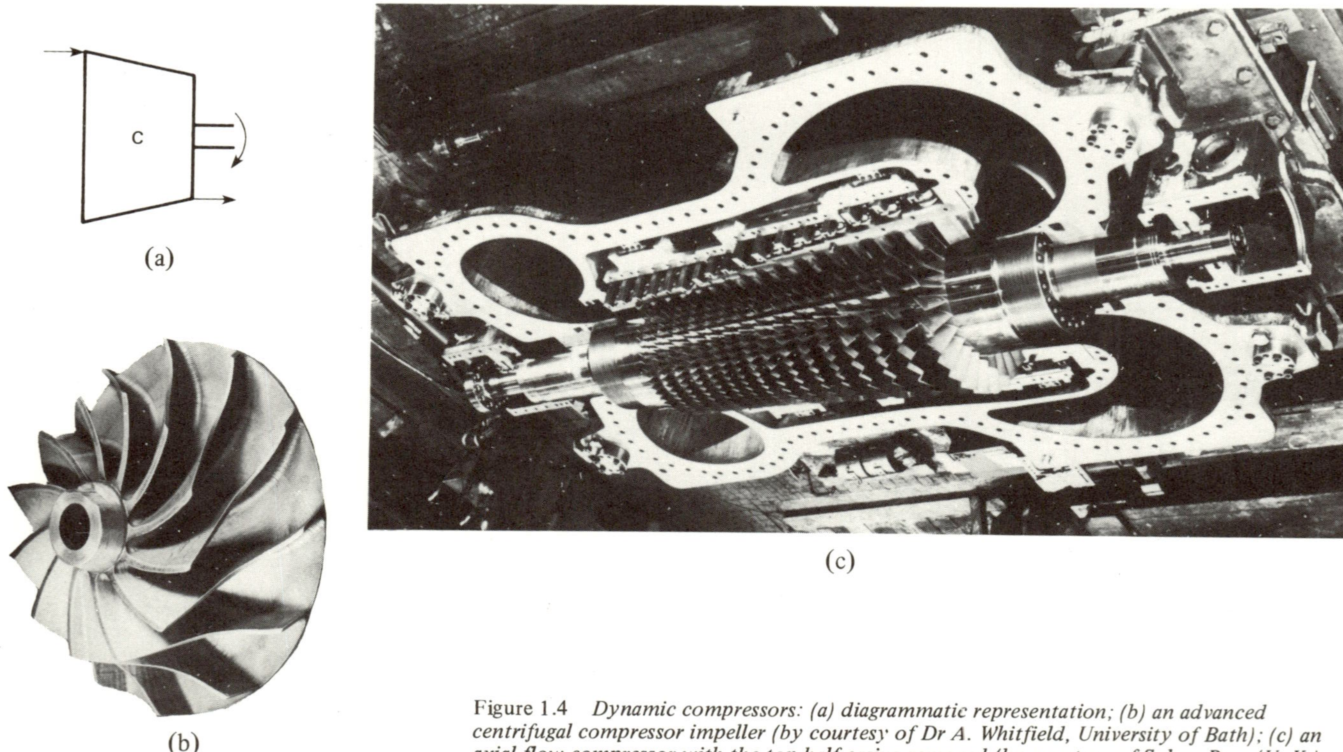

Figure 1.4 *Dynamic compressors: (a) diagrammatic representation; (b) an advanced centrifugal compressor impeller (by courtesy of Dr A. Whitfield, University of Bath); (c) an axial-flow compressor with the top half casing removed (by courtesy of Sulzer Bros (U. K.) Ltd)*

about 5, or higher flows at lower pressure ratio, using a single rotor which inhales axially and exhausts radially (figures 1.4*b* and 5.1). Two or more are sometimes mounted on the same shaft, handling the fluid in series to boost its final pressure.

The *axial compressor* is built up from a number of stages, like a series of fans of reducing flow area (figures 1.4*c* and 6.1). It can handle much larger flows than the centrifugal compressor because its inlet occupies most of its frontal area, and it can generate a pressure ratio of 10 or more if sufficient stages are used. In practice, high pressure ratios are better achieved by running two or more compressor rotors in series, with a net reduction in the over-all number of stages and reduced susceptibility to stall and other problems (section 11.4.6). Shorter compressors offer easier operation and also have higher critical whirl speeds, which can significantly ease the structural design process (section 14.2). In small engines, the optimal means of achieving a high pressure ratio sometimes involves the use of a single rotor with a few axial stages followed by a final centrifugal stage.

Gas turbine *internal combustion* is a continuous process taking place at constant pressure (chapter 7). A steady supply of fuel and air mixes and burns as it flows through a flame zone, which is located in the manner of a cloud in the lee of a mountain. The flame does not touch its container, being stabilised by the inlet airflow pattern which also cools the container walls. Mixtures with a very wide range of fuel/air ratios can be burned, as the mixture in the flame zone is maintained at normal strength in the vicinity of the fuel inlet while any excess air is mixed in downstream of the flame. The combustion process involves very highly developed control of flame stability and can be tuned to emit very low levels of smoke, carbon monoxide, hydrocarbons and oxides of nitrogen. The combustion chamber volume is very small for the rate of heat release, partly because combustion takes place at high pressure: in aircraft engines the volume may be only 5 per cent of that used in a boiler for the same fuel flow. It is conveniently labelled with the symbol B (for burning) in diagrams. Some engines use heat addition by heat exchange from an *external combustion* process, which permits the use of poorer fuels (section 2.7).

The *turbine* is the almost universal means of extracting energy from a high-pressure source of gas. It is represented diagrammatically in figure 1.5*a*, indicating that the gas expands as its pressure falls and it loses energy. Two types are used, the *radial turbine* (figures 1.5*b* and 9.1), which is similar to a centrifugal compressor used in reverse, and the *axial flow turbine* (figures 1.5*c* and 8.1). Typical isentropic efficiency is 0.75 to 0.9 (section C.2).

The axial turbine (chapter 8) is akin to an axial compressor, but its blading is more complicated than a simple reversal of compressor blading (section E.1). It extracts considerable energy, and with one stage can drive a compressor of six to seven stages passing the same mass flow. A single-stage turbine drops the gas stagnation pressure (section B.2) by about 30 to 50 per cent from its absolute value upstream, with a corresponding drop in stagnation temperature of about 8 to 16 per cent. For increased work output, several stages are often combined in

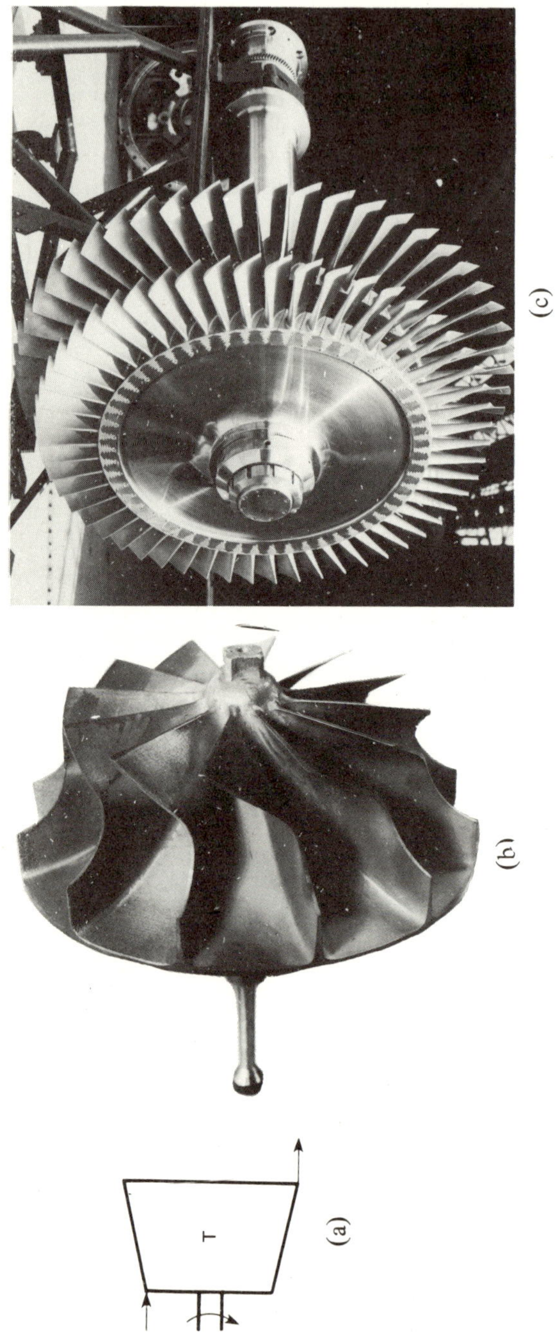

Figure 1.5 *Types of turbine: (a) diagrammatic representation; (b) a radial turbine rotor (by courtesy of Noel Penny Turbines Ltd); (c) an axial turbine rotor (by courtesy of GEC Gas Turbines Ltd)*

the same rotor. The flow velocity through the turbine varies typically above and below Mach 1 (section B.4.3), so the gas imparts its energy very rapidly, with a corresponding rapid fall in temperature and pressure, which can be used to advantage for fast cooling of a gas (section 2.9).

The work capacity of a turbine needs to be carefully matched to the size of compressor it drives (section 11.1). If the turbine is to drive a load other than the engine compressor, so that its speed may vary independently of the rate of supply of gas, then the torque it exerts rises as it is slowed from design speed and may exceed twice the design speed torque when it is stalled (section 9.4).

Regeneration is achieved by two types of hardware. The conventional form is the counter flow *heat exchanger*, in which thin passages containing the hot exhaust gas are interleaved with thin passages passing the colder, compressed air (or gas) in the opposite direction (figure 1.6*a*). The heat transfer occurs in stages; from the hot fluid to the metal walls, through the metal, and into the cold fluid. The metal temperature may not be high enough to prevent deposits forming in the hot passages, reducing the rate of heat transfer, and the unit is bulky.

The later type is the *rotating drum regenerator* (figure 1.6*b*), which was developed mainly for automotive use. It is typically of metal or ceramic honeycomb form, with thin axial passages parallel to the axis. Rotation at about 20 rev/min makes the passages cycle between connecting the hot ducts, then the cold ducts, then the hot ducts again. The material heats up as the hot gas passes, and then gives up this heat to the cold flow. Leakage past the seals between the ducts and the end faces of the drum may lose 5 per cent of the compressed cooler fluid. [6A]

The numbering used for the engine stations in these figures is the conventional sequence from engine inlet to outlet, as shown in figure 1.6*c*. This is a

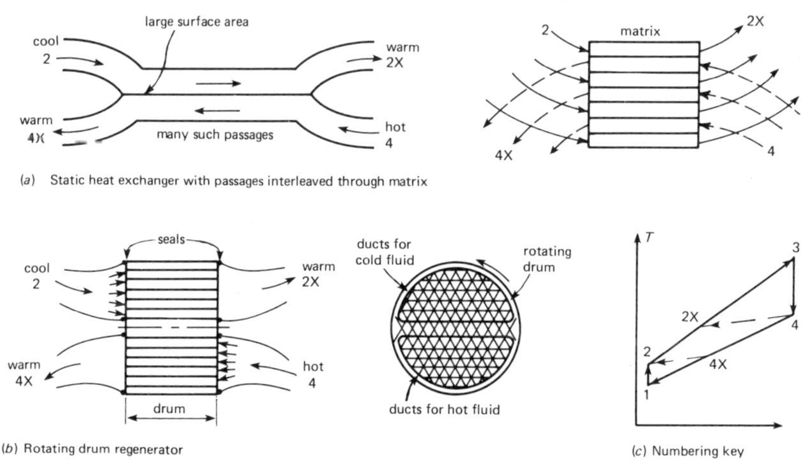

Figure 1.6 *Static and rotating types of regenerator*

simplified version of figure 1.2*c*, but shows the sloping lines from 4 to 2X and 4X to 2 which indicate that not all the heat available is transferred. This is discussed further in section 3.6.3.

1.4 Engine Types and Characteristics

Open-cycle engines (section 1.1) take a number of different forms. The simple engine of figure 1.7*a* generates a high-pressure source of gas which may be accelerated through a nozzle to provide jet propulsion as a *turbojet* engine (section 14.1). In this form, the combustion system is squeezed between the compressor and turbine. Alternatively, a larger turbine may extract more of the exhaust energy, permitting it to drive an external load as well as the compressor as a *turboshaft* engine (figures 1.1*a* and 1.7*b*). Both of these engines are simple cycle, *single-shaft* engines.

A variety of terms is used to describe the configurations of *multi-shaft* engines. The turbojet engine may be coupled to an industrial load by placing an independent turbine in its exhaust stream, rather than making the engine turbine larger. The second turbine is then a free *power turbine*, coupled to the load and running at a different speed from that of the engine. The core engine itself is then called the *gas generator* and the combination of the gas generator with the free power turbine makes a *two-shaft engine* (figure 1.7*c*).

The gas generator may itself have more than one shaft, in split-shaft or compound form. This is usually done to obtain satisfactory compressor

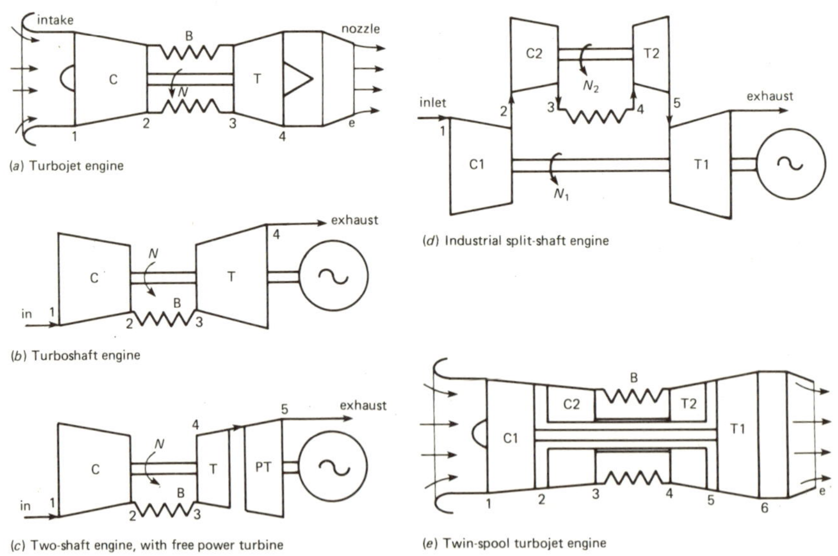

Figure 1.7 *A selection of basic engine arrangements*

operation in high-pressure-ratio engines (section 1.3). Figure 1.7*d* shows such a configuration, in which the load could be coupled to either shaft or to a separate power turbine. There are no unique terms to describe these configurations, but this book will refer to *low pressure* (l-p) and *high pressure* (h-p) components. Thus, the l-p compressor is driven by the l-p turbine and the h-p compressor by the h-p turbine, with the working fluid passing in the sequence shown. The shafts would be located concentrically in the case of an aircraft engine: figure 1.7*e* shows a two-shaft or *twin-spool* turbojet. Some engines have three shafts, in which the middle rotor is the *intermediate pressure* (i-p) system. It is convenient in this book to refer to the highest pressure or core engine rotor as the h-p shaft, even in single-shaft engines, when discussing engine operation.

The shafts of two and three-shaft engines run at their own equilibrium speeds, with no mechanical interconnection. The speed of rotation may range from 2000 to 80 000 rev/min but has little intrinsic significance as regards performance: it varies inversely with rotor diameter to keep the blade tangential velocity at a design speed of about 300 to 450 m/s and to suit the flow Mach number (section B.4.3). The direction of rotation is also of little significance, engines being designed to run in whichever direction suits the blade shape, but a few twin-spool aircraft engines use *contrarotating shafts* to minimise gyroscopic effects in turns and other attitude changes. However, the speed is significant in the case of a turboshaft engine because, depending on the engine size, a gear ratio of up to 20 to 1 may be required after the final turbine to reduce the speed of the output shaft to a conventional speed level for the driven machinery.

In addition to these variations of shaft configuration, there are several means of varying the thermodynamic cycle of industrial engines. The example given in section 1.1 shows that, in compressing gas through a chosen pressure ratio, the work input required varies with the starting temperature. The use of an *intercooler* between compression stages usefully reduces the work required for the h-p compressor of some industrial engines (figure 1.8*a*). Conversely, it is desirable for expansion through the turbine to start at a high temperature, to reduce the pressure drop relative to the work extracted. For this reason, a second combustion process (*reheat*) may be used before the l-p turbine (figure 1.8*b*). Both intercooling and reheating increase the work capacity of a given size of plant

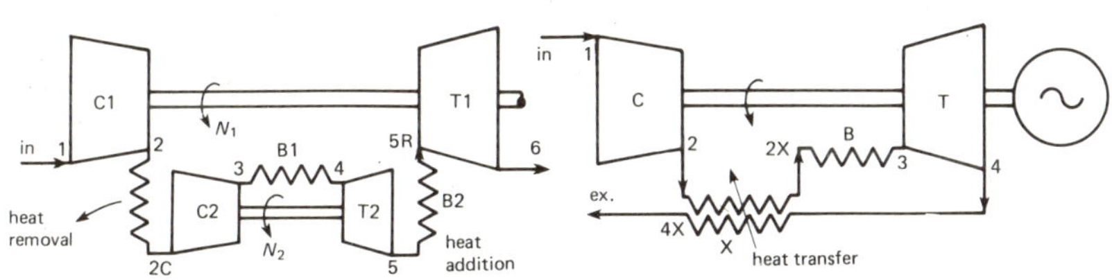

(*a*) Intercooler (*b*) Reheat (*c*) Regeneration by heat exchanger

Figure 1.8 *Basic cycle modifications for industrial engines*

(sections 3.6.1 and 3.6.2), while its efficiency may be increased by *regeneration* (sections 1.1 and 3.6.3), as shown diagrammatically in figure 1.8*c*.

So far, only the operating cycle and the shaft arrangement have been considered. The complete engine also requires a number of systems and some contact with the surroundings. The systems include means to control: the fuel supply, fuel flow rate and shaft speeds (chapter 12); lubrication (section 14.4); heating or cooling of the fuel and oil; air cooling of the discs and shafts with means of sealing to minimise leakage (section 14.2); and compressor operation (section 11.4). Most of the engine accessories which handle these functions are driven from the h-p shaft, though many may be driven separately in industrial applications (section 12.5). External connections may include a fuel inlet; pick-up points for the operator's controls; attachments for slave accessories such as the generator and hydraulic pump on an aircraft engine, and the starter; inlet and exhaust ducting; and mountings to support the engine.

Engine *starting* requires an external power source to rotate and accelerate the h-p shaft, until the combustion system is able to release enough heat to overcome the compressor and turbine losses and make the engine self sustaining. It calls for a particular sequence of actions (section 4.1) which may be controlled by an automatic system (section 12.5.1). The time taken to start, and reach full load, ranges from about fifteen seconds to several minutes. Engine *stopping* is accomplished by shutting off the fuel supply, but a cooling period of several minutes at low running speed is advisable as shutting down from high power may result in internal seizure (section 14.2). In an emergency, however, a fast shutdown is possible (section 12.5.3).

Engine operation may call for variable speed or load (turndown). When the speed is fixed by, for instance, the coupling of an alternator to a power grid, a single-shaft engine permits rapid changes of load simply by changing the input fuel flow (within the stable operating range – see section 3.7.4) but has low efficiency at part load, when the compressor and turbine losses remain high so permit less useful output (section 3.2). A multi-shaft engine is more efficient at low load, when the h-p shaft slows down while the power turbine runs at the speed required by the load, but incurs a time delay in response to changed load requirements because of the time taken for the h-p shaft to change speed.

The gas turbine only delivers useful quantities of power in the upper 10 to 15 per cent of its *speed range* (section 4.2), and its maximum speed is controlled by some operating limit (section 12.2). At any constant speed, the power demands of the compressor and load balance the turbine's power supply. Changes of speed are caused by increasing or reducing the fuel supply at a controlled rate (section 12.3), which directly alters the power output from the turbine and upsets the equilibrium. The torque excess or deficit and the inertia of the rotors then determine the rate of *acceleration* or *deceleration*, which is limited to avoid operational problems (sections 11.5 and 11.6).

The foregoing may apply to gas turbines in any application. The variants used specifically in aircraft are discussed more fully in section 2.1. Industrial engines are considerably more variable, to suit the applications described later in chapter 2. Their *origins* started with heavily built engines drawing to some extent from steam turbine practice. It was found feasible in the late 1950s to use aircraft engines as gas generators. These have evolved to the stage where the blading and combustion remain as developed for flight, but the structure may be made heavier and simpler. One engine alone, or a group of several engines side by side, supply an industrial power turbine. There are now also the medium-weight industrial engines which, like the heavy engines, may use regeneration to provide efficiency levels comparable with the better *aeroderivative* engines. Some comparisons between these types, and between gas turbines and other types of power plant, are made in sections 15.1 and 15.2. Two details, of the difference in turbine blade design philosophy and in engine control design philosophy, are discussed in sections 8.2 and 12.5.

1.5 Gas Turbine Engine Trends

The power output of a gas turbine engine increases relative to its cost by the use of higher turbine inlet temperatures and higher component efficiencies, but the higher temperature demands a correspondingly raised compression ratio (section 3.1).

The temperature and pressure limits are set by the best available materials. The pressure ratio is limited by rotor tip velocity (section 5.2), which implies material stress (section E.8), and the temperature limit is set by turbine blade creep with the loss of strength at elevated temperature (section 8.3). The efficiency is very dependent on the aerodynamic design, and on the quality of sealing to minimise leakage flows in the engine (section 14.2.1) and at the ends of the blades.

The trend of design progress with time is illustrated in table 1.1 by data from a series of engines from one manufacturer. The engine listed as HTDU is the high temperature demonstrator unit described in reference 7. Pressure ratios have risen as aerodynamic understanding has led to transonic and supersonic flow at the compressor inlet. The temperature limit has been avoided by artificial cooling of the blades (section 8.5), which keeps the metal temperature below about 1250 K in all engines, and may in time be met by the introduction of ceramic materials (section 8.3). Turbine efficiency has been increased by shrouding the ends of the blades to improve the sealing, and by detailed improvements to the blade profile and geometry (section E.7): discussion in reference 8 foresees polytropic efficiencies (section C.3) reaching 0.93. In conjunction with other progress [4A] engine efficiencies of 0.4 to 0.5 are foreseen.

While this progress is led by the aircraft industry, it steadily filters into industrial engine designs. A relative reduction in their cost will continue their encroachment into the industrial power field, displacing some traditional power

Table 1.1 Example of Aircraft Engine Design and Development Progress

Engine	Derwent V	AJ 65	Dart 535	Tyne 512	Avon	Spey 506	RB211–22B	HTDU	RB211–524
Design year	1944	1945	1946	1956	1957	1960	1965	1968	1971
Type	Turbojet	Turbojet	Turbojet	Turbojet	Turbojet	Bypass jet	Turbofan	Experimental	Turbofan
Compressor	Centrifugal	Axial	Centrifugal	Axial	Axial	Axial	Axial	Axial	Axial
Shafts	1	1	1	2	1	2	3	1	3
No. of stages	1	12	2	6 + 9 = 15	17	4 + 12 = 16	1 + 7 + 6 = 14	7	1 + 7 + 6 = 14
Mass flow (kg/s)	28.1	50.0	10.6	21.1	83.7	92.0	628	54.0	686
Bypass ratio	–	–	–	–	–	1.0	4.7	–	4.5
Pressure ratio	4.2	6.3	5.75	13.9	10.3	17.1 (2.6 x 6.6)	25.5 (1.36 x 4.6 x 4.1)	7.0	28.4 (1.49 x 4.5 x 4.2)
Turbine stages	1	2	2	1 + 3 = 4	3	2 + 2 = 4	1 + 1 + 3 = 5	2	1 + 1 + 3 = 5
Shrouded, cooled	no, no	yes, no	yes, no	yes, yes	yes, yes	yes, yes	yes, yes	yes, yes	yes, yes
TET[1] (max)	1100	1150	1160	1275	1250	1350	1493[2]	1800	1583[2]
TET (cruise)	–	–	1135	1180		1190	1337[2]	–	1386[2]
Thrust (kN)	14.23	20.02	–	–	54.27	46.26	186.83	–	222.41
Thrust (lbf)	3200	6500	2280[3]	5500[3]	12200	10400	42000	–	50000
s.f.c.[4](cruise)	0.125	0.106	–	–	0.094	0.077	0.065	–	0.066
(lb/lbf h)	(1.23)	(1.04)	(0.735)[5]	(0.455)[5]	(0.92)	(0.76)	(0.64)	–	(0.65)
Spec. wt.(lb/lbf)	0.36	0.33	–	–	0.32	0.22	0.21		0.20

1 **Turbine entry temperature (K).** 2 Nozzle exit temperature (K). 3 Total equivalent horse power. 4 Specific fuel consumption in kg (fuel)/N (thrust) at 0.8M, **9.15 km altitude.** 5 Units are lb (fuel)/(shaft horse power)/hour. Data courtesy of Rolls-Royce Ltd

plants (section 15.1). Further success is assured by current serious research work into the use of coal-derived fuel (section 7.6), and heat transfer from external combustion (section 2.7) to reduce turbine erosion (section 8.4) when using low-grade fuels.

2

Applications for Gas Turbine Engines

Gas turbine engines are most generally known for their use in aircraft (section 2.1)[9] but they are also used in surface transportation (section 2.2) and increasingly in stationary applications. The latter field encompasses the greatest variety of configurations and purposes and is still expanding, whereas the aviation field is consolidating.

The stationary uses are discussed here in the sequence: turboshaft applications in electricity generation (section 2.3) and the oil and gas industry (section 2.4); the use of exhaust heat for combined cycles (section 2.5) and other purposes (section 2.6); the use of other sources of heat and external combustion (section 2.7), particularly in closed cycle engines (section 2.8); chemical and process industry applications (section 2.9) and some unusual arrangements (section 2.10).

Information on many such applications, including which makers are involved in which application, can be found in catalogues such as reference 10 and the others listed in section 15.2.3. Some further information to act as a guide in the selection of an engine for a required duty is given in chapter 15.

2.1 Engines for Aircraft

Aircraft engines exert a forward thrust on the airframe in reaction to the rearward acceleration imparted to part of the passing airflow. This is achieved by means of a propeller, or by generating a high-pressure jet which emerges through a nozzle (section 14.1). The propeller may be turned by any type of engine, but the gas turbine is by far the most effective means of providing jet propulsion.

The choice of a propeller or jet propulsion is determined by the required *flight speed*. Propulsive efficiency (section C.6) is lost if the outlet flow from the propeller or jet has significant rearward velocity, which represents wasted kinetic energy. The efflux velocity from a propeller is low, which suits low flight velocities: medium and high jet velocities suit medium and high-speed flight.

The jet engine thus overcomes the speed limitation of the propeller, and is also well suited to operation at different *altitudes*. Its thrust varies with atmospheric pressure (section B.5), being high at low altitudes where the aircraft drag is high, and falling at high altitude where the lower air density also reduces the drag. Its net thrust (section 14.1) falls only slightly as the flight speed increases, permitting very high speeds. Nevertheless, jet aircraft fly at a range of speeds, which calls for a range of engine configurations, as discussed later.

Given the required thrust in an efficient engine configuration, the next most important criterion for a heavier-than-air machine is its *weight*. A heavier engine requires a heavier supporting structure, and more fuel to carry it throughout the aircraft's life. The extra fuel needs a bigger, heavier tank which, in turn, needs more fuel and a more powerful engine to lift it. This design spiral is so significant that, in 1962 for example, the economic rule-of-thumb was that an airline would pay an extra £150 to save a pound of engine weight. This could then save a further 5 lb of aircraft weight.

Its high power-to-weight ratio and its jet favour the gas turbine over the piston engine in all but the smallest aircraft. It also has a longer life and is more reliable. The low weight calls for very sophisticated design, and the lightness of the rotating parts confers good handling properties by permitting low acceleration times (section 1.4). The high power is achieved by high internal pressures and temperatures (sections 1.5 and 3.5).

Aircraft gas turbine engines are now made by about fifty companies, [11] mostly in a few, well-known *configurations*. They may be considered as variations of the single or multispool turbojet (figures 1.7*a* and *e*), in which the combustion system is squeezed between the compressor and turbine to minimise weight and frontal area (cf. figure 1.1*a*). In turboshaft form (section 1.4), its shaft may drive the rotors of a helicopter. If the reduction gear is built into the engine to drive a propeller, it becomes a *turboprop* or *prop-jet* engine (figure 2.1*a*), from which the exhaust gas exits through a nozzle to give a small residual jet thrust.

Enlargement of the l-p compressor (C1 in figure 1.7*e*) makes a *bypass-jet* engine if the excess l-p air is ducted to mix with the exhaust (figure 2.1*b*), or a *fan-jet* or *turbofan* if it emerges separately (figure 2.1*c*). The choice of l-p air outlet position significantly affects engine matching (section 11.3) and noise (section 13.4). This flow bypasses the h-p compressor, combustion system and turbines. The *bypass ratio* is the ratio of this flow to the flow through the h-p system, and is typically 0.3 to 1.5 for the bypass engine, or up to 6 for the fan engine. The higher bypass ratio engine provides better propulsive and over-all efficiency (section C.6) for subsonic flight, because it generates its thrust by accelerating a large mass flow to a low velocity (section 3.5.2).

The fan or l-p compressor may have one or several stages of blading, and is driven by the l-p turbine in a two or three-shaft engine. There is little balance of advantage for either number of shafts: the three-shaft engine [7] may be a little

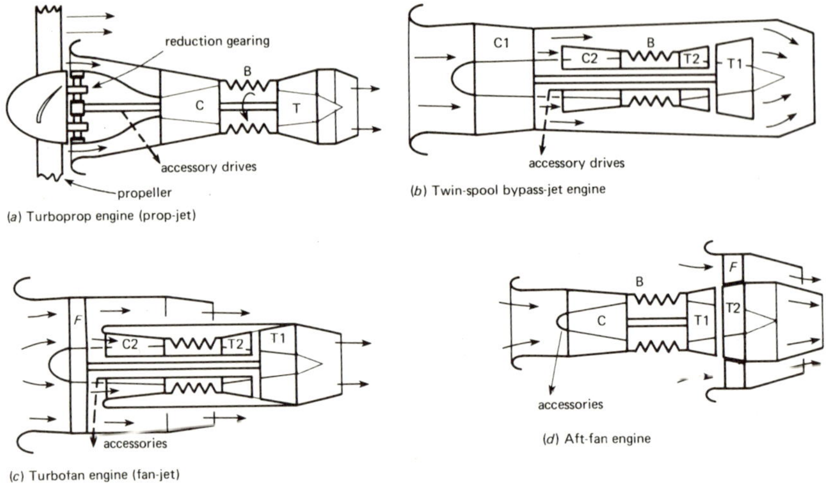

Figure 2.1 *Various aircraft engine configurations*

more expensive and heavy, but may offer higher efficiency and better handling response (sections 11.6.4 and 11.6.5). The fan could instead be mounted as an *aft-fan* (figure 2.1*d*) behind an existing engine, to improve its efficiency, thrust and noise. This avoids the need for a shaft: the turbine blades or buckets are extended to form fan blades protruding into a large duct outside the nacelle, the one-piece combined blades being called 'bluckets'.

Fan engines also make good *lift engines* for vertical (VTOL) or short take-off and landing (STOL). They may be installed vertically for direct lift, possibly rotating to horizontal for forward flight. The l-p air may be ducted to a lifting wing [12] for STOL, or may be taken directly through vectoring nozzles (figure 2.2) for VTOL.

Supersonic flight requires a considerably increased jet velocity. The *afterburner* (sections 3.5.3 and 7.4) reheats the exhaust gas after the turbines, permitting it to accelerate to an appropriate level above the flight velocity and boosting the thrust to overcome the increased drag.

The last aircraft engine type is the *auxiliary power unit* (APU). It is a small turboshaft engine which provides air-conditioning and electric and hydraulic power while the main engines are stopped, but its main function is to start the main engines without external assistance. The APU is usually started from batteries by an electric motor. When it reaches its operating speed, a substantial flow of air is bled from its compressor outlet and is ducted to drive the air turbine starters on the main engines. The fuel flow is increased when the turbine air supply is reduced by the air bleed, to provide the energy required for compression. These engines are also found on ground carts, which may be

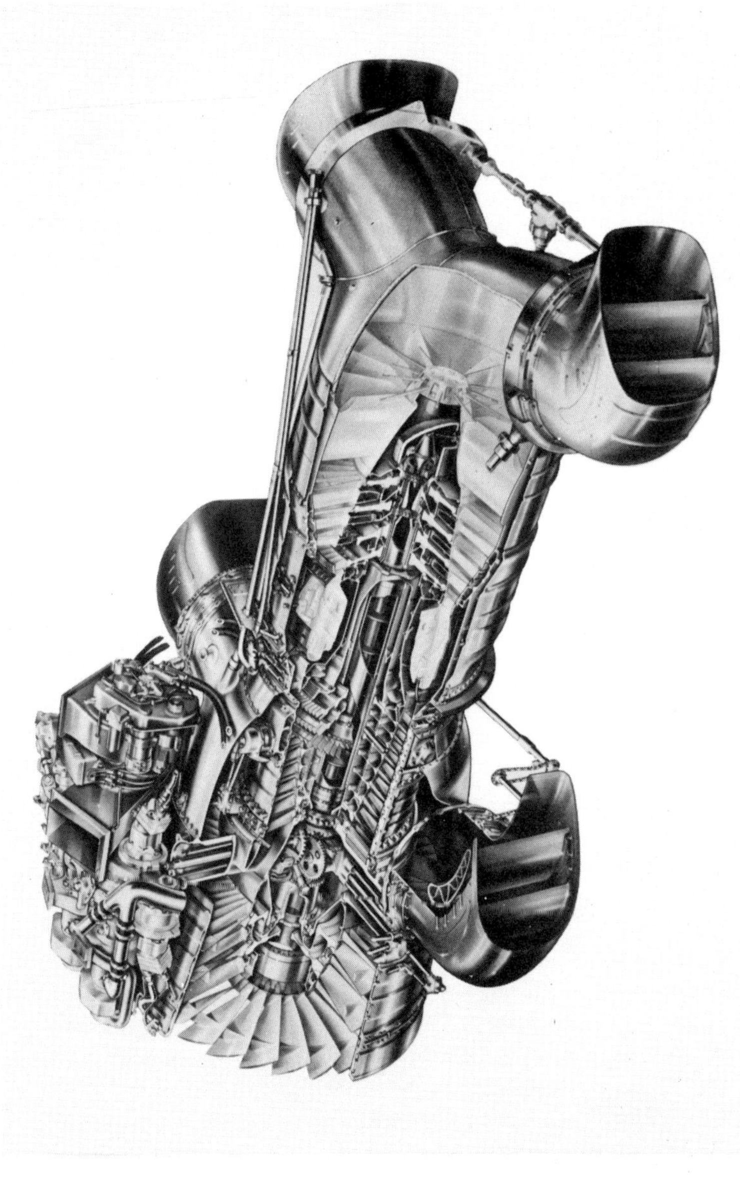

Figure 2.2 *Sectioned view of the Rolls-Royce (Bristol) Pegasus engine with vectoring nozzles for the VTOL Harrier aircraft. The h-p and l-p shafts rotate in opposite directions (by courtesy of Flight International, from the issue of 19 October 1972)*

temporarily connected to an aircraft to service it. They may also have uses in industrial plant requiring air at 3 or 4 bar.

Aircraft engine operating practice is described briefly in section 4.4, and its design trends in references 4A and 4B.

2.2 Engines for Surface Transportation

This category includes engines for rail, road and off-road transport, some of which may be for military purposes, and marine use. The low weight and volume are assets in all cases, but the relatively high fuel consumption is more acceptable in military use. The engines are of turboshaft type with a free power turbine: they can burn a variety of liquid or gaseous fuels and need no liquid cooling system.

Railway applications are rare, because of the poor efficiency relative to diesel–electric or all-electric systems. Regeneration is essential to make the fuel consumption acceptable. The transmission may be mechanical, electrical or hydraulic, and the task is made easier by the use of gentle gradients and long journeys.

Road use uncovers two more drawbacks. The engine has slow torque response to accelerator pedal movement, and inadequate torque to cope with gradients if geared for level running. The slow torque response stems from the loss of gas generator rotor speed at part throttle, and its long acceleration time. Some makers use variable-angle nozzle guide vanes (VNGVs) (section 9.2) to maintain a high gas generator speed. Variable power turbine NGVs may also be used, even to the extent of providing reverse rotation when required. Volvo use a turbine to drive differentially both the compressor shaft and the output shaft: at low road speeds the compressor speed remains high while, at higher speeds, the power is passed mainly to the output shaft.[13] This also overcomes the second drawback: generally, the modest rate of torque gain with reduced speed (section 9.4) necessitates the use of a multiratio gearbox or automatic transmission. Considerable reduction gearing is needed between the high turbine speed and low wheel speed, and regeneration is also essential.

Despite all these disadvantages, the automotive gas turbine still has good prospects for the future because of its low levels of pollution (section 7.3). With the Stirling cycle reciprocating engine it may offer the solution to the very tight emission control requirements of some countries. Work is in progress to introduce ceramic turbines (section 8.3), permitting higher temperatures and improving the cycle efficiency.

Off-road applications are usually military and may include landing craft, armoured cars, tanks, hovercraft and trucks.[14] They may also use regeneration, and inlet air filtration is essential (section 15.4).

The gas turbine is well suited to *marine* use as it matches the fluid nature of the supporting and propelling medium. The shaft drives the propeller or water jet unit or, in the case of hovercraft, the lifting fan and air propeller.[4C] The engines are usually aeroderivatives and are the almost universal power plant for new naval ships. Regeneration is not used because of its bulk and weight: it also adds to the complexity of changing to a spare engine if problems arise. The more efficient later engines are ousting the diesel and steam plant, which were sometimes used for economical cruise, and supplemented by gas turbines for 'sprint' use. Gas turbines are also finding favour in liquefied natural gas (LNG) tankers, in which they use as fuel the small 'boil-off' quantity which evaporates to keep the main bulk cool. The boil-off can be increased by exhaust heat if necessary.

The main environmental problem is *corrosion*. It may affect all parts – discs, shafts, casings, blades and blade roots and external parts. This can be controlled on steel components by the use of sacrificial coatings, often containing aluminium, iron and chromium, or by using nickel or cadmium (on cold parts only). A more severe problem is 'hot corrosion' (section 8.4) of nickel and cobalt turbine blade alloys, resulting from combination of sodium from the salt atmosphere with sulphur from the fuel. This problem concerns operation at sea, and also land-based operation near the sea. A second problem is compressor fouling or erosion, by salt or sand or dirt deposits, causing a loss of performance (section 6.4.1). Both of these problems call for extensive filtration plant (section 15.4).

2.3 Applications in Electricity Generation

A turboshaft engine is ideal for driving an alternator, and most manufacturers will supply the engine matched to the electrical equipment as a turbogenerator set. Such units come from all origins (section 1.4) and may have one or more shafts. The single-shaft engine in this application may be designed for a single operating speed and the shaft speed may be governed by the alternator or may be controlled to provide very accurate *frequency control* for specialist tasks. An example of this is the power supply for computers.

These units are suitable for use by national electricity generating authorities. Many are used on *base load*, where their efficiency is little less than that of other plant (section 15.1). The advantage of such units is that they can be distributed throughout the country in medium sizes, being located near the demand with no regard for supplies of cooling water, and thereby minimising transmission losses.

Local electricity boards and companies use turbogenerators for *peak lopping*. It is common practice for all the electricity supplied in a given period to be priced in proportion to the maximum rate of usage at any time during that period. If the peaks are lopped by, for instance, using additional local generating capacity to absorb the evening meal cooking load, then the savings to the local board, which buys power from the main grid, may pay for the peak lopping engine within a few days. However, the engine life is reduced by such duty (section

15.2.1) and may be shorter still on *reserve peak* duty: each engine run may last only a few minutes, which gives a high proportion of starts to hours. Aero-derivative engines are often chosen for this duty.

2.4 The Oil and Gas Industry

Turboshaft engines are extensively used by the oil and gas industry, which can afford to burn its own fuel at, sometimes, relatively low efficiency.

A major application is *pumping oil and gas* through pipelines over long distances, through deserts, arctic lands, sea and other remote places. As the pressure falls owing to frictional losses in the pipeline, pumping stations are built at intervals along the line to boost the pressure. Gas is transmitted at a typical pressure of 68 bar (1000 lb/in^2), but the pressure varies considerably as the pipeline serves the secondary function of storage. Two-shaft engines are suitable for this duty, as the gas generator may then change speed with only moderate loss of efficiency (section 1.4) to cater for the load variations. The engine is usually fuelled by tapping off a small amount of the pumped oil or gas, and may operate on a remote site for weeks or months, without supervision. It is controlled from a main centre, started, stopped and monitored as required (section 12.5.1).

Gas turbines are also used extensively on off-shore *oil drilling rigs*. There, they generate electricity, supply shaft power for pumping duties, and may heat the rig with their exhaust gas (section 2.5). The low weight and bulk, and supply as a prepackaged and tested unit are of benefit. The danger to personnel in this environment is often reduced by the provision of considerable excess engine capacity (section 15.3).

The oil industry also uses gas turbine components separately, as discussed in section 2.9.

2.5 Combined Cycles and Cogeneration

The relatively low efficiency of the turboshaft engine results from the high temperature remaining in the exhaust gas when the pressure is too low to provide useful work (section 1.1). The heat may be used to raise the efficiency by regeneration, heating the compressed air before the combustion process takes place. In the appropriate plant, the heat may be used more effectively to raise steam by feeding the engine exhaust gas into a boiler, to drive a steam turbine and generate more power. A suitably designed plant may permit the gas and steam sections to be operated independently as well as together [15], while old steam plant may be repowered in this form.[16] An unusual variant is discussed in section 2.10.

Cogeneration gives higher *efficiency* than either the gas turbine or steam plant by itself. Typical gas turbine efficiency is 0.2 to 0.3, but the higher pressure ratio, aeroderivative engines may exceed 0.35. Steam plant used for electricity generation alone can achieve peak levels of 0.35 to 0.38 in very large plant sizes (over 200 MW), while smaller units (under 20 MW) may peak at only 0.3 (section 15.1).

In an *unfired* cogeneration plant, where no additional fuel is burned in the boiler, the over-all plant efficiency may lie between 0.4 and 0.5 (section 3.6.7). In a *fired* plant, the additional fuel may be of poorer quality and will add to the power output but, in the limit, as the gas turbine fuel input falls to only a small proportion of the plant total consumption, the efficiency falls back towards that of the steam plant alone. The plant then begins to resemble an energy topping application.

Energy topping is a method of improving the cycle efficiency of existing or proposed steam heating plant. It involves mounting a small gas turbine engine to use the boiler's air intake flow before discharging it into the boiler (figure 2.3). As the gas turbine burns more fuel, the boiler burns less fuel. While the engine runs at idle, supplying no output power, all the fuel energy input appears as exhaust heat, which reduces the boiler fuel input correspondingly. When shaft power is drawn off, perhaps 30 per cent of the fuel energy input may be used. The total plant fuel input must be increased to supply that power in addition to the normal quantity of steam. In fact, the total fuel input may be increased by 8 to 10 per cent, as also is the air mass flow: all of that 8 or 10 per cent emerges as shaft power, an incremental efficiency of 1.0.

This calculation is demonstrated in general terms in section 3.7.1, and an example of the energy balance at full load is shown in figure 2.3*b*. It assumes a boiler capacity of 10 000 kg/h and an engine with efficiency of 0.3 and compression ratio of 12.5 to 1. Note that some of the energy in the gas passing through the combustor is recycled through the turbine, shaft and compressor to

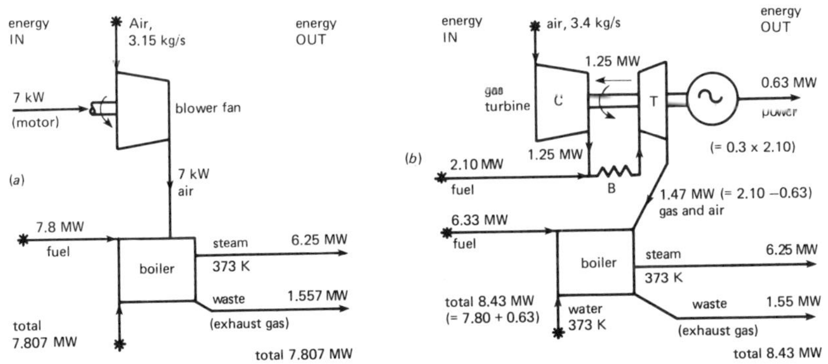

Figure 2.3 *Gas turbine topping of process steam plant: (a) the standard boiler generating 10 000 kg/h of steam; (b) the changed energy flows with the engine using the air first*

re-enter the combustor. This application offers a uniquely efficient means of obtaining power for the lighting or other loads in a plant, with two qualifications. The engine is likely to need higher quality and more expensive fuel than is normally used in the boiler. Secondly, the engine taking only 3.4 kg/s of air in this example is a very small engine and its power output is quite low: the smallest engine available would probably suit a boiler of 4000 kg/h but its compression ratio may be only 4 to 1, limiting the power output to 170 kW. The principle could be applicable to completely different types of fuel (section 2.9).

The most efficient combined cycle is the *total energy* package. Efficiencies of up to 0.85 are feasible, where the steam exhaust from the turbine is used for process work or heating. The steam turbine may then be run to a chosen back pressure, which provides the process temperature required. The residual exhaust gas heat may also be usable, but the cost of recovering all the waste heat may be prohibitive (section 15.3). A number of factories use these units for all their power and heat requirements, controlling this balance by varying the amount of supplementary firing.[17]

A pioneering combination of the gas turbine with steam was the Brown, Boveri *Velox boiler* of the 1930s (figure 2.4). The combustion chamber of the engine is the flame tube of the boiler which, by virtue of the high gas pressure, is considerably more compact than a boiler having its combustion at atmospheric pressure. This made it suitable for rail and marine use as well as stationary use, but its efficiency could not match the use of exhaust heat for steam generation. It is described in reference 18.

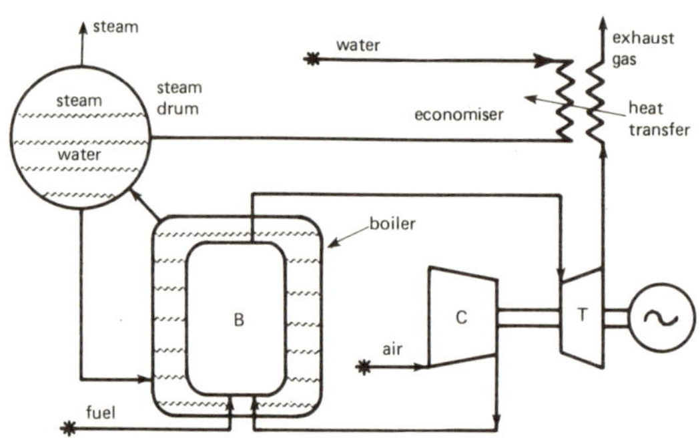

Figure 2.4 *Simplified diagram of Brown, Boveri Velox boiler*

2.6 Uses for the Exhaust Gases

The exhaust gas heat may have other uses than steam raising. It may be used for *drying processes*, with cold air mixed in to reduce the temperature as required. Applications in industry include kiln-drying of *cement*, or *tobacco*, or *blood* from a slaughterhouse.

The combination of heat and power is used extensively for the *desalination* of sea-water, particularly in the Arabian Gulf and other desert regions.

The partial use of the oxygen from the intake air permits the exhaust gas to form the basis of an *inert gas.* More fuel must be burned in it to use up the remaining oxygen and it will then need cooling before it can be used. A significant use for it is *purging* empty oil tanks, removing the oxygen to eliminate the risk of explosion. With the large mass flows involved, even a small engine provides plenty of gas.

In arctic or other regions, the gas from a portable gas turbine may be used to *thaw* material and equipment. It serves the purpose of a blow torch, but the use of the compressor and turbine provides considerably greater mass flow at more moderate temperatures.

A more powerful version uses the blast from the jet in addition to the heat, for *snow clearing* from airfield taxiways and runways. Old turbojet engines are used for this: the high fuel consumption is offset by the time saving, or even the ability to open the airfield at all on any given day. It could also be used on roads and railways. A fishtail nozzle is used to provide a wide but shallow jet, with its area trimmed to retain the correct engine matching (section 11.1). The jet needs to be located near the ground, pointing slightly downwards, and the engine is mounted on a heavy vehicle with adequate brakes and traction. An air mass flow of 10 to 20 kg/s at a low power setting is very effective. People should be kept clear of the blast, and the blown snow and ice.

2.7 External Combustion and other Heat Sources

Most of the well-known fuels are unsuitable for gas turbine engines because they are solid, or slow burning, or corrosive, or form ash which damages the turbine (section 8.4). Research work is in hand to modify or gasify coal, to make it satisfactory (section 7.6).

Many cheap fuels will be usable when suitable *heat exchangers* are made to transfer their heat of combustion into the compressed working fluid. The heat exchanger may take the form of a water tube boiler, or a regenerator (section 1.3). It may require cleaning, but the turbine will be exposed only to the clean working fluid. Moreover, being static and without the high rotational stresses of the turbine, the heat exchanger could be run hotter than turbine blades usually are. The working fluid can never reach the temperature of the combustion products, which would limit the performance unless ceramic material is used for the heat exchanger.

There are other sources of heat apart from combustion products. Two options are solar heat, and the heat of a chemical reaction. There is also nuclear heat, which is capable of being handled by closed cycle gas turbines (section 2.8).

2.8 Closed-cycle Gas Turbines

The closed-cycle gas turbine uses the same working fluid repeatedly. The compressed gas is heated by external combustion or other heat source (section 2.7), and the expanded gas is cooled by a process which can make use of its heat (section 2.5 or 2.6). Given a need for this heat, which is often used for district heating in Germany and Eastern Europe, efficiencies of 0.85 are feasible.

Closed-cycle engines appear to compete well with steam plant in sizes from 6 MW upwards. [19] Its capital cost is greater for size below 100 MW, but its efficiency, unlike that of steam plant, does not fall with size. Its efficiency also remains high at part load, and a wide variation of power can be achieved at constant speed by varying the total mass (and pressure) of working fluid in the system.

There is also the option of using different working fluids. Air works well, and incurs no cost of replenishment to compensate for leakage losses. Other gases are used in specific applications where their properties are advantageous. For instance, *helium* is considered better than air in generation plants over 30 to 50 MW,[19] where the expense of the fluid is more than offset by the reduced size of the turbomachinery (section E.9). *Carbon dioxide* is another relatively inert gas which is a viable working fluid, and might even be replenished from the combustion gases. *Nitrogen* is used in the regasification of LNG, which provides the cooling necessary before recompression: in this application, the regenerative cycle with external combustion also provides useful shaft power for electricity generation. The use of *hydrogen* is another possibility, as discussed in section 2.10.

External combustion experience is expected to lead to the use of helium closed-cycle gas turbines as the means of extracting power from advanced, gas-cooled *nuclear reactors*. The helium would be heated in its passage between spheres of active material, and would then flow through the turbomachinery, as shown in figure 2.5. The avoidance of the need for intermediate fluids, with heat exchange from the reactor coolant to the turbomachinery circuit, yields the highest possible temperature at turbine entry. The field is covered in reference 20 and in those listed in reference 19.

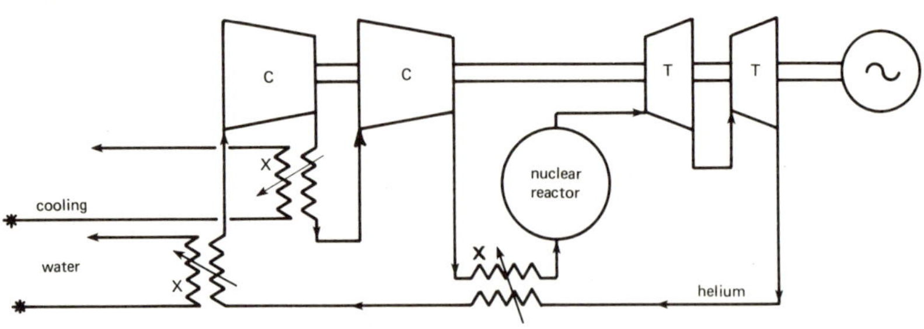

Figure 2.5 *Gas turbine in a helium-cooled reactor*

2.9 Chemical and Process Industry Applications

The ability to compress or expand large quantities of air or gas is used in many plants to adjust between the different temperature and pressure levels of the process. This may be done by compressors and turbines acting separately, or together, or as part of an engine. Normally, a dynamic compressor handles so much air and absorbs so much power that the turbine is the essential driver: it is fortunate that a compression stage is usually accompanied by an expansion process, making the units compatible. The turbine may alternatively be coupled to an external load if the source of pressure arises by other means.

In all cases, the turbo machinery works without contact between its lubricant and the working fluid. The chemical process may therefore remain *uncontaminated* by oil, an advantage relative to most other types of compressor.

The earliest chemical industry application was the *Houdry process* of 1937, which converted a wide range of hydrocarbons by catalytic cracking to relatively uniform, high-grade gasoline.[21] Three beds of catalyst (clay and other materials) were used sequentially for 10 minutes each, by which time they were coated with carbon. Each bed was then purged for 5 minutes to recover the oil and vapour content, regenerated for 10 minutes with hot compressed air from the gas turbine, and purged for 5 more minutes before its next working period. The gas turbine section of the plant is shown in figure 2.6. This process has been superseded by catalyst in fluidised beds, partly because the turbine blades suffered erosion from catalyst dust.

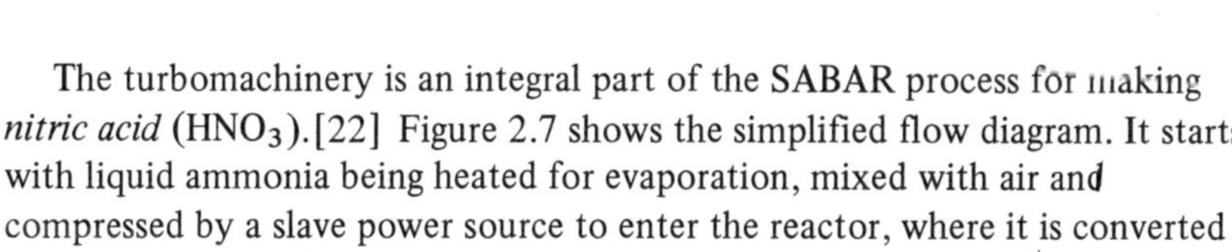

Figure 2.6 *The Houdry process for catalytic cracking of gasoline*

The turbomachinery is an integral part of the SABAR process for making *nitric acid* (HNO_3).[22] Figure 2.7 shows the simplified flow diagram. It starts with liquid ammonia being heated for evaporation, mixed with air and compressed by a slave power source to enter the reactor, where it is converted catalytically to water vapour and various oxides of nitrogen (NO_x). The heat of reaction is taken for steam generation by a boiler and other heat exchange processes (not all shown), and the steam turbine provides power to compress the nitrous gases for the absorption processes. The tail gas, with very little NO_x remaining, is expanded to help the compression process.

Energy topping (section 2.5) could be applied to industrial *combustion processes*, which burn waste products or materials not normally regarded as fuels.

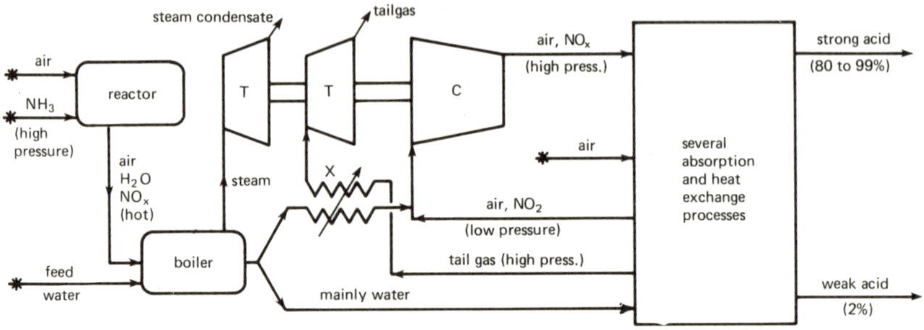

Figure 2.7 *The SABAR process for making nitric acid (HNO_3)*

Waste products include *methane* gas, produced from sewage or manure or the digestion of offal, or *blastfurnace gas* in which the carbon monoxide content warrants further combustion (section 7.6). There may be a requirement for such fuels to be filtered, decontaminated to protect the turbine (section 8.4), and dry (section 15.3).

A less usual fuel is *sulphur*, which is burned in large quantities to make sulphuric acid via sulphur dioxide (SO_2) and trioxide (SO_3) stages. The heat released is sometimes dissipated unused, sometimes partially recovered in steam plant to generate electricity for the plant and the local community. It is shown in reference 23 that topping such a plant could increase its electricity output by over 60 per cent at very moderate capital cost. Over 50 per cent of the fuel may be burned in the engine, compared with 25 per cent in figure 2.3, at zero fuel cost since the fuel would be burned anyway. The problems of handling the fuel in liquid form are well known, and the corrosion problem may well be negligible even though sulphur is undesirable in normal fuels. Many other unusual materials, which are converted by combustion processes, could similarly be burned in gas turbine engines. Corrosion, erosion and combustion intensity (section 7.2) would need to be satisfactory, along with numerous other factors. If in doubt, the use of an external combustion engine (section 2.7) would increase the operational feasibility including, perhaps, that of burning sulphur.

The *turbo-expander*, shown in figure 2.8, which was derived from reference 24, is used variously to separate carbon dioxide, hydrogen sulphide or nitrogen from natural gas. The gas often comes at high temperature and pressure from the ground, thermally degraded (CO_2 or N_2) or soured (H_2S). A turbine extracts work, cooling the gas to permit demethanisation after partial condensation. The turbine drives a compressor to repressurise and heat the gas for delivery, and a steam turbine may additionally be required. Older plants often used refrigeration.

A turbine may also be used for *ultra-fast cooling* processes, and may cool a flow of gas at a rate as high as 500 000 K/s (section 3.7.2). This capability could be of use where the high-temperature equilibrium condition of a gaseous mixture is required at lower temperature, where the equilibrium would normally be

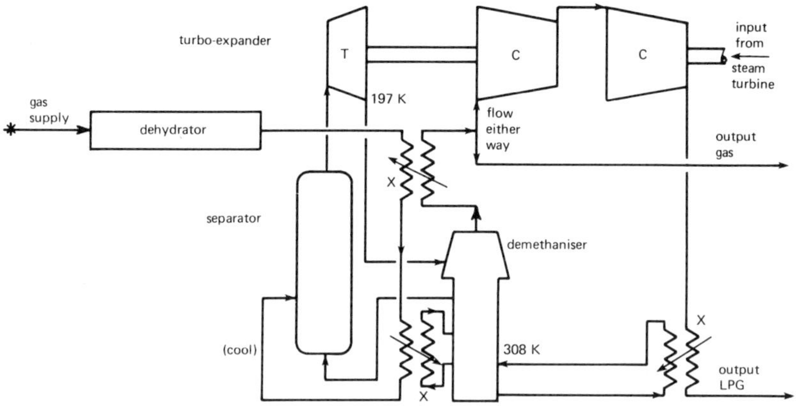

Figure 2.8 *Turbo-expander used to produce LPG*

different. The gases involved would need to be checked for compatibility with the turbine material; a reducing or corrosive atmosphere may not be suitable (section 8.4), and the temperature at turbine entry could not normally exceed 1200 K.

2.10 Unusual Applications

There are several miscellaneous configurations which are considered feasible, and some are now in operation.

Electrical load variations during the day may be smoothed by the *air storage* gas turbine, which absorbs energy from the grid at low demand periods and feeds the grid at times of peak demand. It involves an electric motor driving a compressor, to pressurise overnight a large underground rock cavern.[25] When required, the air is released through a combustion system to drive a turbine and alternator. Further heat may be recovered from the exhaust gas.

Another possibility is *air-cycle refrigeration*. The simple arrangement of figure 2.9 cools the compressed air, which is then expanded to a very low temperature. The turbine extracts less energy than the compressor requires, so another engine or electric motor is required to supply the deficit. This unit would require pre-

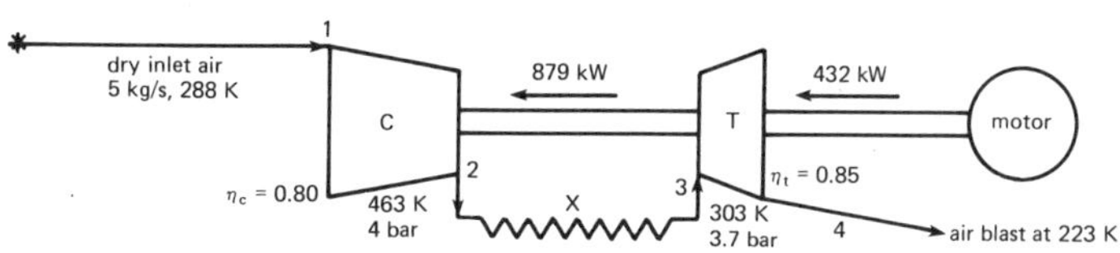

Figure 2.9 *Air-cycle refrigeration system for blast freezing (simple)*

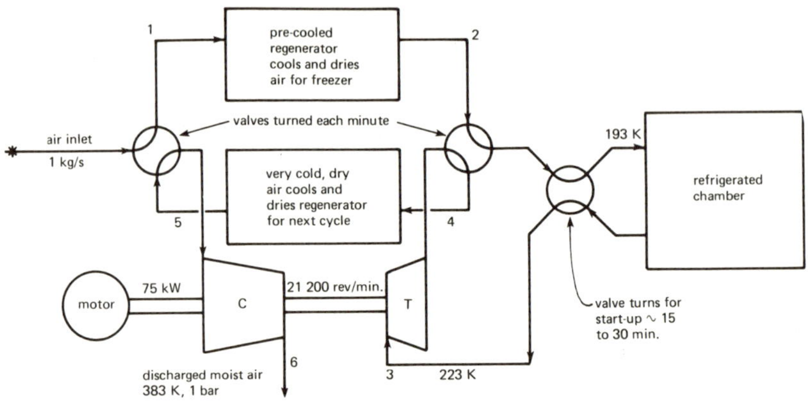

Figure 2.10 *Air-cycle refrigeration system (practical)*

dried air: the cycle of a unit handling humid air (figure 2.10) is discussed in section 3.7.3.[26] Neither has a very high coefficient of performance, but they offer a portable and very powerful source of cold air for blast freezing, available almost at the instant of start-up, and without the need for freon or ammonia. The subject is covered more fully in reference 27.

Waste heat at low pressure, which accounts for the inefficiency of many thermal cycles, may be partially recovered by *sub-atmospheric pressure cogeneration.* This could involve a turbine expanding the hot exhaust gas from a boiler down to a low pressure, cooling in a heat exchanger, and then recompression to exhaust to the atmosphere. The turbomachinery would be quite small, though bigger than the energy topping engine which works at above atmospheric pressure. A high boiler exhaust temperature and high expansion and recompression ratios are required to obtain a worthwhile output.

A study [28] has shown the feasibility of a high-efficiency power cycle which minimises the usage of high-grade energy for compression. It involves a closed cycle in which the hydrogen working fluid is compressed by an *absorption/ desorption* cycle, using metal hydride ($FeTiH_x$) beds, as shown in figure 2.11. It

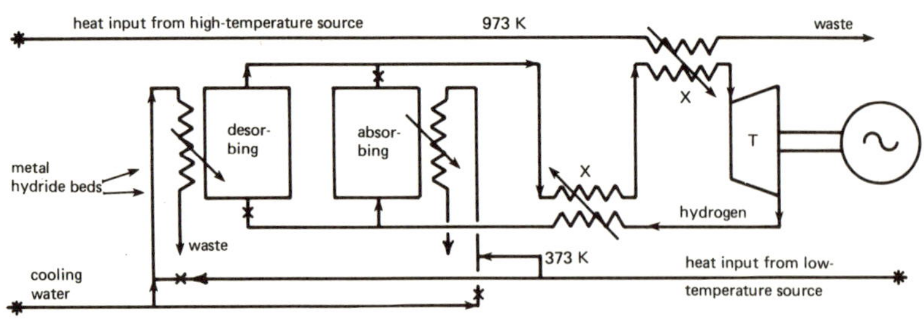

Figure 2.11 *Gas turbine with absorption/desorption compressor*

is based on the fact that, as the gas is absorbed, it gives out heat while, to desorb, heat must be fed in. Two or more hydride beds are used, with inlet or outlet valving sequenced to let one absorb as the other desorbs. Absorption takes place at low pressure, with the heat removed by water. To desorb into the high pressure ducting, heat is added at about 373 K from a solar or geothermal source. After regeneration, the high-pressure gas is further heated with high-grade energy from a source at about 973 K. The turbine then drives only the electrical load. It is anticipated that about 90 per cent of the high-grade energy is recovered as electricity, while about three times as much low-grade heat is also required. The x in $FeTiH_x$ varies during the cycle from about 0.2 to 0.6.

3

Cycle Calculations: Design Point Performance

Engine performance can be calculated at a multitude of different levels. It may be assessed at the very elementary level introduced in section 1.1, or it may make allowance for component inefficiences, pressure losses, non-ideal behaviour of gases and numerous other effects. All these affect the performance at the design point; that is, the normal operating condition of a plant or the cruise condition of an aircraft engine.

Then there are the complications of operation at other speeds, off-design. Here, the effects of the component characteristics add to the complication, involving the matching between the turbine and compressor characteristics at many operating speeds (chapter 10). Further allowance may be required for non-standard atmospheric conditions, such as the raised inlet pressure and temperature with forward flight for aircraft engines or the reduced pressure and temperature for all engines at higher altitudes (section B.4.5). Then, the working fluid may not be air (section E.9), having a different set of properties as listed for a few gases in appendix D. Finally, the calculations may get down to the specific aerodynamic behaviour of the blade shapes chosen, as introduced in appendix E.

The scope of this chapter is necessarily limited in several ways. It looks at design point performance only, and in such a way that the performance of a proposed gas turbine installation could be studied to assess its viability. Typical values of efficiency and other parameters are used, and rule-of-thumb equations are presented in place of standard graphs to determine the value of a few parameters. The method of calculation of a basic cycle is presented, with the effects of a number of standard variations. A general guide to cycle analysis may also be found in reference 29, and a much more detailed guide to the initial design process is given in reference 30. Other books giving fuller background to the design and performance estimation are references 31 and 32A. Beyond these, the manufacturers have their own, unpublished data and computer programs.

A very helpful simplification is the use of non-dimensional, or dimensionless, performance parameters (section B.5). These convert pressures and temperatures to the form of ratios rather than actual values. Thus, the pressure ratio across a compressor may be 2.5, whereas its delivery pressure would be 2.533 bar if the inlet condition was the International Standard Atmosphere, Sea Level, Static condition (ISASLS). Other parameters also have dimensionless forms. They permit the engine performance throughout the dimensionless speed range to be plotted on simple graphs (section 4.2), from which the actual values of any parameter can be found by inserting the ambient atmospheric conditions as the reference inlet and outlet data.

3.1 The Ideal Operating Cycle

As introduced in section 1.1, the gas turbine engine works ideally to the Joule or Brayton cycle, involving compression and expansion at constant entropy (isentropic, section C.2) and heat addition and rejection at constant pressure. This cycle is different from those of the various reciprocating engine types, which work ideally to the Diesel, Otto (petrol engines), Stirling or other cycles. These use various combinations of processes involving constant entropy, pressure, temperature and/or volume.

The diagrams in figure 1.2 show the lines of constant pressure as straight lines, with no particular values to the horizontal scale. Cycle diagrams are more commonly drawn on a plot of temperature against *entropy*, or pressure against volume, as in figure 3.1. The use of a linear entropy scale causes the lines of constant pressure to bend as shown, but this does not alter the nature of the argument in section 1.1. Entropy is discussed briefly in sections B.3 and C.2 as a measure of wasted energy: changes of entropy must therefore always be positive,

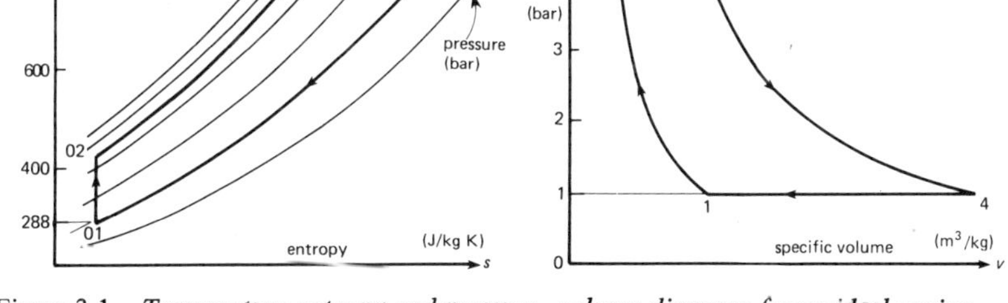

Figure 3.1 *Temperature-entropy and pressure–volume diagrams for an ideal engine*

yet a negative change is indicated during heat rejection from station 4 to station 1 (figure 3.1). The explanation of this apparent anomaly is that, while the working fluid itself loses heat, the surrounding atmosphere gains heat. Furthermore, the working fluid is hotter than the atmosphere, so the net change in entropy is still positive, as $(+dQ/T_a)$ is greater than $(-dQ/T_{41})$ (section B.3).

The relationship between the pressure ratio and temperature ratio in isentropic compression or expansion is given by equation B.12 (section B.3)

$$r_p = \frac{p_{02}}{p_{01}} = \left[\frac{T_{02}}{T_{01}}\right]^{\gamma/(\gamma-1)}$$

Here, the suffix '0' indicates that stagnation or total measurements are used (section B.2), and the '1' and '2' refer to the beginning and end of the process. In terms of figure 3.1, the numbers could be changed to '3' and '4' for the expansion, but '1' and '2' can be used generally for either process. The value of γ for air is taken as 1.4, although in practice it reduces as temperature rises (section 3.2). Thus the power $\gamma/(\gamma-1)$ has the value 3.5, and the pressure ratios corresponding to the two temperature ratios used in section 1.1 are: $1.5^{3.5} = 4.13$ and $2^{3.5} = 11.30$.

The term temperature ratio should be used carefully, as it is sometimes taken as the ratio of turbine entry temperature (the maximum temperature of the cycle) to the inlet temperature. To distinguish the two, in this book, the latter will be called the *over-all temperature ratio*, which is T_{03}/T_{01} in figure 3.1.

The conclusions to be drawn from the ideal cycle (section 1.1) are that its efficiency varies with the temperature ratio of the compression process, which is related to its pressure ratio, and that the power output depends on the amount of heat added after compression and the compression ratio. The optimum amount of heat added is also related to the compression ratio. If the pressure ratio is so high that T_{02} reaches the allowable level of T_{03}, then no heat can be added and no work extracted. This is when $r_p = (T_{03}/T_{01})^{\gamma/(\gamma-1)}$. It is shown in section C.5 that the maximum work output is obtained when $r_p = (T_{03}/T_{01})^{\gamma/2(\gamma-1)}$.

These conclusions have to be modified when the real cycle is considered, because the pressure ratio and over-all temperature ratio both interact with efficiency and work output.

3.2 The Real Operating Cycle

The effects that will be considered are as follows.

3.2.1 Inefficiency of Compression and Expansion

The paths followed move to the right because the entropy increases (section C.2).

Thus, the compressor work demand is increased while the turbine work capacity is reduced. These are shown in figure C.2, and in figure 3.2*a* in conjunction with other effects as they apply to industrial engine plants.

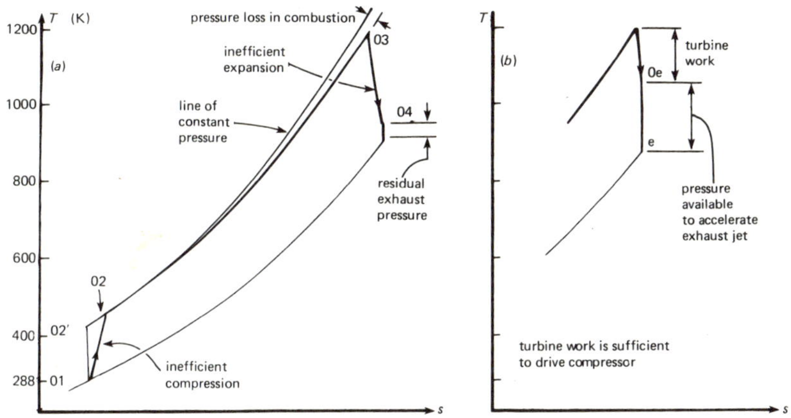

Figure 3.2 *Temperature–entropy diagrams for a real engine, in configurations producing: (a) shaft power output; (b) jet exhaust*

3.2.2 Loss of Pressure during Heat Addition and Rejection

The loss of pressure, as the compressed air flows through and reacts with the fuel in the combustion system (section 7.5), can be represented by showing p_{03} a little lower than p_{02}. It reduces the available isentropic temperature drop relative to dc (figure 1.2*a*). Pressure losses in regeneration, reheat and inter-cooling (section 3.6) also have to be considered.

In a closed-cycle engine, the drop from p_{04} to p_{01} could be explained by friction in the cooling system. In an open-cycle engine, the gas has very little ducting after the turbine so might expect to avoid this loss. In practice, the exhaust must leave with some velocity, so that it can get clear of the engine before its kinetic energy is dissipated in the atmosphere. Its static pressure cannot fall below the ambient static pressure, p_a, which is p_{01} in the case of zero inlet pressure loss except in circumstances of forward flight. Therefore p_{04} is higher than p_a, and a typical pressure ratio would be $p_{04}/p_a = 1.1$ to 1.2.

The case of a jet engine is shown in figure 3.2*b*. Here, the inefficient turbine expansion is limited to the work required to drive the compressor. The expansion through the nozzle is substantially isentropic, the total and static conditions at the exhaust outlet being shown by Oe and e respectively.

3.2.3 Variation of Specific Heat

The specific heat of most working fluids varies with temperature (appendix D). With air, the effect is increased by the addition of the products of combustion. The value of C_p is related to γ by the equation (B.6): $C_p = R\gamma/(\gamma - 1)$. Often, the value of γ is taken simply as 1.4 for air, and 1.33 for gas with combustion products. The corresponding values for C_p would be 1005 and 1148 J/kg K, the value of R being 287 J/kg K (appendix D).

A more accurate value for air and combustion gases can be obtained from the equation

$$C_p = 950 + 0.21 T_{01} \tag{3.1}$$

where T_{01} is taken at the start of the process, whether it is a compression or expansion. This equation gives a reasonable value for the mean specific heat through the process, and it is derived from curves given in NGTE report No. R74 and other books. For absolute accuracy, the full curves should be used and a mean value between the end points used for each process. The value 1005 is usually used for cold air.

The effect of varying C_p is that the turbine needs a smaller temperature drop than the temperature rise in the compressor. The full expression for power output is

$$P = \dot{m} C_p \Delta T_0 \tag{3.2}$$

where P is the power or rate of doing work, $\dot{m}$ is the mass flow rate of the working fluid, and ΔT_0 is the total change in temperature. Obviously, as C_p is higher in the turbine than the compressor, $\Delta T_{0\mathrm{T}}$ is less than $\Delta T_{0\mathrm{C}}$. The product $C_p \Delta T_0$ is therefore often used, as enthalpy, and is given the symbol H_0 (section B.3).

3.2.4 Variable Mass Flow Rate: Engine Cooling Flows, etc.

One of the ways of cooling the internal engine components, to avoid loss of strength with excessive temperature, is by bleeding air from the compressor and letting it flow between the discs and past the shaft and casings. Air taken from a low-pressure source ends up being vented to atmosphere and lost. Air from a high-pressure source may be able to return to the main flow at a point of lower pressure in the turbine. Some of this air arises from leakage past labyrinth seals (section 14.2.1) which is controlled in quantity to serve its cooling function. In that it starts as compressed air, some energy is lost in these cooling flows.

More recent engines operate at too high a turbine entry temperature for the blade material, with the object of increasing the work output. This requires an additional cooling supply of air, which flows through passages in the blades to cool them (section 8.5). The effect on performance is demonstrated in section

3.5.1. This cooling air returns to the working fluid for the later turbine stages, increasing their work capacity.

The fuel input also increases the mass flow, as does water injection (section 3.5.3), which is used to boost performance on hot days. The cycle of the APU (section 2.1) involves a deliberate, substantial loss of compressed air as part of the engine's duty. Its turbine temperature drop is boosted to compensate.

Figure 3.2 is not sufficiently detailed to show any of these effects.

3.2.5 Combustion Inefficiency

If the chemical process of burning the fuel fails to reach completion, then the full energy content of the fuel is not released (section 7.3). Although this does not show on the *T–s* diagram, the additional fuel needed to reach the required temperature represents a loss of thermal efficiency, η_{th}, or an increase in specific fuel consumption, s.f.c. (fuel flow divided by output power).

3.2.6 Engine Accessory Loads

Aircraft engines have accessories such as fuel and oil pumps for internal use, and to provide services such as air, electricity and hydraulics for the aircraft. These are driven by the h-p or core shaft, and add to the compressor load which has to be supplied by the turbine. Industrial engines may well use externally powered slave systems which do not affect the engine cycle, but which subtract from the net output of the plant (figure 2.1).

These items cover the main departures from the simple, ideal cycle, and are used to draw figure 3.2. The effect of items 3.2.1, 3.2.2 and 3.2.6 is to reduce the useful output available from the turbine, which therefore also reduces the cycle efficiency relative to the given heat addition. The effect of variable specific heat is beneficial (3.2.3), while the loss of air for cooling is slightly detrimental (section 3.5).

Figure 3.3 shows the effects of varying the pressure ratio and over-all temperature ratio in a real engine. The assumptions made were a range of efficiencies for the compressor and turbine, both having the same efficiency in each case; specific heat varying according to equation 3.1; no allowance for variations in mass flow, even fuel; no pressure loss in the combustion system but an exhaust pressure ratio of $p_{04}/p_a = 1.1$ to compensate. Three sample compression ratios were used: 4, 10 and 20 to 1. Five turbine entry temperatures were used: 800, 1000, 1200, 1400 and 1600 K, with no allowance for blade cooling. The air mass flow was 1 kg/s, the inlet pressure p_{01} was 1 bar and T_{01} was 288 K.

These carpet plots show the substantially constant thermal efficiency when

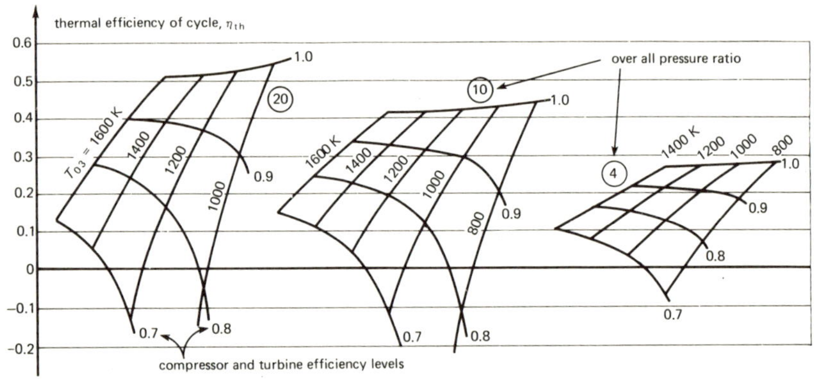

Figure 3.3 *The effect of cycle pressure ratio and component efficiencies on cycle efficiency (The turbine expansion ratio is assumed to be 90 per cent of the over-all compression ratio)*

the compressor and turbine have no losses. The thermal efficiency falls with flame temperature, particularly as the rotor efficiencies fall and particularly in the higher pressure ratio engines. The high-pressure ratio must clearly be matched by the higher temperatures, as the engine's performance becomes meaningless at only 800 K. This is because the value of T_{02} is 721.5 K at a compressor efficiency of 0.9, and 775 K at 0.8, leaving insufficient margin for a worthwhile heat addition. The sensitivity of the engine output to component losses may thus be seen as the variability of the small difference between two large quantities, the turbine output and compressor demand.

3.3 Basic Engine Cycle Calculation

This will show the calculation method used for each component in turn, using the following data

inlet conditions ISASLS: T_a = 288 K, p_a = 1.013 bar

intake: loss 1.3 per cent of absolute pressure

mass flow $\dot{m}_1$ = 50 kg/s

compressor: pressure ratio $r_{p12} = p_{02}/p_{01}$ = 5.0

isentropic efficiency η_{12} = 0.84

combustion: chamber pressure loss = 5 per cent ($p_{03} = 0.95\, p_{02}$)

efficiency η_B = 1.0

fuel energy input = 42 000 kJ/kg (lower calorific value, section 7.1)

fuel air ratio: see under (2) below

turbine: entry temperature = 1200 K, blades uncooled

isentropic efficiency $\eta_{34} = 0.89$

air offtake to cool discs and shafts, and dump = 0.25 kg/s from $\dot{m}_2$

power offtake to drive accessories = 40 kW

specific heat varies as per equation 3.4 ($C_p = 950 + 0.21T$)

R = gas constant for air = 287 J/kg K

The calculation step by step is then as follows.

(0) *Intake* $T_{01} = 288$ K, $p_{01} = 1$ bar, $\dot{m}_1 = 50$ kg/s

(1) *Compressor* $p_{02} = r_{p12} \times p_{01} = 5$ bar

C_{p12} for cold air = 1005 J/kg K

$\gamma/(\gamma - 1) = C_p/R$ (equation B.6) = 1005/287 = 3.5

isentropic $T_{02}'/T_{01} = r_{p12}{}^{1/3.5} = 1.583$

$T_{02}' - T_{01} = 1.583 \times 288 - 288 = 0.583 \times 288 = 168$ °C

Note that absolute temperatures are expressed in kelvins, changes in degrees Celsius.

$T_{02} - T_{01} = (T_{02}' - T_{01})/\eta_{12} = 168/0.84$ (equation C.1)
$= 200$ °C

$P_{12} = 50 \times 1.005 \times 200$ (equation 3.2) = 10058 kW

$T_{02} = 200 + 288 = T_{0A}$ for combustion = 488 K

(2) *Combustion* $\dot{m}_2 = \dot{m}_1 - 0.25$ for cooling = 49.75 kg/s

T_{03}', given symbol T_{0B} for equation below = 1200 K

$T_{0B} - T_{0A}$ = combustion temperature rise = 712 °C

The fuel/air ratio is found from equation 7.6, which is appropriate to fuel of the given lower calorific value (LCV) and which suits the required temperature rise.

$$f/a = \frac{1.1(T_{0B} - T_{0A} - 50)(1 + T_{0A}/3250)}{42000\eta_B}$$

$f/a = 1.1[(712 - 50)(1 + 488/3250)/42000 = 0.01994$

$\dot{m}_f$ = fuel flow = $0.01994 \times 49.75 = 0.992$ kg/s

$p_{03} = 0.95 \times 5$ bar = 4.75 bar

(3) *Turbine* $\dot{m}_3 = \dot{m}_2 + \dot{m}_f = 49.75 + 0.992 = 50.742$ kg/s

$$P_{34} = P_{12} + 40 \text{ (accessories)} = \dot{m}_3 C_{p34}(T_{03} - T_{04}) = 10098 \text{ kW}$$

$$C_{p34} = 950 + 0.21 \times 1200 = 1202 \text{ J/kg K}$$

$$\gamma/(\gamma - 1) = 1202/287 = 4.190$$

$$T_{03} - T_{04} = 10098/(50.742 \times 1202) = 165.6\ ^\circ\text{C}$$

$$T_{03} - T_{04}' = (T_{03} - T_{04})/\eta_{34} \text{ (equation C.2)} = 165.6/0.89 = 186.0\ ^\circ\text{C}$$

$$\frac{p_{03}}{p_{04}} = \left(\frac{T_{03}}{T_{04}'}\right)^{\gamma/(\gamma-1)} = \left(\frac{1200}{1200 - 186.0}\right)^{4.19} = 2.026$$

$$p_{04} = 4.75/2.026 = 2.345 \text{ bar}$$

$$T_{04} = 1200 - 165.6 = 1034.4 \text{ K}$$

p_{04} is the pressure available for expansion through a power turbine to do useful work, or to expand through a nozzle for jet thrust.

(4) *Power turbine* p_{05}/p_{01} (from 3.2.1, a typical value) = 1.150

$$p_{04}/p_{05} = 2.345/1.150 = 2.039$$

$$C_{p45} = 950 + 0.21 \times 1034.4 = 1167 \text{ J/kg K}$$

$$\gamma/(\gamma - 1) = 1167/287 = 4.067$$

$$T_{04}/T_{05}' = 2.039^{1/4.067} = 1.1915$$

$$T_{05}' = 1034.4/1.1915 = 868.1 \text{ K}$$

$$T_{04} - T_{05}' = 1034.4 - 868.1 = 166.3\ ^\circ\text{C}$$

$$T_{04} - T_{05} = 0.89 \times 166.3 \text{ (assuming } \eta_{45} = \eta_{34}) = 148.0\ ^\circ\text{C}$$

$$T_{05} = 1034.4 - 148.0 = 886.4 \text{ K}$$

$$P_{45} = 50.742 \times 1167 \times 148.0 = 8764 \text{ kW}$$

Note that the power turbine calculation is different from that for the engine turbine. The former (power turbine) is based on a known pressure ratio whereas the latter is based on known work and the pressure ratio has to be determined.

(5) *Efficiency*

$$\eta_{th} = \text{thermal efficiency} = \text{power out/fuel energy input}$$
$$= P_{45}/\dot{m}_f \times \text{LCV} = 8764/0.992 \times 42\,000 = 0.21$$

The alternative calculation, of thrust from a nozzle, is as follows

(6) *Jet exhaust* $p_{0e} = p_{04}$ (no loss in nozzle) = 2.345 bar

$$C_{pe} = C_{p45} \text{ as above} = 1167 \text{ J/kg K}$$

$$T_{0e} = T_{04} \text{ (no work done in nozzle)} = 1034.4 \text{ K}$$

Check to determine whether the nozzle is choked (section B.4.3).

$$\gamma/(\gamma - 1) = 4.067 \text{ so } \gamma = 4.067/3.067 = 1.326$$

$$p_{0e}/p_e = \left(\frac{\gamma + 1}{2}\right)^{\gamma/(\gamma-1)} = 1.163^{4.067} = 1.848$$

This is less than p_{0e}/p_a (2.345/1.013 = 2.314). Therefore

$$p_e = 2.345/1.848 = 1.269 \text{ bar}$$

That is, the static pressure in the nozzle is greater than the ambient pressure outside the engine (section 14.1). Using equation B.25

$$\frac{\dot{m}\sqrt{T_{0e}}}{A_e p_{0e}} = \gamma\sqrt{\left[\left(\frac{2}{\gamma + 1}\right)^{(\gamma+1)/(\gamma-1)} \Big/ C_p(\gamma - 1)\right]}$$

where the pressure is in the consistent units N/m^2 (appendix A).

$$\frac{50.742\sqrt{1034.4}}{A_e \times 234500} = 1.326\sqrt{\left[\left(\frac{2}{2.326}\right)^{2.326/0.326} \Big/ 1167 \times 0.326\right]}$$
$$= 0.3967$$

$$A_e = \frac{0.00696}{0.03967} = 0.1753 \text{ m}^2$$

This is the effective area of the nozzle. The *thrust* is then found from equation B.26

$$\frac{F_e}{A_e p_a} = 2\left(\frac{2}{\gamma + 1}\right)^{1/(\gamma-1)} \frac{p_{0e}}{p_a} - 1 = 1.2585 \times 2.345 - 1 = 1.951$$

$$F_e = 1.951 \times 0.1753 \times 101325 = 34654 \text{ N}$$

The *specific fuel consumption* is

$$\text{s.f.c.} = 3600\, \dot{m}_f/F_e = 3600 \times 0.992/34654 = 0.1030 \text{ kg/N h}$$

In older units, this converts to

$$0.1030 \times 9.807 = 1.0106 \text{ kg of fuel/kg thrust h} = 1.0106 \text{ lb/lbf h}$$

The s.f.c. is effectively an inversion of the thermal efficiency, but it is not dimensionless like the efficiency. A low value of s.f.c. is desirable.

(7) *Rotational speed and engine diameter*

This performance calculation has been made independently of the shaft speed and with no indication of the engine size. The figures could be calculated as follows. Assume the inlet Mach number = 0.4 for an industrial engine. Equation B.24 yields: $\dot{m}_1\sqrt{T_{01}}/A_1 p_{01} = 0.0254$. Using $p_{01} = 1$ bar, $T_{01} = 288$ K and $\dot{m}_1 = 50$ kg/s, the inlet area is therefore 0.333 m^2. Using a blade hub/tip diameter ratio of 0.4 (section E.6), because the central hub blocks part of the area, the compressor outside diameter would be 0.71 m. Using a typical blade tip speed of 400 m/s (section 1.4), the rotational speed must be about 1125 rad/s, or 179 rev/s (10740 rev/min).

The engine would be smaller if a higher inlet Mach number were used. For instance, at Mach 0.5, the speed would be 195 rev/s and the diameter 0.65 m. The choice of tip speed will also affect the rotational speed and blade design.

3.4 Basic Twin-spool Cycle Calculation

The methods used above may be applied to a split-shaft engine, which will form the basis of many other engine types. The numbering of the stations through the engine is now as shown in figure 1.7*d*. The previous example will be treated as the h-p shaft system for this engine, and the data assumed for the l-p system and other new inputs will be as follows.

inlet data (section 3.3): $T_{01} = 288$ K, $p_{01} = 1$ bar, $\dot{m}_1 = 50$ kg/s

l-p compressor: pressure ratio $r_{P12} = 2.5$

isentropic efficiency $\eta_{12} = 0.87$

h-p compressor, as in section 3.3: $r_{P23} = 5$, $\eta_{23} = 0.84$

combustion, as in section 3.3: $p_{04} = 0.95\, p_{03}$, $\eta_B = 1.0$

h-p turbine, as in section 3.3: $T_{04} = 1200$ K, $\eta_{45} = 0.89$

l-p turbine: isentropic efficiency $\eta_{56} = 0.89$

air offtakes: from 2, to cool shaft and bearings, dumped: 0.12 kg/s

from 3, to cool h-p turbine discs, return at 5: 0.60 kg/s

from 3, to cool l-p turbine discs, return at 6: 0.40 kg/s

from 3, seal leakage to l-p system, dumped: 0.24 kg/s

The word 'dumped' indicates that the air pressure at outlet from the cooling passage is too low to permit mixing back into the main flow, so the cooling air is dumped overboard.

The l-p compressor efficiency is higher than the h-p compressor because it involves fewer stages and works to a lower pressure ratio. The calculation follows the method in section 3.3.

(1) *L-p compressor*

$$T_{02} - T_{01} = 99.1\ ^\circ\text{C},\ T_{02} = 387.1\ \text{K},\ p_{02} = 2.5\ \text{bar},\ P_{12} = 4978\ \text{kW}$$

(2) *H-p compressor*

$$\dot{m}_2 = 49.88\ \text{kg/s},\ p_{03} = 12.50\ \text{bar},\ C_{p23} = 1031.3$$

$$T_{03} = 647.5\ \text{K},\ P_{34} = 13\ 394\ \text{kW}$$

The noticeably higher power requirement of this compressor compared with the earlier calculation (section 3.3) results from its increased starting temperature, as discussed in section 1.1. If this over-all compression ratio of 12.5 to 1 had been generated by a single compressor, at a similar over-all efficiency, the final temperature would have been the same as with two rotors. However, the compressor would have many more stages, as indicated in table 1.1, with the attendant problems (section 1.3).

(3) *Combustion*

$$\dot{m}_3 = 48.64\ \text{kg/s},\ \text{fuel/air ratio} = 0.01578,\ \dot{m}_\text{f} = 0.768\ \text{kg/s}$$

$$p_{04} = 11.88\ \text{bar}$$

(4) *H-p turbine*

$$\dot{m}_4 = 49.41\ \text{kg/s},\ P_{45} = 13434\ \text{kW},\ C_{p45} = 1.202,\ T_{05} = 973.8\ \text{K}$$

$$p_{05} = 4.384\ \text{bar}$$

After this turbine, 0.6 kg/s of cooling air re-enters the flow. The heat it gains is balanced by a loss of heat from the flow of station 5. Although the mean values of C_p at this stage may not be exactly as shown, this calculation gives an adequate indication of the mixed condition at 5c.

$$\dot{m}_4 C_{\text{p}56}(T_{05} - T_{05\text{c}}) = \dot{m}_\text{c} C_{\text{p}34}(T_{05\text{c}} - T_{03})$$

$$49.41 \times 1154(973.8 - T_{05\text{c}}) = 0.6 \times 1086 \times (T_{05\text{c}} - 647.5)$$

so

$$T_{05\text{c}} = 970.1\ \text{K}$$

(5) *L-p turbine*

$$\dot{m}_{5\text{c}} = 50.01\ \text{kg/s},\ P_{56} = 4978\ \text{kW},\ C_{p5\text{c}6} = 1154,\ T_{06} = 883.8\ \text{K}$$

$$p_{06} = 2.979\ \text{bar}$$

mixing in 0.4 kg/s cooling air

$$\dot{m}_{6\text{c}} = 50.41\ \text{kg/s},\ T_{06\text{c}} = 882.0\ \text{K}$$

(6) *Power turbine*

$$p_{06\text{c}}/p_{07} = 2.590,\ C_{p67} = 1135,\ T_{07} = 714.1\ \text{K},\ P_{67} = 9604\ \text{kW}$$

over-all thermal efficiency = 0.298

(7) *Jet exhaust*

$$\dot{m}\sqrt{T_{0e}}/A_e p_{0e} = 0.03982, A_e = 0.1262 \text{ m}^2, F_e = 35210 \text{ N}$$

$$\text{s.f.c.} = 0.0785 \text{ kg/N h}$$

As a turboshaft engine, the increased pressure ratio permits a slight increase in useful output with a drastic reduction in fuel input, the flame temperature being maintained constant. The efficiency is thereby greatly improved. The s.f.c. of the turbojet engine is also improved, but it gains only a trace of thrust because the exhaust gas is much cooler so its velocity is lower. The engine in twin-spool form therefore needs a higher flame temperature to justify its extra hardware and pressure ratio.

3.5 Cycle Variations for Aircraft Engines

The single shaft and twin-spool engines calculated above could well be practical turbojet engines. They will be used as a basis for comparison with a number of variations to the cycle but, before that, one level of simplication will be introduced. The internal cooling air flows add to the complication yet are by no means of a standard magnitude or nature. The comparison will therefore be made without the allowance for these flow adjustments (section 3.2.4).

It will be convenient to make the comparisons by tabulation of the main parameters through the engine. Lines 1 and 2 in table 3.1 are the engines of sections 3.3 and 3.4, revised to eliminate the allowance for cooling flows. Comparison of the cooled and uncooled engines shows that the effect of taking x per cent of the inlet air for internal cooling reduces thermal efficiency by about x per cent, the thrust of a turbojet engine by about $2x$ per cent and the power turbine output by about $4x$ per cent.

The assumptions for the later calculations in the table will be the same as the original assumptions, except where introduced differently in each case. For simplicity, the single shaft engine is shown with the station numbering for a h-p shaft.

3.5.1 Raise Turbine Entry Temperature

Increasing the fuel flow will increase the value of T_{04}. This permits the compressors to be driven with a reduced loss of pressure. The percentage increase in thrust and s.f.c. is similar to the percentage increase in T_{04}. If the engine is used to drive a power turbine, the power output increase is about three times the percentage increase in temperature, and the efficiency is slightly increased. In comparison with the single-shaft engine, the thrust and the s.f.c. are greatly improved by the increased pressure and temperature (line 3.5a, table 3.1).

The increased temperature involves a penalty. The tabulated figures allow for

Table 3.1 Calculated Data for Representative Engine Cycles

Line	m_1 (kg/s)	p_{02} (bar)	T_{02} (K)	P_{12} (kW)	p_{03} (bar)	T_{03} (K)	P_{23} (kW)	$\dot{m}_f$ (kg/s)	p_{04} (bar)	T_{04} (K)	p_{05} (bar)	T_{05} (K)	p_{06} (bar)	T_{06} (K)	turboshaft output: p_{07} (bar)	turboshaft output: T_{07} (K)	turboshaft output: P_{67} (kW)	turboshaft output: η_{th}	turbojet: A_e (m^2)	turbojet: F_e (N)	turbojet: s.f.c. (kg/N h)
1	50	–	–	–	5.00	488.0	10058	0.967	4.75	1200	2.353	1035.2	–	–	1.15	886.4	8850	.212	0.1753	34150	0.1020
2	50	2.5	387.1	4978	12.50	647.5	13426	0.766	11.875	1200	4.504	979.3	2.982	894.4	1.15	724.4	9822	.298	0.1280	35120	0.0785
3.3 (a)	50	2.5	387.1	4978	12.50	647.5	13426	1.223	11.875	*1500*	5.449	1283.7 5c 1261.5	3.985	1181.5	1.15	910.8	16616	.315	0.1117	44660	0.0986
3.5 (b)	50 250	2.5 1.6	387.1 334.0	4978 11555	12.50	647.5	13426	1.223	11.875	*1500*	5.449	1261.5	1.738 1.600	995.8 334.0	–	–	–	–	–	27520 71725	0.0444
3.6 (a)	50	2.5 2.375	387.1 288.0	4978	11.875	488.0	10058	0.967	11.281	1200	5.566	1034.3	3.777	950.6	1.15	733.6	12226	.293	CICBTTT		
3.6 (b)	50	2.5	387.1	4978	12.50	647.5	13426	0.766 R 0.319	11.875	1200	4.504 4.279	979.3 12000	3.074	1118.9	1.15	907.9	12773	.273	CCBTRTT		
3.6 (c)	50	2.5	387.1	4978	12.50 11.875	647.5 719.8	13426	0.668	11.281	1200	4.249	978.0	2.804	892.6	1.15	732.6 668.0	9210	.322	CCBTTTX		
3.6 (d)	50	2.5	387.1	4978	12.5 11.875	647.5 869.8	13426	0.465 R 0.319	11.281	1200	4.249 4.037	978.0 1200	2.893	1118.4	1.15	909.0 697.4	11988	.355	CCBTRTTX		
3.6 (e)	50	2.5 2.375	387.1 288.0	4978	11.875 11.28	488.0 877.0	10058	0.442 R 0.354	10.717	1200	5.272	1033.6	3.563 3.385	949.1 1200	1.15	957.4 598.7	14827	.432	CICBTTRTX		
3.6 (f)	50	2.5 2.375	387.1 288.0	4978	11.875 11.28	488.0 861.6	10058	0.909	10.717	*1500*	6.171	1336.8 5c 1308.1	4.454	1228.3	1.15	927.6 603.9	18497	.472	CICBTTTX		

4 per cent of the h-p compressor delivery air to cool the h-p turbine blading, mixing in again at station 5. For this and the next calculation, the value of T_{05} is taken after the mixing of this air.

3.5.2 Conversion to a Turbofan Engine

This is one of several possible conversions, two others being to turboprop or bypass-jet form (section 2.1). In this case, the power turbine is linked to the l-p turbine as extra stages, and the extra power is used to drive a large fan with its flow outside the l-p compressor. It could be in the form of very long blades at the first stage of the 1-p compressor. The inner 50 kg/s passes through the engine as before, but the outer 250 kg/s is pressurised to 1.6 bar at 90 per cent efficiency and allowed to exhaust directly. This represents a bypass ratio of 5 to 1. See figure 2.1*c*.

The engine has the high temperature of the previous example. Both the hot and cold nozzles are unchoked; that is, the total to ambient pressure ratios are less than about 1.85 and 1.89 respectively. The thrust calculation is handled as follows.

Outlet from	*Fan (cold)*	*Turbine (hot)*
p_{0e}/p_a	1.580	1.715
T_{0e}	334 K	995.8 K
C_{pe}	1020.1	1159.1
$\gamma/(\gamma-1)$	3.554	4.039
$T_{0e}/T_e = (P_{0e}/P_a)^{(\gamma-1)/\gamma}$	1.137	1.143
T_e	293.7 K	871.3 K
$V_e = \sqrt{[2C_p(T_{0e} - T_e)]}$	286.9 m/s	537.2 m/s
$\dot{m}$	250 kg/s	51.223 kg/s
$F_e = \dot{m}V_e$ (mass flow x velocity)	71725 N	27520 N

The total thrust, therefore, is 71725 + 27520 = 99245 N

$$\text{s.f.c.} = 3600 \times 1.223/99245 = 0.0444 \text{ kg/N h}$$

Using the gas pressure to drive the fan brings such great improvements as to explain the widespread use of the turbofan engine for subsonic applications. At high-altitude cruise, the fan thrust falls very significantly: the cruise value of s.f.c. is typically 0.065 to 0.07 kg/N h and is at its lowest value when the specific thrust ($F_n/\dot{m}_1$) is about 100 N s/kg.[33]

3.5.3 Other Aircraft Engine Cycle Variations

Water, or water–methanol mixture, is sometimes injected into engines used in hot climates. It may be injected at the compressor inlet or the combustion system. As it evaporates and is then heated to the mean turbine inlet temperature, it cools the gases. More fuel is therefore injected to restore the maximum permissible temperature level. The thrust is increased approximately in proportion to the total increase in mass flow (water and fuel), but the s.f.c. is significantly increased also. The effect of day temperature on thrust levels is shown in section 12.1.3. The injection of water can also significantly reduce the formation of polluting nitrogen oxides (section 7.3).

Reheat, or afterburning in the exhaust from the last turbine, provides a substantial increase in exhaust temperature, velocity and thrust (section 14.1): the value of $V_e/\sqrt{T_{0e}}$ (equation B.22) is almost constant even if T_{0e} is doubled, so V_e is increased by about 41 per cent. This suits military use, boosting the thrust/weight ratio for a substantial s.f.c. penalty, and is essential to supersonic flight to maintain an appropriate ratio of exhaust velocity to flight velocity (section C.6). Means of improving fuel economy are discussed in greater detail in references 8, 11, 33 and 34.

3.6 Industrial Turboshaft Engine Cycles

Industrial engines may be fitted with compressor intercoolers, regenerating heat exchangers and reheating burners to tune the cycle efficiency and power output as required. The weight and space requirement of these components rule them out for aircraft use. When a number of these units are fitted, it is convenient to use a code to describe the cycle, as used in reference 29. Compressors, combustion and turbines get the letters C, B and T as used before. Intercoolers are I, reheaters are R, and the heat exchanger X indicates that heat is fed to the compressed air immediately before combustion. The engines considered so far have been to the simple CBT cycle. The data for the following cycles is also shown in table 3.1.

3.6.1 Engine with Intercooling CICBTTT

This is the basic twin-spool engine (section 3.4) with the air cooled at station 2 to 288 K for the h-p compressor. The lower temperature means that the h-p compressor would be a smaller actual unit than that of the non-intercooled engine. The flow remains as before but a pressure loss of 5 per cent is assumed. The three T's indicate the h-p, l-p and power turbines, but the last two could be combined if the l-p shaft runs at the right speed for the load. The cycle is shown in figure 3.4*a*.

The h-p compressor power requirement is reduced to that of the single-shaft engine, thereby leaving a higher pressure and temperature after the h-p and l-p turbines. The power turbine is therefore able to extract much greater power

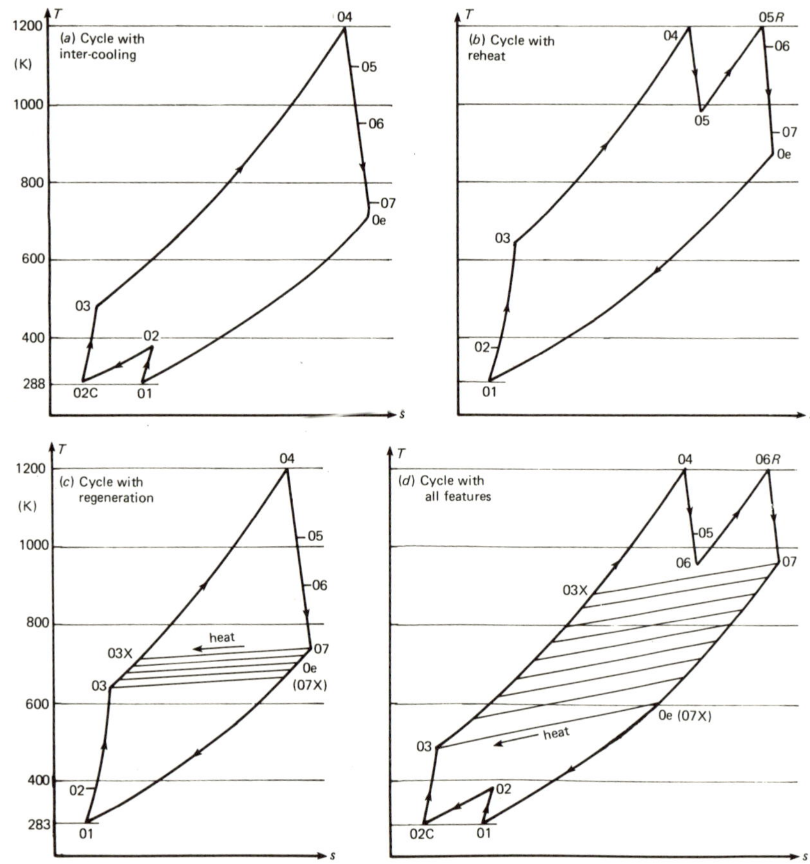

Figure 3.4 *Temperature–entropy diagrams for various cycles*

from the given size of engine. The reduced h-p compressor delivery temperature means that substantially more fuel is required to attain the maximum temperature of 1200 K, but the over-all efficiency is hardly affected. This calculation may be optimistic in the extent of the intercooling, which may not be able to lower the temperature as far as 288 K.

3.6.2 Engine with Reheat, CCBTRTT

The reheating combustion system is placed before the l-p turbine in this case, leaving the temperature and pressure at station 6 still high enough for the power turbine: see figure 3.4*b*. If the reheat were used after the l-p turbine, the pressure may have fallen so far that the expansion ratio would be too low and the exhaust too hot.

In comparison with line 2, the fuel input is the same at B (0.766 kg/s) but a further 0.319 kg/s is put in at R to raise the temperature from 979.3 K to 1200 K. A further 5 per cent pressure loss is assumed, reducing p_{05} at entry to the l-p turbine to 4.279 bar. The work output is increased while the efficiency is slightly reduced.

3.6.3 Engine with Regeneration, CCBTTTX

This cycle incorporates a heat exchanger, taking the gas from the power turbine outlet to heat the compressed air before combustion. The gas enters the hot passages at 732.6 K and the compressed air enters the cold passages from the other end at 647.5 K (figure 1.6*a*). The air cannot reach the level of 732.6 K, but may achieve 85 per cent of the temperature difference, as shown in figure 3.4*c*. Compare this with the 100 per cent of figure 1.2*c*.

$$0.85 \times (732.6 - 647.5) = 0.85 \times 85.1 = 72.3\ ^{\circ}\mathrm{C}$$

The temperature at the burner inlet is then 647.5 + 72.3 = 719.8 K. The reduced combustion temperature rise reduces the fuel input to 0.668 kg/s. The temperature drop to station e is 64.6 °C, less than the 72.3 °C temperature rise because the mass flow and specific heat are greater at station 7 than at station 3.

The increase in thermal efficiency relative to line 2 is disappointingly small in this case, for several reasons. The temperature difference between the exhaust and the compressed air is small, so only a small saving in fuel input is possible. The power output is acutally reduced, because the heat exchange reduces the pressure before combustion and at all subsequent stages: it could fall further still as the pressure loss after the 1-p turbine has been omitted. The regenerator may also have an adverse effect on exhaust pollution (section 7.3).

Figure 3.5 shows for a particular case how increasing the heat exchanger effectiveness reduces the pressure ratio and, thereby, the amount of machinery required to obtain the maximum efficiency at any given turbine inlet temperature.

3.6.4 Engine with Reheat and Regeneration, CCBTRTTX

This cycle improves on the last by increasing the temperature for the last two turbines, and then transferring heat to the compressed air during regeneration. There is a major increase in power output with a slight gain in efficiency. The power output and efficiency would be similar if the reheater were placed after the l-p turbine.

3.6.5 Engine with Intercooler, Reheat and Regeneration, CICBTTRTX

The addition of intercooling leaves T_{05} higher than in case 3.6.4, so that the

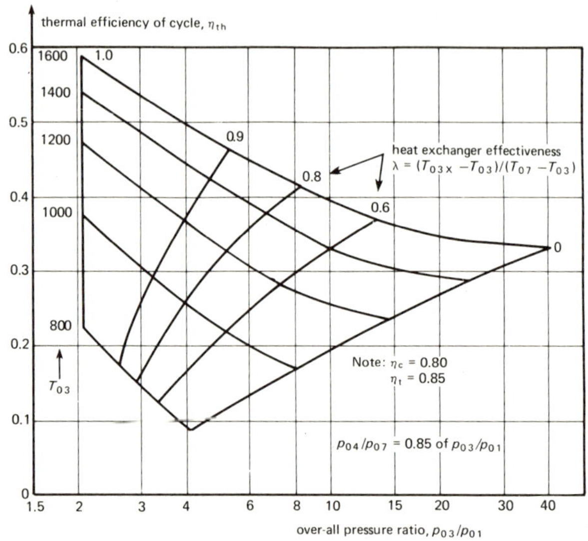

Figure 3.5 *The pressure ratio at which the maximum thermal efficiency is obtained, at various levels of turbine entry temperature and heat exchanger effectiveness. Note: engine as figure 3.4(c) with assumed compressor efficiency 0.8, turbine efficiency 0.85. Expansion ratio is 85 per cent of compression ratio*

available temperature increase would hardly justify the burner installation. This burner is therefore located before the power turbine. This cycle is the best so far, producing a greater work output and efficiency and leaving quite a cool exhaust: see figure 3.4*d*.

3.6.6 Engine with Intercooling, Regeneration, Raised Flame Temperature, CICBTTTX

This is very similar to cycle 3.6.5, except that the turbine entry temperature of 1500 K leaves T_{06} too high to reheat, at 1263.4 K. Whereas the h-p turbine may have expensive cooled blades, this is not warranted in the power turbine. Nevertheless, the power output and over-all efficiency are boosted further still, to levels which make this cycle very competitive in any power generation system. This represents an ideal, as continuous operation at such temperatures will not be possible for many years to come.

3.6.7 Engine with Exhaust Heat Recovery

This application is discussed in section 2.5. Using, as a rule of thumb, a boiler exhaust gas temperature of 430 K and a combined boiler and steam turbine efficiency of 0.20, the increase in over-all efficiency can be determined as follows. The available energy in the exhaust is given by $\dot{m}C_{p7}(T_{07} - 430)$. The steam turbine output is 20 per cent of this energy, say P_s. Then, the combined cycle efficiency is

$$\eta_{cc} = \eta_{th}(P_s + P_{67})/P_{67}$$

Examples of this calculation, using the cycles in table 3.1, are

cycle 2 :	$P_s = 3294$ kW	$\eta_{cc} = 0.397$
cycle 3.5a :	$P_s = 5610$ kW	$\eta_{cc} = 0.422$
cycle 3.6b :	$P_s = 5570$ kW	$\eta_{cc} = 0.392$
cycle 3.6c :	$P_s = 1843$ kW	$\eta_{cc} = 0.486$

Thus, efficiencies approaching 0.5 are feasible.

3.7 Miscellaneous Cycles

3.7.1 Energy-topping Cycle

This cycle goes a stage beyond 3.6g, in which the engine was primary and the boiler boosted the power output and efficiency.

In energy-topping applications, the boiler is the primary component, being integrated with a plant. If a gas turbine engine is mounted to exhaust into the boiler, then the boiler may continue to operate normally while some of the fuel is burned in the engine to produce useful power at high efficiency (section 2.5).

The calculation for the general case is as follows. Assuming that the over-all fuel/air ratio in the boiler remains as before, the additional fuel burned will be matched by extra air to the extent of $(1 + q)$ times the original flows. If fraction $1/r$ of the total fuel input is used by the engine at full load, and if the new engine and boiler fuel flows and the original boiler fuel flow are denoted by e, b and o respectively, then

$$(e + b) = (1 + q)o \tag{3.3}$$

$$re = (1 + q)o \tag{3.4}$$

Assuming that the heat input to the boiler remains as before

$$(1 - \eta_{th})e + b = o \tag{3.5}$$

Subtracting equation 3.5 from 3.3, dividing the result by 3.4 and rearranging yields

$$q = \eta_{th}/(r - \eta_{th}) \tag{3.6}$$

Using the typical values $\eta_{th} = 0.3$ and $r = 4$, then $q = 0.081$. Thus, the air and fuel flows are increased by 8.1 per cent and the output shaft power equals the additional energy input – an incremental efficiency of 100 per cent (section 2.5).

Different values apply in the case of the process combustion topping application (section 2.9). In plants where the steam is usually expanded through a turbine, the steam temperature is probably below 500 K even though the combustion temperature is 2500 K. The reason why the gas turbine recovers so much more energy from the same fuel input is that it extracts energy with a starting temperature of 1200 K, permitting a greater temperature drop (section 1.1).

3.7.2 Cooling Effect of the Turbine

The cooling in the turbine (section 2.9) can be calculated using typical data from section 3.6. The h-p turbine of the twin-spool engine extracts 13 500 kW from a flow of about 50 kg/s, with a temperature drop of 225 °C. The physical size of the turbine could be diameter, 0.7 m and length, about 0.15 m. Using a mean blade speed U of 400 m/s and an axial velocity of about $0.8U$(section 8.2), the time taken for the flow to pass through the turbine is 0.15/320 = 0.000469 s. The rate of loss of temperature is 225 x 320/0.15 = 480 000 °C/s. Normal variations of these parameters could make the cooling rate vary by 25 per cent up or down.

3.7.3 Air Cycle Refrigeration

The refrigeration systems described in section 2.10 are depicted in the T–s diagram of figure 3.6. The simple system could be used for blast freezing and

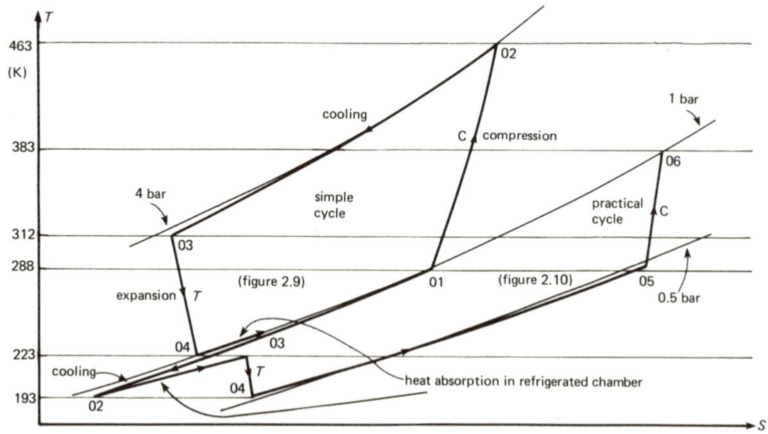

Figure 3.6 *Temperature–entropy diagram (for simple and practical air-cycle refrigeration systems*

works at above atmospheric pressure, while the practical system for maintaining a cold store works at below 1 bar.

The hypothetical simple cycle is calculated as follows, assuming a compression ratio of 4, compressor and turbine isentropic efficiencies of 0.8 and 0.9 a 5 per cent pressure loss in the cooler, and constant C_p = 1005 J/kg K. Taking in dry air at 288 K, the compression temperature rise is 175 °C. The air is then cooled to 312 K so the turbine may exhaust at 223 K (–50 °C). The effect of lowering the temperature by 65 °C from station 1 to 4 is equivalent to a power of 1005 x 65 = 65.4 kW (3720 Btu/min) per unit mass flow (kg/s). An electric motor or other engine is required to supply the difference between the compressor and turbine power, 1005 x (175 – 89) = 86.5 kW per unit mass flow, to compensate for the rotor inefficiency and the fact that compression takes place at a higher temperature than the expansion. The coefficient of performance (COP) of this cycle is

$$\text{COP} = \text{cooling effect/work input} = 65.4/86.5 = 0.756$$

This is a low value, compared with the level of 1 to 2 for standard refrigeration plant. It can be improved if the ambient inlet temperature is raised or the air delivery temperature is reduced. By heating water and providing a strong blast of air, it serves two plant functions which are normally separate from the refrigeration system. In a standard plant, fans for blast freezing may use as much power again as the refrigerating system.

These systems have several advantages over Freon or ammonia systems of the same cooling power. They are portable, much lighter and more compact and therefore cheaper. They can be started and stopped very quickly, and are independent of the supply and danger of refrigerating fluids.

3.7.4 Variable Mass Flow in a Closed Cycle

The usual way of varying the power output from the engine is to vary the fuel flow, and let the engine speed vary accordingly. Changing the speed may reduce the efficiency of the compressor and turbine, and may cause compressor operational problems (chapter 10). Alternatively, an engine which is maintained at constant speed by an alternator or other load (section 1.4) may not have the required turndown ratio, because its combustion must be maintained well within the rich and weak combustion stability limits (section 7.2) at constant air flow.

A closed-cycle engine may well achieve a greater turndown ratio and the power output range could be considerably broadened at constant shaft speed by varying the mass of working fluid in the system. Injection of more gas will raise the pressure, mass flow and power at virtually identical internal flow conditions and efficiencies. Such a proposal has been made in the past by Esscher Wyss.

4

Engine Operation

There is much more to engine operation than the design point performance covered in chapter 3. Some introduction to the broader aspects was given in section 1.4, and this is extended in this chapter. Later chapters will study the major components and the deeper aspects of operation and operating characteristics.

4.1 Engine Starting

The starting procedure for the engine has to take into account the starting procedure for the plant to which the engine is attached. The plant requirements, whether the correct setting of air valves and switches, a long period of preheat, correct phasing of an alternator with the local grid or other specialised activity may take precedence over the simple task of starting the gas turbine. A large installation may also have engine parameters which require monitoring during start-up, such as control of cooling water circuits and temperatures; the rate of engine temperature increase, which affects thermal stresses in bulky metal components and differential rates of expansion; the exhaust gas temperature; or the manual control of compressor bleed valves (section 11.4.3). The requirements vary from one installation to another but the principles involved apply generally: an example is discussed in section 12.5.

The variation of net engine shaft torque with rotational speed is shown in figure 4.1 for the acceleration from rest to idle. The negative torque at below the self-sustaining speed indicates that the compressor requires more torque than the turbine is able to exert. Both units are very inefficient at low speeds, when the airflow angles and the rate of contraction or expansion of passage areas do not match the rate of compression or expansion of the gas. Also, the low flow velocities limit the momentum and the work capacity of the flow.

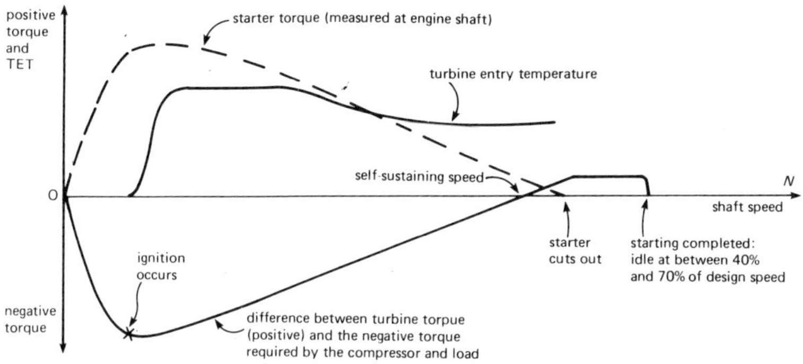

Figure 4.1 *The basic engine starting process*

The torque deficit is made up by the starter, which also provides additional torque to accelerate the shaft. In multi-shaft engines, the starter turns only the h-p shaft, so may be a smaller unit than is required by an equivalent single-shaft engine. The starter itself is commonly an electric or hydraulic motor or an air-driven turbine geared to the main engine shaft, or an air jet directed on to the engine's turbine blades. The former types normally disconnect by a freewheel device at greater than self-sustaining speed, commonly between 40 and 50 per cent of the h-p shaft's design speed. The air supply may come from another engine or from compressed-air bottles or a cartridge, in which case it is sometimes passed through a combustor to burn fuel and increase the starter turbine power. An aircraft engine which loses combustion in flight may be restarted using the windmilling effect to turn the shafts.

The starting procedure can significantly affect the engine turbine's life. The ignition plugs should be firing before the fuel is admitted, to ignite the fuel immediately and prevent fires and overheating in the turbine and downstream ducting. The turbines should be rotating before the fuel is admitted, to avoid burning the blades downstream of the initial ignition zone; the higher the speed before light-up the better. Standard checks include monitoring the temperature and measuring the time to light-up from fuel admittance. Above self-sustaining speed the starter system cuts out and the engine continues to accelerate to its idle speed: the time to this stage is also recorded. Failure of the starter to disconnect automatically at the appropriate speed is the last starting risk, which can result in severe damage if not detected before the engine is accelerated to operating speed. After completion of the start, the ignition system is usually switched off unless bad weather is considered likely to cause a flame-out by water ingestion. This procedure may be carried out and monitored automatically for some engines (section 12.4).

Two conditions may prevent successful completion of the start. Inadequate starter torque or fuel supply, or failure of the combustion system to light round (section 7.4) may result in a *hung start*, in which the engine fails to reach self-sustaining speed. Increasing the fuel supply may alternatively cause a *hot start*, involving compressor malfunction (section 6.4.3) and turbine overheat. An engine with inadequate surge margin (section 11.1) for low-speed acceleration (see sections 11.5 and 11.6.1) may hang when starting from cold, while its structure is absorbing heat from the flow, but overheat when starting while still hot from a previous run. No single control adjustment could help this engine, but the hot start may be avoidable by blowing off some air from the h-p compressor delivery. The main solution to starting problems is the use of an adequate starter, with good connections to ensure an adequate supply pressure or voltage.

The *idle speed* is chosen to balance several factors. It is a safe margin above the self-sustaining speed to avoid unexpected run-down, but not high enough to use fuel excessively or generate thrust for an aircraft. It is located beyond the peak of exhaust gas temperature, which results from severe inefficiencies, but an axial compressor may be in a rotating stall condition (section 10.4) until it reaches a speed well above idle, with resultant high vibrational stresses in the blades (section F.3). Finally, the idle speed may be set to avoid a speed causing blade resonance.

4.2 Performance through the Speed Range, Off-design

The engine is designed to operate most efficiently at one condition, usually on load for an industrial engine or at cruise in an aircraft. Regular operation at other speeds is also quite common and could be calculated, as at design point, with knowledge of the component characteristics. For brevity, only the typical results of such calculations will be presented: they could equally well be derived from engine testing (section 4.3).

The variation of the measured parameters against h-p shaft speed is plotted for a typical split-shaft engine in figure 4.2, the measurements being taken at steady speeds and plotted non-dimensionally (chapter 3). This engine has separate hot and cold flow outlets, but they could have been mixed before a common nozzle with little change to the appearance of the curves. It is apparent in this example that at h-p speeds below about 80 per cent the useful output is negligible, and the performance increases rapidly with speed above 90 per cent. One reason for the low performance at speeds below 80 per cent is that the h-p compressor mid-stage bleed valve is open (section 11.4), without which the compressor would be in surge. The kink point in the relationship between the two shaft speeds occurs where the variable h-p compressor inlet guide vanes (section 6.1) reach their fully open stop at 675 $N_2/\sqrt{T_{01}}$, the geometry being fixed above that speed (section 11.4).

A common alternative form of performance graph involves plotting parameters in dimensional form, but corrected to the values that would be measured on an

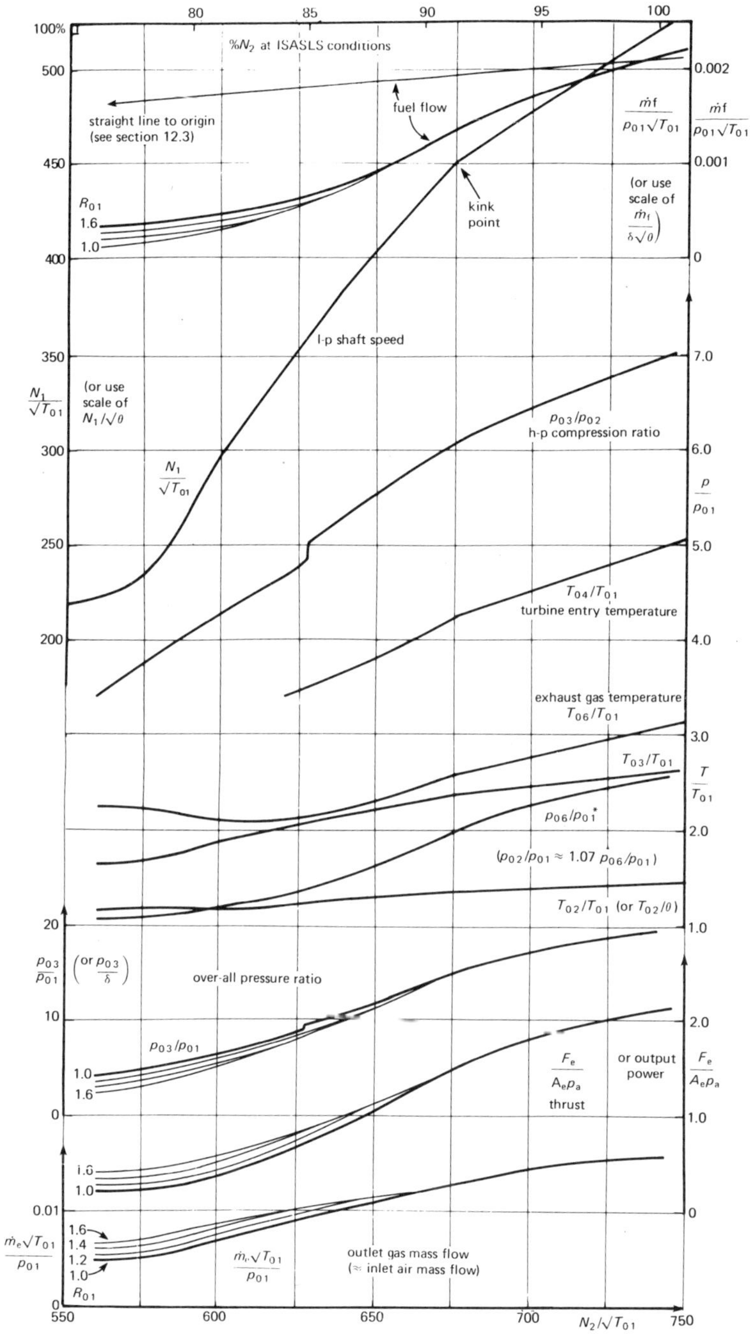

Figure 4.2 *Typical performance curves for twin-spool, bypass aircraft engine*

I.S.A. day. Pressure and temperature *correction factors* are used in the form $\delta = p_{01}/1.013$ and $\theta = T_{01}/288$. Pressures and temperatures are then plotted as p_{0n}/δ and T_{0n}/θ, and the speed as $N/\sqrt{\theta}$. The performance is therefore presented as standard values rather than the ratios used in this book.

Two particular aspects of the use of dimensionless curves deserve mentioning. Consider operation of an aircraft engine at 10 000 m altitude and Mach 0.8. The engine inlet conditions may be determined from section B.4.5 and equation B.16 as

$$\text{static temperature } T_a = 288.15 - 65 = 223.15 \text{ K}$$

$$\text{static pressure, } p_a = 1.013\,(223.15/288.15)^{5.25} = 0.265 \text{ bar}$$

$$\text{ram temperature ratio, } T_{01}/T_a = 1 + 0.2 \times 0.8 \times 0.8 = 1.128$$

$$\text{ram pressure ratio, } R_{01} = p_{01}/p_a = 1.128^{3.5} = 1.524$$

The first aspect is the change in *relative shaft speeds.* At the value $N_2/\sqrt{T_{01}} = 680$, the ISASLS value of N_2 is 11540 rev/min. The corresponding l-p values are 455.5 $N_1/\sqrt{T_{01}}$ and 7730 rev/min. At altitude, where $T_{01} = 251.7$ K, the value of $N_2/\sqrt{T_{01}}$ at 11540 rev/min is 727.4. From figure 4.2, the corresponding value of $N_1/\sqrt{T_{01}}$ is 506.2 so the l-p shaft speed is 8030 rev/min. Thus the l-p shaft speed rises relatively to h-p speed at altitude for this typical engine, and it does so because this part of the speed match characteristic is aligned with a point below the origin.

The second aspect is that of the *feathers*, the spreading of the curves at lower engine speeds. It occurs where the engine no longer behaves in a dimensionless fashion, below the speed at which the final nozzle chokes. At ISASLS conditions, this occurs where $p_{0e}/p_a = 1.852$, $N_2/\sqrt{T_1} = 87$ per cent. If the engine is slowed to $N_2/\sqrt{T_{01}} = 77.7$ per cent when flying as above, then the nozzle is still choked because its pressure ratio is $R_{01} \times p_{0e}/p_a = 1.524 \times 1.215 = 1.852$. Thus, *ram pressure* at low engine speeds increases the final nozzle pressure ratio and its non-dimensional mass flow. In contrast, when the nozzle is already choked, as at higher engine speeds, increasing ram pressure leaves the critical pressure ratio p_{0e}/p_e constant, so the non-dimensional mass flow $\dot{m}\sqrt{T_{0e}}/Ap_{0e}$ remains constant (section B.5).

It can be shown that, when running at constant engine pressure ratios, increasing ram pressure with an unchoked nozzle increases the actual and non-dimensional air mass flow; increases fuel flow in proportion to air flow but decreases non-dimensional fuel flow; and increases the nozzle thrust F_e (equation 14.3). There is also a slight increase in ram temperature, but the windmilling effect makes a greater increase in shaft speed so that the non-dimensional speed also rises: at constant $N/\sqrt{T_{01}}$, therefore, increasing ram pressure slightly reduces the internal pressure ratios. Figure 4.2 shows the feathers with exaggerated vertical scales: the thrust curve could be reduced to a single line by plotting the parameter $(1 + F_e/A_e p_a)/R_{01}$ (equation B.29).

Other parameters could be used instead of h-p shaft speed along the x-axis, as a base for plotting the performance. A common one for aircraft engines is thrust, the parameter for which the engine is sold and the most important requirement for the performance of the aircraft. An equivalent for a turboshaft engine would be its shaft power, which varies with speed in a similar manner to thrust.

4.3 Engine Testing and Correction of Data

Engines are tested mainly to demonstrate that they have been built properly and that they achieve the specified performance objectives. In heavy installations, this is part of the plant commissioning process and may be the first occasion on which a new engine design is run, as the engine is not assembled until it is located at its intended site. Smaller engine types are run extensively at the manufacturer's plant before delivery to the customer and also after installation in his equipment. At this stage, the testing ensures that all systems are working and that the engine is satisfactorily connected to the fuel supply, controls, ducting, etc. It may be repeated intermittently to check for faults or make periodic control adjustments.

In many cases, special development engines that will not be sold are built for bench testing, to prove the engine type. This permits acceleration of the fault-finding process, using engines built to assess different facets such as performance; vibrations and stress levels; oil, fuel and air systems; the life, reliability and operation of the major components; handling characteristics; the customer's ancillary equipment; etc. Engines for passenger transportation, particularly in aircraft, are required by local authorities to pass a *type test* to demonstrate satisfactory operation in all circumstances. A typical type test lasts 150 hours, with steady running at several speeds throughout the speed range, fast accelerations and decelerations and extensive running at maximum speed (section 14.6): engines subsequently built to the same standard are then cleared for use.

The testing may take place on site or installed in the open air, or in special test beds. Standard precautions are taken before a test run to check the safety and completion of all connections; to check and calibrate instrumentation, to inspect the intake and exhaust for signs of damage or debris; to check for leaks; to sweep the floor or ground in the cell or test area and to remove larger loose objects such as tools and flashlights which could be ingested by the engine (section 6.4.2). The dangerous region behind the engine must also be cleared for open-air running.

A typical test cell is shown in figure 4.3. Air is drawn in through splitters, made of acoustically absorbent material and with crooked passages to prevent direct noise transmission. Some of the air passes through the engine as working fluid, while more air flows past the engine to be entrained by the exhaust gas and cool it for the quietening system. A detuner is shown, which acts as a large

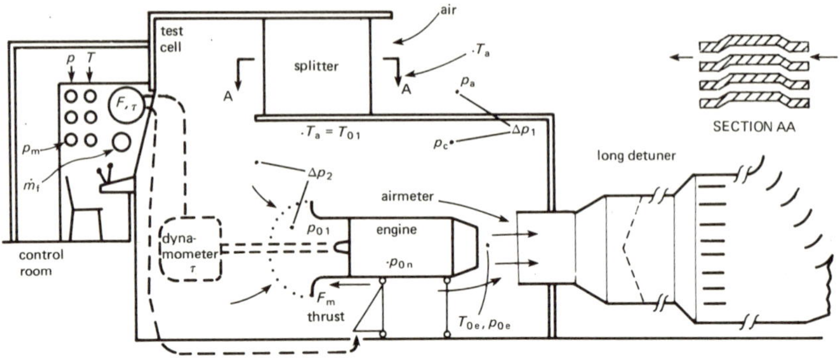

Figure 4.3 *Typical test-cell arrangement, with optional dynamometer shown dotted*

muffler, but splitters similar to those at the intake may also be used, particularly with turbofan or turboprop engines (section 13.5).

The engine is bolted to a floating cradle on the test stand and is connected if applicable to a dynamometer or test propeller. The test stand for aircraft engines is often suspended from the ceiling, leaving the floor clear for access. The floating cradle moves in response to engine thrust or torque, and is restrained by a weighgear linkage or strain-gauged link which transmits the signal to the thrust or torque indicator in the control room. This room is located for good visibility but not in the plane of the main rotors: its window is made of armoured glass to exclude pieces of a failed engine.

Downtime in the cell may be reduced by pre-rigging; that is, connecting an engine to a test frame in a separate room while another engine is on test. When the new engine goes in for test, the frame is easily attached and the instrumentation wiring and plumbing joined up with plug-in connections. This permits a large proportion of the test-cell time to be used in actual testing.

The test cell itself has to be calibrated, to determine correction factors which will adjust the measured performance. Two main corrections are required, the first allowing for cell depression. The air entering through the intake suffers a pressure loss Δp_1, which may amount to 10 or 20 mm of water gauge. This means the engine runs in a slightly low atmospheric pressure, and the engine inlet pressure may be still lower (by Δp_2) if a wire-mesh screen is fitted over the airmeter. The second is that the air flowing past the engine and test cradle creates an aerodynamic rearward drag, F_d, which reduces the thrust load on the weighgear. This drag is somewhat proportional to test cell depression, as a greater depression implies a greater flow past the engine.

The correction of test-cell measurements is established by running a calibration engine on an open-air test stand on a still day, and then again in the cell. The differences are then applied to the measured data as follows, noting that the pressure gauges or manometers in the control room read relative to the outside ambient pressure.

The *pressure ratio* at any station n is given by p_{0n}/p_{01} and is found from the pressure measurement p_m by

$$\frac{p_{0n}}{p_{01}} = \frac{p_m + p_a}{p_a - \Delta p_1 - \Delta p_2} = \frac{p_m + p_a}{p_c - \Delta p_2} = \frac{p_m + p_a}{p_{01}} \tag{4.1}$$

The *thrust* is corrected to

$$F = F_m \times p_a/p_c + F_d \tag{4.2}$$

The non-dimensional form of pressure ratio permits it to be plotted against *speed* in the form of $N_m/\sqrt{T_a}$, or $N_m \times \sqrt{(288/T_a)}$. The unique curve of pressure ratio against non-dimensional speed may then be applied to any operating environment. The standard, dimensionless forms for thrust, fuel and and air mass flows, and temperature are given in section B.5. Further information on testing is included in references 35 and 36.

4.4 Aircraft Engine Operation

Airline operations follow well-established patterns and the engine history is required to be well controlled and documented. Engine use is recorded in terms of hours and of flight cycles, one cycle typically involving the starting; taxi out; take-off at high engine speed; climb, cruise descent and landing at successively lower speed levels; reverse thrust at high speed again; and then taxi-in and shut-down.

In operation, the engine deteriorates as a result of both the hours and cycles. The exposure of the compressor to dirt and grit (section 6.4.1) and of the turbine to prolonged heat effects (sections 8.3 and 8.4) are largely a matter of time, but thermal fatigue aspects of turbine deterioration result from the number of times the temperature is changed significantly. As the average flight length for different operators may range from 30 min to 8 h, the short haul engine spends a greater proportion of its life at high temperature and its hourly life is therefore low. Although the operating life may be very short on new engine types, it is extended with experience and with engine development until most engines are capable of reaching about 4000 hours or flights, whichever comes first.

The other major life limiter is the permitted *cyclic life* of the main rotating components, discs and shafts. Weight limitations dictate that these be stressed well above their material fatigue limit at take-off speeds, so the design life is typically between 10 000 and 20 000 flights.[37] Engines new to service may have some components with much lower cycle lives. When stripped for repair and overhaul, those components which will reach their life limit before the combustion system or turbine is expected to require further attention are rejected and new components fitted. Occasional engines are built with short life components to use them up but, when *time expired* (T-X), the components are scrapped. The determination of the safe cyclic life is discussed in section 14.5.

Apart from scheduled removal for T-X components, most engines are run on condition, with frequent inspection of the rate of deterioration of known areas. The hourly parts cost is minimised by well-judged removal times, and severely boosted by the secondary damage resulting from a failure. Unscheduled removals result from various types of failure, including blade or disc rim breakage, turbine nozzle deterioration, high oil consumption (section 14.4), bearing and seal failures, and accessory problems which result in faulty operation (chapter 12). Some operating problems may be corrected by overnight maintenance, but may cause delay or cancellation of a flight. Some require an engine to be shut down in flight. A typical objective is to keep the shut-down rate and the unscheduled removal rate both below 0.5/1000 h, averaged across the fleet.

Cost control is essential, in operation and at overhaul. Fuel consumption is reduced by flying at the highest altitude possible and at a lower speed than could be achieved. Engine efficiency is restored by periodic internal cleaning (section 6.4.1). The turbine life is prolonged by the use of flexible take-off power, that is, the minimum power level consistent with the conditions of aircraft weight, runway altitude and length, and ambient temperature. A 20 K reduction in turbine entry temperature could even double the h-p blade life, and component cyclic lives could be increased.

The airline is restricted by the requirements of the local aviation authority, and is required to work to the manufacturer's extensive maintenance, repair, overhaul and operating manuals. Numerous modifications are introduced as a result of the manufacturer's development programme, a very few of which may be considered essential for safety to the extent that the fleet is grounded until the modification is embodied. Most modifications are incorporated at repair or overhaul and some are not taken up, being production improvements and rationalisation on later engines.

5

The Centrifugal (Radial) Compressor

The centrifugal compressor is one type of dynamic compressor, another being the axial compressor which is covered in chapter 6. As introduced in section 1.3, the dynamic compressor works by conversion of velocity to pressure in a continuous flow. This distinguishes it from the reciprocating compressor, which compresses the fluid intermittently in separate portions, or the rotary compressor of vane or screw or other type which compresses continuously in portions separated by moving walls or boundaries.

In the dynamic compressor, the passages through which the fluid flows are continuous and contain no barriers across the flow path. The centrifugal compressor consists of a simple rotor which scoops in the working fluid and flings it into a peripheral ring of fixed stator passages. The divergence (increasing cross-sectional area) of these passages diffuses (slows) the flow to a lower velocity and higher static pressure (section B.2). Having no barriers, the flow can reverse its direction in extreme circumstances (section 10.1) if the outlet pressure rises too high for the operating speed and flow.

The features of the centrifugal compressor are the result of very intensive study and development, making discussion of the design and operation inseparable (section 5.1). The deeper aspects of performance and flow in the impeller and diffuser are covered separately in sections 5.2 and 5.3, and the operating characteristic is covered in chapter 10 along with the axial compressor. The centrifugal compressor gives little trouble in service, although the performance may suffer from dirt deposits (section 6.4.1). The other major problem is breakage resulting from vibrations (section 5.4).

5.1 Design and Operation

Figure 5.1 illustrates a typical centrifugal compressor. The *impeller* (rotor) has a

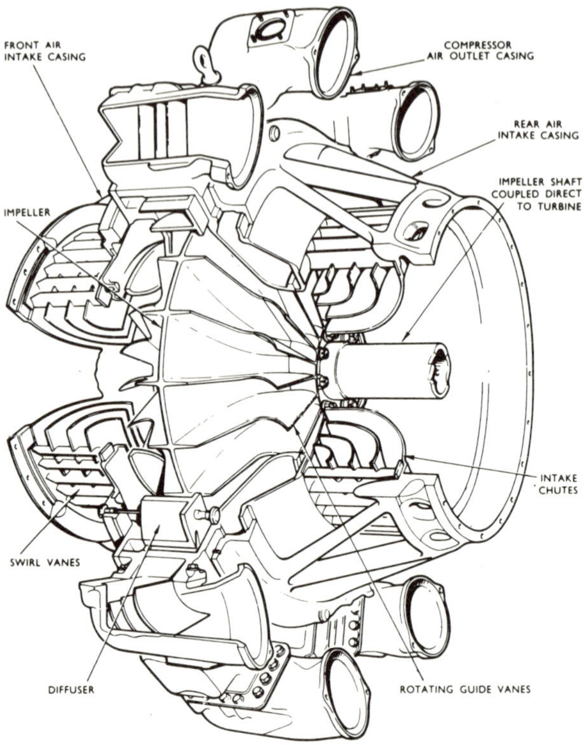

Figure 5.1 *A double-sided centrifugal compressor (by courtesy of Rolls-Royce Ltd.)*

series of vanes which take an axial flow of fluid and expel it radially. At the eye (inlet), the vanes are curved to induce the flow: this axial portion is called the inducer or rotating guide vanes and may be integral with or made separately from the main impeller. The impeller vanes are aligned radially, but may be curved near the tip (section 5.2 and figure 1.4*b*). The impeller may be double sided as shown, to pass the maximum flow relative to its diameter and to balance the stresses, or single sided as in figure 5.2.

The vanes are cast, or forged, or machined integrally with the impeller hub, which is the name given to the curved disc extending from the eye root to the outlet. The number of vanes is usually a prime number, typically from 19 to 37, to avoid vibration problems (section 5.4). Half vanes are sometimes used towards the tip, as shown in figure 5.2, to improve the flow pattern where the full vanes are widely spaced (section 5.2). The outer curve of the vanes is sealed by the shroud, which may be part of the stationary structure or may rotate with the rotor. The impeller material is often aluminium, with titanium or steel for smaller, high duty machines. Typical rotor proportions are that the eye root diameter is about half the eye tip diameter, while the outlet diameter is nearly twice the eye tip diameter (section 5.2).

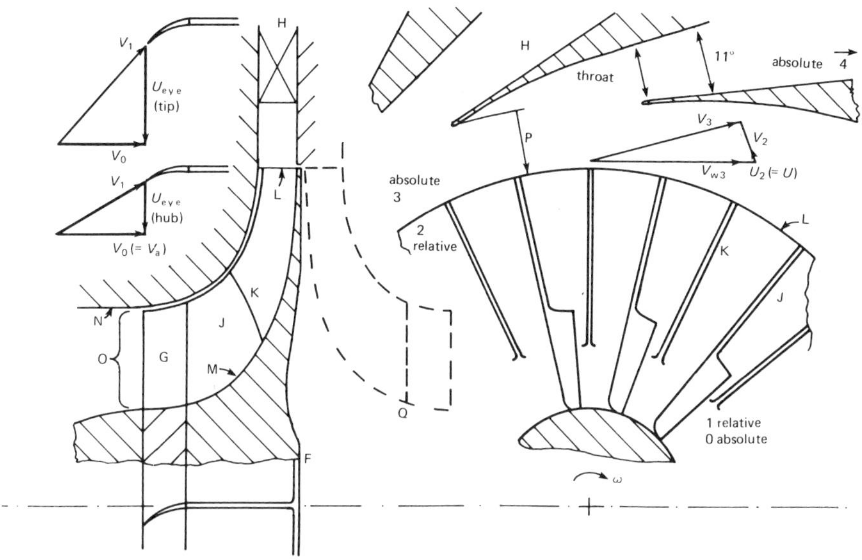

Figure 5.2 *Nomenclature and velocity triangles for a single-sided compressor with channel-type diffuser. F impeller; G inducer (rotating guide vanes); H diffuser; J impeller vane; K half vane; L impeller tip; M impeller hub; N shroud (casing); O impeller eye; P vaneless gap; Q addition of second side to rotor*

The *velocity triangles* for a typical impeller are shown also in figure 5.2. The flow at approach to the eye may be axial, V_a, with a typical velocity of 150 m/s for air, or may have a swirl angle (prewhirl). The flow velocity adds vectorially to the tangential impeller velocity U to give the relative velocity V_1, which varies in magnitude and direction from eye hub to tip because U varies with radius. This requires the inducer vanes to be twisted, to align with the flow at all radii. It is normal for the relative Mach number (section B.4.3), at the tip to be below 0.9, and for it to fall to below 0.5 for most of the subsequent passage length to minimise frictional losses

The tip tangential velocity U_2 at outlet is slightly greater than the whirl velocity V_{w3} of the flow, the difference being called the *slip* (section 5.2). This causes the relative velocity to be angled backwards relative to the vanes. Adding V_2 vectorially to U_2 gives the absolute outlet velocity V_3, which is greater than V_0 and represents the effect of the work input to the rotor shaft (section 5.2).

The fixed structure involves the diffuser, shroud, casing and bearings. Swirl vanes may be fitted in the inlet duct, and may swivel about radial axes to vary the swirl angle with speed. They keep V_1 aligned with the inducer leading edge, resulting in minimum incidence or shock angle (section E.4). Their effect is to widen the operating speed range, increase efficiency and reduce the relative Mach number at the eye tip, but they also reduce the mass flow by reducing the inlet area.

Several types of *diffuser* are used, of which the channel type is shown in figure 5.2. Others include the vaneless, cascade, vaned and pipe diffusers, which are illustrated in figure 5.3*a* to *d* respectively. All diffusers have an initial vaneless gap outboard of the impeller, in which the flow settles. The vaneless diffuser is bulky and can be inefficient relative to the other types. The cascade may involve one, two or three rows of vanes, with performance akin to that of axial cascades (section E.3). Vaned diffusers may use curved or straight vanes, having a longer path length than the cascade diffuser. Both these types may lead into a volute-type collector. The pipe and channel diffusers collect the flow in separate passages, sometimes with corner vanes at their outlet. The passages diverge slowly, the pipe consisting of a conical hole and the channel type being rectangular in section with one pair of opposed walls diverging at 10 to 11°, the others being parallel. The pipe and channel diffusers may feed separate combustors, or may merge downstream. Of all these types, the vaned diffuser may be made variable, the vanes pivoting about an axis such as that shown in figure 5.3*c* to accommodate different incident flow angles through the speed range. This also permits the flow capacity of the impeller and diffuser to be matched through the whole speed range (section 5.3). The choice of diffuser type has a significant effect on the compressor noise (section 13.5).

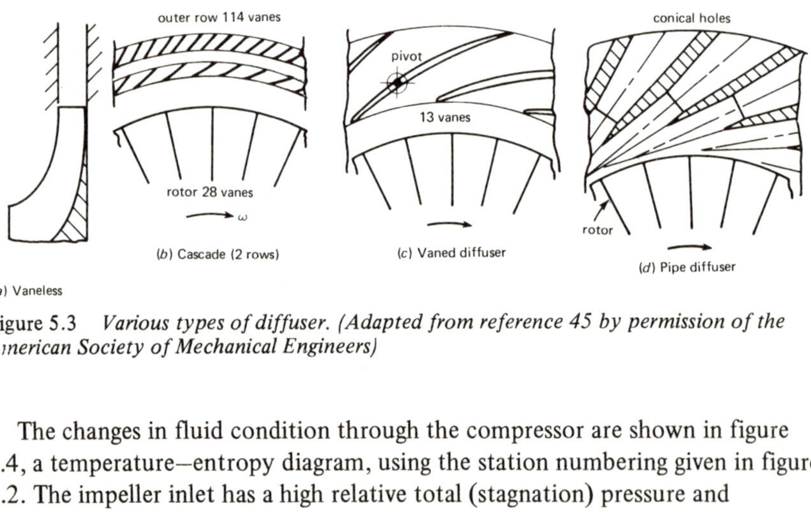

Figure 5.3 *Various types of diffuser. (Adapted from reference 45 by permission of the American Society of Mechanical Engineers)*

The changes in fluid condition through the compressor are shown in figure 5.4, a temperature–entropy diagram, using the station numbering given in figure 5.2. The impeller inlet has a high relative total (stagnation) pressure and temperature, but only the same static conditions as the approach flow. The difference between the total and static temperatures is the velocity equivalent, $V_1{}^2/2C_p$, from equation B.13. At impeller outlet, the static conditions are the same for the relative and absolute velocities but the absolute total conditions are higher. This total temperature, T_{03}, is the highest level reached and equals T_{04} as no further work is put in. However, the diffuser losses add to the entropy (sections B.3 and C.2) and reduce the pressure to its outlet value.

Nominally, the rotor puts energy into the fluid in the form of velocity while the stator converts it to pressure energy. In practice, the impeller also generates some of the static pressure rise. It can be shown that, if the relative velocity in the impeller remained constant throughout, the static temperature rise in the impeller would be about half the total temperature rise – a condition of approximately 50 per cent reaction (section E.3). As the velocity reduces through the impeller (section 5.2), the centrifugal compressor has a higher degree of reaction than 50 per cent.

The stagnation pressure ratio of a centrifugal compressor, shown as p_{04}/p_{00} in figure 5.4, is usually between 3 and 5 to 1 for normal good designs. The upper limit applies particularly with aluminium impellers, because the material stress limits the permissible tip speed U_2 to about 450 m/s (section 5.2). The efficiency (section C.2) is usually between 0.7 and 0.8, although a few are higher [6B]. Advanced designs for small, high duty applications use stronger materials, higher tip speeds and precise control of supersonic flow, with shock waves to generate pressure ratios of over 10 to 1 at similar efficiency levels. Typical flow Mach numbers are 1.1 to 1.3 at station 1, and 1.0 to 1.4 at station 3 [38A].

The design process remains a mixture of art and science, many good compressors having been designed using empirical factors and a good eye for shape. Whittle [2] achieved dramatic improvements over pre-1939 levels of efficiency by attacking the several causes of loss, and early work at NACA on axially long, parabolically curved inducer vanes continued this phase of development. Intricate flow patterns occur in the impeller which defy full mathematical analysis, although several methods of calculation are summarised in reference 6B. The comparatively recent development of high performance, advanced compressors indicates that the flow patterns can at least be catered for, if not calculated.

5.2 Impeller Performance and Flow Patterns

Much of the understanding of the flow patterns in the impeller derives from Fowler's work with a large low-speed rotor, with which he rotated inside the transparent hub.[39] A summary of the flow patterns is given at the end of this section but only the simple effect of impeller rotation will be considered initially.

The flow is in nominally straight, parallel streamlines as it approaches the impeller eye. Having no initial spin or rotation, it remains substantially irrotational as it passes through the impeller, while the passage rotates about it (figure 5.5). Relative to the passage, this appears as a circulating flow which, when superimposed on the nominally even radial flow, results in an uneven velocity gradient and pressure gradient across the passage, as shown exaggeratedly in figure 5.6. Although this is distinctly simplified at this stage, the effect of the circulation backwards relative to the impeller motion at its outlet (tip) reduces the whirl

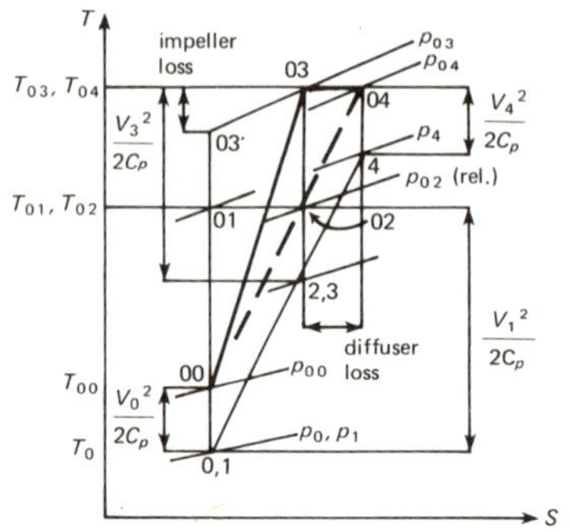

Figure 5.4 *Temperature–entropy diagram for a centrifugal compressor*

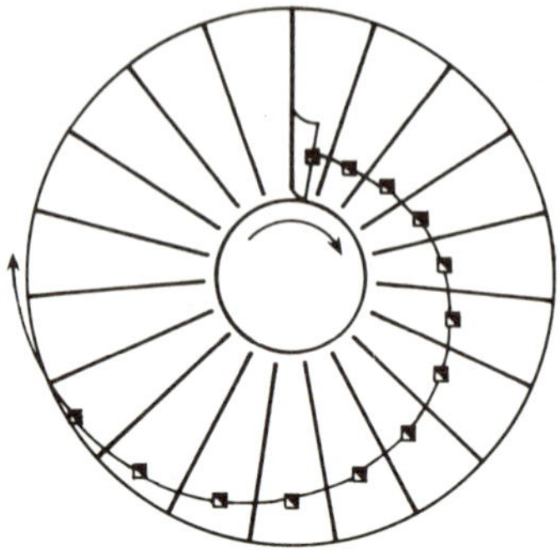

Figure 5.5 *Successive orientations of an element of the flow as it moves through a rotating passage*

velocity of the flow relative to the rotor and results in the slip. This is often quantified by the *slip factor*, σ, which is defined as

$$\sigma = V_{w3}/U_2 \tag{5.1}$$

The value of σ can be related to the number of vanes, n, in the impeller: the most appropriate value remains that determined by Buseman's tests [6B] and expressed empirically as $\sigma = 1 - (2.4/n)$. Slip has also been found to vary with mass flow [40], as discussed later. The effect of using half vanes is dubious but the value of n would double, yielding an appropriately lower slip.

The driving torque applied to the rotor generates a circumferential pressure gradient across each passage (figure 5.6*d*), as each vane pushes the first layer of molecules, which pushes the next layer, etc. The pressure gradient across adjacent passages causes a pressure difference across each vane, which exerts a tangential force in reaction to the applied torque. The impeller puts work into the flow by accelerating each flowing element of the fluid in a tangential direction until it reaches its final whirl velocity V_{w3} at outlet. Radial vanes are unable to apply

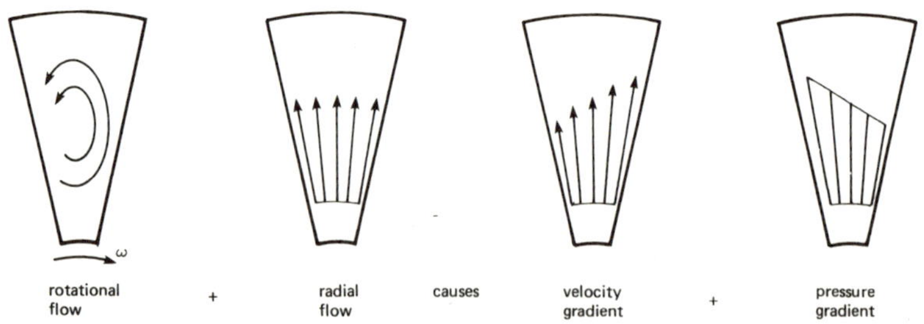

Figure 5.6 *Simplified indication of velocity and static pressure gradients across an impeller passage*

any radial force on the flow except for a small inward force caused by boundary layer drag.

The tangential acceleration is named after *Coriolis* and is the inevitable result of rotating a flow vector about its source. The velocity diagram in figure 5.7 illustrates this, the source O supplying elements of the flow such as that at P. As the element moves outwards to a nearby position at Q, there is a conflict between the higher whirl velocity imposed by the increased radius and the lower whirl velocity it would assume in a natural free vortex condition. (Whirl velocity times radius is constant, making the angular momentum constant at all radii.)

Consider the case where OQ is a 1 per cent greater radius than OP, and rotation is at ω rad/s. The whirl velocity at P is ωR: at Q it is 1 per cent greater where it would prefer to be 1 per cent less. The forced velocity increase is therefore approximately 2 per cent of ωR. The time taken to move from P to Q at a mean radial velocity of V_r is the distance divided by the velocity, $0.01R/V_r$. The Coriolis acceleration is the change in velocity divided by the time, that is: $(0.02\omega R) \div (0.01R/V_r) = 2\omega V_r$. It acts in the direction of rotation if the flow is outwards, or against the direction of rotation if the flow is inwards (section 9.1).

The torque applied to drive the impeller is quantified by the change in angular momentum between inlet and outlet: the SI units involved are discussed in appendix A. Assuming no inlet swirl (prewhirl) so that $V_{w1} = 0$, the torque is

$$\tau = \dot{m}(V_{w2}R_2 - V_{w1}R_1) = \dot{m}V_{w2}R_2$$

This is the maximum torque, as it is not reduced by any inlet swirl. The *power input* is given by torque times rotational speed

$$P = \tau\omega = \dot{m}V_{w2}U_2 \qquad (R\omega = U)$$

Introducing the slip factor from equation 4.1 and a *power input factor* ϕ (discussed next) yields

$$P = \dot{m}\phi\sigma U_2{}^2 = \dot{m}C_p(T_{03} - T_{00}) \tag{5.2}$$

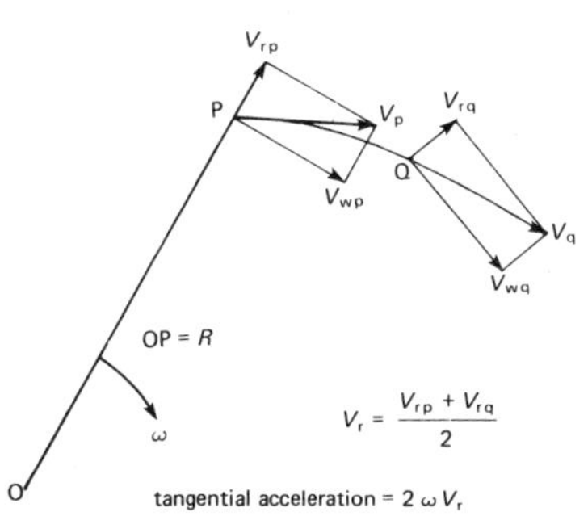

Figure 5.7 *Tangential (Coriolis) acceleration of a particle*

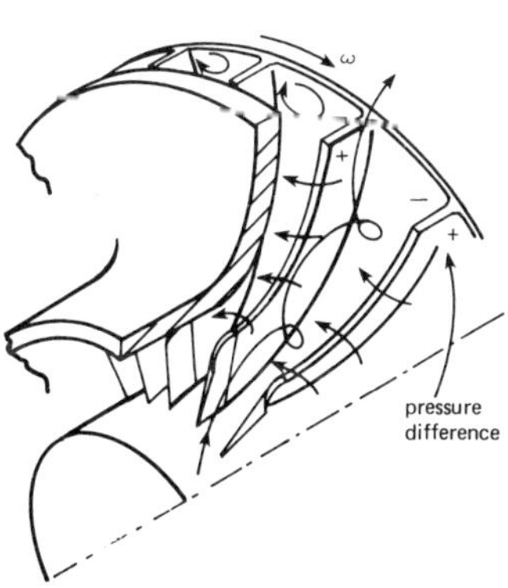

Figure 5.8 *The main flow patterns within a centrifugal compressor*

The numbering for stations 3 and 0 applies to figure 5.2. In consideration of the compressor as a unit, between engine stations 1 and 2, the last term in equation 5.2 would be $\dot{m}C_p(T_{02} - T_{01})$, which will be used in subsequent equations.

Whereas the slip factor is typically about 0.9, less than unity, the power input factor is usually about 1.035 to 1.04. It arises from the power wastage which drives some secondary flows. The main one is a helical flow up each passage (figure 5.8), caused by drag from the tangential motion of the fixed shroud relative to the impeller channel. There are also vortices from the inducer, and leakage flows through the clearance between the impeller vanes and the shroud. Reference 40 gives formulae for calculating the value of ϕ. Fowler [39] found that, at an optimum value of clearance, the leakage flow improves the radial velocity distribution in the next passage: a suitable clearance was found to be about 5 per cent of the axial width of the passage at the impeller tip. It has been shown that the clearance could rise satisfactorily to as high as 10 per cent of the passage width.[41]

It can be seen that, apart from the correction factors ϕ and σ, the power input can be expressed simply in terms of impeller tip velocity and flow. A standard alternative expression is the *work done*, which is the power input per unit mass flow (using engine station numbering)

$$W = P/\dot{m} = C_p(T_{02} - T_{01}) = \phi\sigma U_2{}^2 \tag{5.3}$$

The stagnation temperature rise leads to the over-all stagnation pressure ratio generated by the machine

$$r_p = \frac{p_{02}}{p_{01}} = \left[1 + \frac{\eta_c(T_{02} - T_{01})}{T_{01}}\right]^{\gamma/(\gamma-1)} \tag{5.4}$$

η_c is the isentropic efficiency (section C.2) which allows for the losses in the impeller and diffuser (section 5.3). It is the ratio of useful energy input to total energy input: the difference between these is the wasted energy in the various circulating flows of figures 5.6*a* and 5.8, and in the diffuser losses. It has been shown theoretically how the impeller efficiency may well exceed 0.9 if prewhirl and splitter (half) vanes are used; it is lower without these features and it also falls as the hub/tip ratio is increased.[41]

The over-all impeller flow pattern can now be considered. Fowler [39] showed that the relative flow velocity profile at the impeller outlet was uneven, as shown in figure 5.9*a*: no indication of the flow direction was obtained, but the velocity profile contradicts that of figure 5.6*c*. In later work [42], he tested the outlet flow from a channel with the neighbouring channels blocked, and measured a very even velocity profile, as shown in figure 5.9*b*. This clearly shows that interaction between the passages is significant.

Whitfield [40] linked the slip to the part of the passage which is usefully filled by the flow, dividing the flow into the high-velocity jet and the turbulent separated wake from the suction side of each vane. These regions are shown in

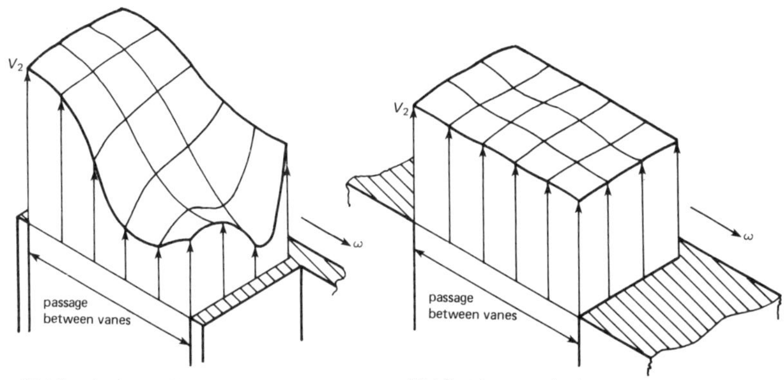

(a) Adjacent passages open (b) Adjacent passages closed

Figure 5.9 *Velocity profile at outlet from one passage, showing the effect of the adjacent passages. (Adapted from reference 42 by permission of the National Research Council of Canada)*

figure 5.10, in which adjacent passages are represented as straightened out to include the inducer and outlet in one plane. A small portion of the flow crosses past the tip of the vane and flows back into the low-pressure zone of the adjacent passage, because of the strong pressure difference. It is this flow past the tip and the reverse flow across the wake which represents the slip, rather than the simpler explanation based on figure 5.6*a*. The recirculating flow of this figure may exist in the region of high-speed flow, but applies more relevantly to a circulating flow without the work input of the centrifugal impeller. The pressure distribution of figure 5.6*d* remains substantially correct.

The likelihood of flow separation depends on the diffusion ratio V_1/V_2, which accounts for the loss of relative velocity in the flow passage. It is akin to the degree of reaction and is related to the area ratio A_2/A_1. Separation is considered very likely if V_1/V_2 exceeds 1.8, and typical values in practice are 1.3 to 1.4. The

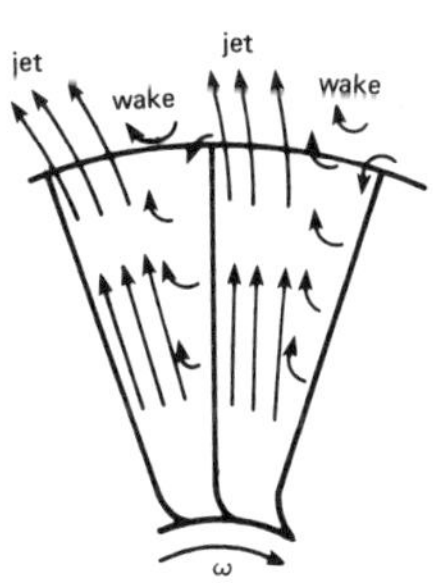

Figure 5.10 *The flow division into smooth jet flow and disturbed wake*

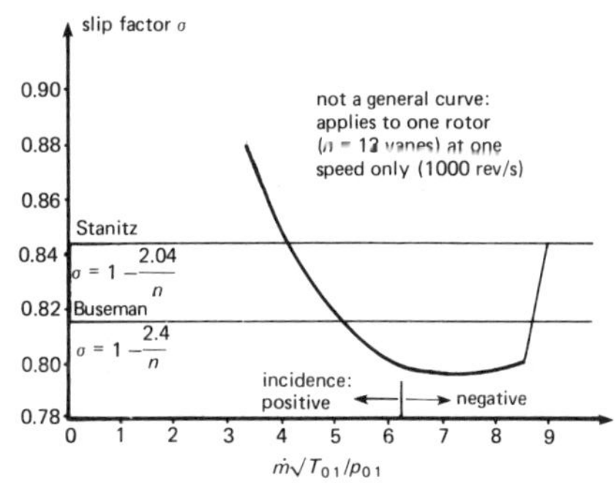

Figure 5.11 *The variation of centrifugal compressor slip with mass flow; adapted from reference 40 by permission of the Council of the Institution of Mechanical Engineers)*

diffusing action at the convex side of the inducer vanes may well help to trigger the separation: in figure 5.11, the increase of slip with mass flow is suddenly reduced when the incidence angle (section E.4) at the inducer entry becomes negative. The slip factors of Buseman and Stanitz are included for comparison.

The disturbed nature of the flow in the passage is a good reason for avoiding half vanes, whose misalignment at flow entry could lose more than the reduction of slip recovers. The flow at outlet is also confused, by the variable total pressure and velocity and by the slip and secondary flow patterns. The flow delivered to the diffuser is thus anything but ideal but its turbulence can be reduced by backward curved vane tips (figure 1.4*b*), which increase the slip but reduce the pressure variations. It also increases the mass flow range at any speed.[6B]

5.3 The Diffuser

The object of the diffuser is to recover as much as possible of the impeller kinetic energy input and convert it to a flow with high static pressure, at a Mach number of about 0.2. Losses in the diffuser will convert it to heat at low pressure if uncontrolled, with excessive loss of total pressure. The diffuser must accept a confused inlet flow and may generate its own flow problems, making its design more difficult than that of the impeller.

Diffusion is difficult, as it requires the flow to enter a region of increased pressure. Low energy boundary layers build up, effectively blocking part of the passage area. If the passage divergence angle or roughness is excessive, reverse flow occurs at the walls and the pressure recovery is low (figure 5.12).

The vaneless diffuser (figure 5.3*a*) has an increasing area with radius, which causes the flow to decelerate on an approximately logarithmic spiral path. The path length required is long, making it bulky, while flow breakdown may develop causing recirculation, as shown in figure 5.13. This is given the name rotating stall, because the recirculating flow is normally in two opposed regions which rotate slowly at about one-sixth of impeller speed.[43] The stall bands may

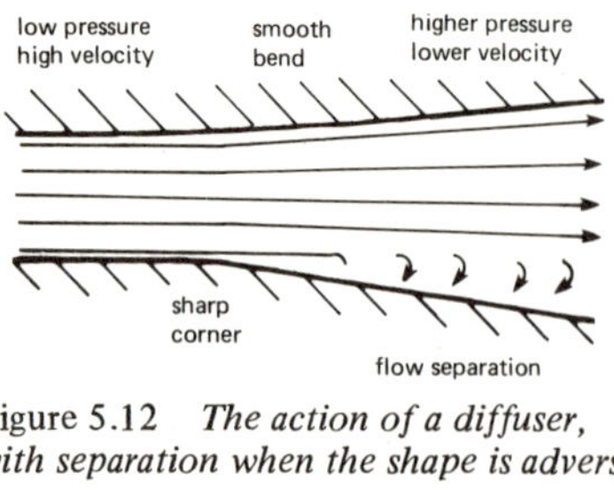

Figure 5.12 *The action of a diffuser, with separation when the shape is adverse*

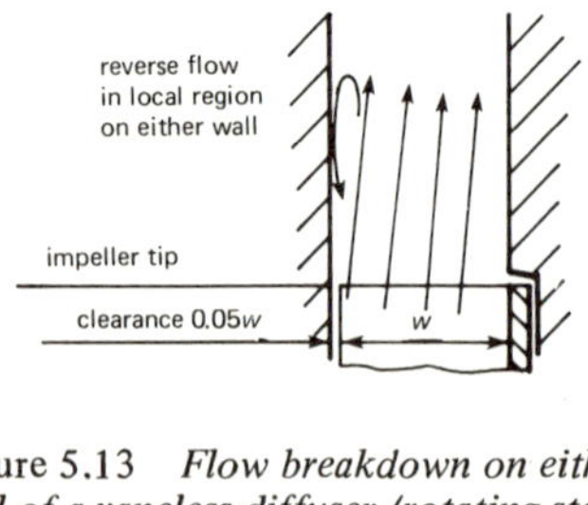

Figure 5.13 *Flow breakdown on either wall of a vaneless diffuser (rotating stall)*

attach to either wall of the diffuser and are quite stable, causing a cyclic variation of radial velocity. It develops particularly if the spiral angle is too low (radial velocity low relative to whirl velocity): this cause can be reduced by a narrower diffuser passage width. These flows have been analysed to determine the design parameters required to avoid them.[44]

The other types of diffuser shorten the diffusion path length by the use of vanes and other features, as introduced in section 5.1 and as discussed in reference 45. The radius at which the vanes or passage entrances start is variable: it may be large in a subsonic diffuser to give the impeller leaving-flow time to slow to subsonic velocity and for its turbulence to settle, or there may be only a radially small vaneless gap to minimise rotating stall and permit a supersonic diffuser entry to benefit.

The number of vanes or passages is important. In earlier work when the knowledge of flow patterns was less, it was found necessary to have only a few vanes. Cheshire [43] found that the flow variation from one impeller passage entering three or four cascade passages at a time caused very different flow patterns between adjacent cascade passages. With the rapid, cyclic variation of flow with impeller rotation, the diffuser was inclined to choke and cause compressor surge (section 10.3). In contrast, one diffuser passage is much more easily able to take the averaged output of two or three impeller passages. More recent designs of cascade diffuser may be considerably more successful.[41] Of the other types, the pipe diffuser [45] appears most satisfactory because it can be machined by simple though precise drilling and boring techniques and is made in one strong piece.

The angle of the leading edge of the diffuser passages should ideally be aligned with the flow, which requires the inside radius of a vane or channel upstream of the throat to match the spiral flow path. Blunt vane leading edges create large stagnation zones, with static pressure variations which excite vibrations of the impeller (section 5.4), but they are more tolerant of varying incidence angle than the sharp leading edge of pipe diffusers (figure 5.3*d*). The spiral flow angle is substantially constant through the speed range, except when the outlet pressure is externally affected: an increased back pressure reduces V_{r2} and causes negative incidence, while reduced back pressure in a subsonic design will cause positive incidence (figure 5.14).

The diffuser efficiency may be assessed in terms of outlet total pressure relative to inlet total pressure, p_{04}/p_{03}. The total pressure loss relative to inlet pressure, $(p_{03} - p_{04})/p_{03}$ is plotted in reference 45 for several types of diffuser, a typical loss characteristic looking like figure 5.15. It also shows the effect of varying incidence angle: when too positive the loss increases drastically, with separation and blockage on the convex side of the vanes. The diffuser tolerates negative incidence until rotating stall of the vanes starts, akin to rotating stall of the axial compressor (10.4). Stalling the diffuser blocks the flow outlet so it reverses direction or surges back through the impeller (section 10.1). An alternative diffusion philosophy is discussed in section 7.4.

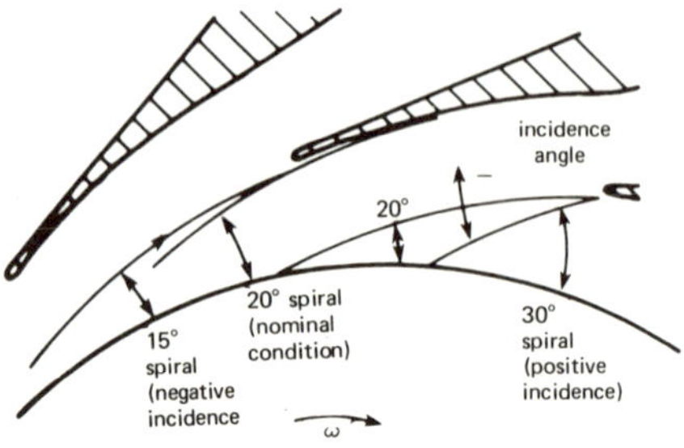

Figure 5.14 *Variation of the spiral angle in the vaneless gap, causing positive or negative incidence*

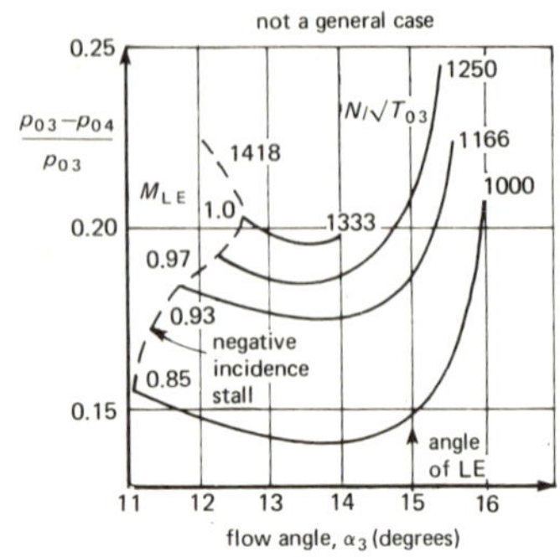

Figure 5.15 *The diffuser characteristics, showing the loss of total pressure over a range of inlet conditions (Redrawn from reference 45 by permission of the American Society of Mechanical Engineers)*

5.4 Vibration Problems

The centrifugal compressor is a very reliable unit but in earlier times was prone to fatigue failures, losing either the tip of a vane or a complete vane from the hub, with severe secondary damage to the other vanes (figure 5.16*a* and *b*).

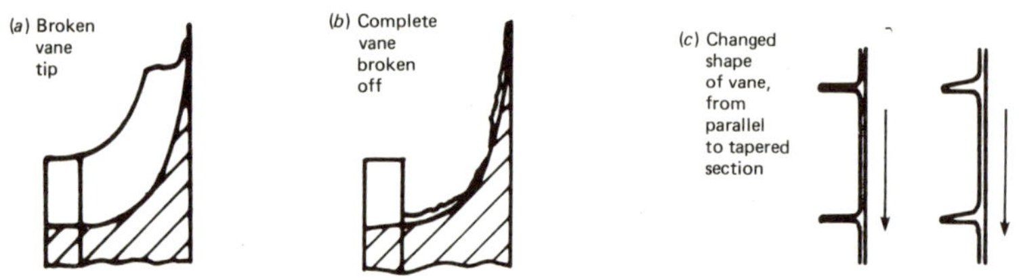

Figure 5.16 *Types of failure of thin impeller vanes, with the buttress section which overcomes the problem*

It happened largely because the vanes were too thin. The flap vibration mode's natural frequency may match the excitation arising from passing the given number of diffuser vanes at a critical rotational speed: appendix F explains this type of excitation mechanism. The design change of thickening the impeller vanes, particularly near the tip to resist the input signal from the diffuser vanes, overcame the problem (figure 5.16*c*). It also reduced the compressor efficiency.[46

6

The Axial Flow Compressor

The axial flow compressor is the second type of dynamic compressor, whose general nature and capabilities were introduced in section 1.3. It absorbs energy from an external power source, raising the pressure of a continuous flow of working fluid by means of its blading geometry and motion.

The fluid flows through an annular passage surrounding the central shaft, entering and leaving at opposite ends of the rotor. As the flow usually has constant axial velocity through the machine, the cross-sectional area of the annular passage reduces towards the outlet. The contraction ratio, outlet area/inlet area, may not always match the flow ideally as the rotational speed and outlet pressure vary: this causes a variety of problems when operating away from the design point, as discussed with other operating problems in section 6.4.

Axial compressors also have a variety of vibration problems. These are discussed later in terms of rotor imbalance (section 14.3) and blade vibrations (appendix F).

6.1 Construction

The axial compressor is built up of a series of stages, each consisting of a disc of rotor blades followed by a ring of stator vanes. The manner in which these accelerate and diffuse the flow to generate pressure is shown in section 6.2. An additional ring of stator vanes may be fitted in front of the first stage as inlet guide vanes (IGVs), to direct the flow to suit the first-stage rotor. The final ring of stator vanes may be called the outlet guide vanes (OGVs), as they guide the flow to the axial direction to suit the compressor outlet.

A typical blading arrangement is shown in figure 6.1, which depicts an eight-stage compressor of constant mean radius. The shape of the annulus and typical blade proportions are discussed in section E.6. The rotor construction shown is an imaginary composite of several methods of construction. Such a mixture would never be used in practice, but it serves to explain a variety of design

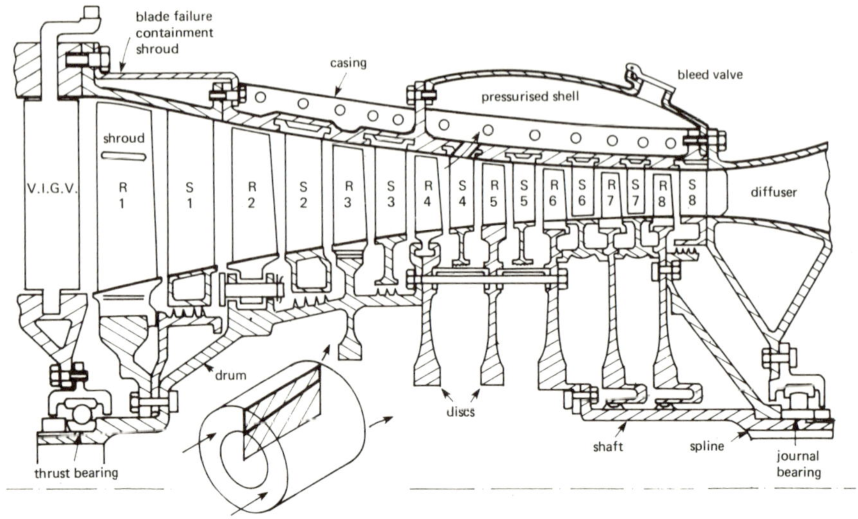

Figure 6.1 *A section through a composite axial compressor showing a variety of different types of design feature*

philosophies. In addition to some references already listed, further illustrations of construction can be found in reference 36, with a description of some manufacturing methods.

Figure 6.1 shows the rotor blades attached by, in sequence: a dovetail root, pin fixing, fir tree, and straddle T root. At stage 5 the blades are cast integrally with the disc, while stages 6 to 8 use dovetails. These are common in aircraft engines, while the straddle T is of heavier construction and suits industrial applications best. The first three types of root fixing are illustrated in end view in figure F.1. Compressor rotor blades have unshrouded ends, unlike high duty turbine blades, but may use intermediate shrouds to counter vibration problems (section F.5). Some looseness in the fit of the blade roots may be beneficial in damping or detuning blade vibrations, but it may also lead to frettage problems if the centrifugal blade load does not apply a sufficient frictional grip.

The rotor design philosophies shown are: a disc (stage 1), a drum (stages 2 and 3), through-bolted discs (stages 4 to 6), and clamped discs splined to a shaft (stages 7 and 8). The drum construction is appropriate for low speeds and light blades, and is often a welded assembly of a series of rings: it is of light weight, has a high whirl speed and retains its balance well. Where the blades are heavier, or the disc or drum material strength is reduced by a high operating temperature, an inner hub is required to hold the rim together, as shown at stages 3 to 8: without the hub, the rim alone could not withstand the centrifugal blade load and would burst. The hub stretches less than the rim as speed increases, so pulls the rim in with a radial tensile stress while the inner thickening of the hub keeps the peak stress level uniform.

The through-bolted discs are of this type, and such construction may extend the whole length of the rotor. The bolts are a very neat fit in all the holes in the discs and spacers, making it a relatively expensive construction. The clamped discs appear simpler, the spline having a hairpin fixing so that radial growth of the hub does not loosen the grip of the splines on the shaft. However, this construction can be more prone to frettage, a major long-term problem in compressors which loosens the fits and may initiate fatigue cracking of the blade roots (section F.5). The discs and spacers may also fret, as torsional deflection of the shaft with speed changes calls for slight circumferential slip at the outer diameters, and vibration of the blades causes slight axial movements. The rotor construction incorporates some form of triangulation to transmit the axial blade loads to the shaft with minimal deflection. This is shown by the conical front end of the drum and conical rear seal. Welded rotor construction gives easier balancing than the assembled types (section 14.3).

The stator vanes are often shrouded at their inner ends, for strength and to seal against leakage flows forward from the higher pressure behind the vanes. The shroud may also reduce losses caused by circulating flows in the cavity between adjacent discs, if it is large enough to fill the cavity as shown at stages 1 and 2. The vanes may be T-slotted into a casing, which is split across a diameter for access and inspection, or may form complete rings as at stage 1. A welded ring of this type may be used to support a rotor bearing in aircraft engines, but industrial practice more often supports the bearings in the casing structure. Some engines use vanes which swivel about a radial axis to vary the flow angles at low speeds (section 11.4.2); only *variable inlet guide vanes* (VIGVs) of this type are shown but stator vanes of this type are also sometimes used for the first few stages. The seals shown are of labyrinth type, as discussed in section 14.2.

The casings are shown as a casting at inlet (left side of figure 6.1), a ring with stators at stage 1, split casings bolted together through diametrically opposed flanges at stages 2 to 8, and a lightweight diffuser casing to slow the flow for the combustion system. The compressor casing may be reinforced (stage 1) or thickened (stage 2) to ensure that debris from a blade failure is contained (section 14.2). The last five stages are shown enclosed by a spherical casing, an industrial practice to reduce the casing loading. In this case, it is pressurised internally by air entering through slots between the stage 4 stator vanes. A valve is shown which opens at low speed, to bleed the compressor and maintain satisfactory operation (11.4). If this were an h-p compressor in a bypass or fan engine, this bleed flow could well mix with the delivery from the l-p compressor to avoid total wastage of its energy content.

6.2 Operation and Performance

In simple terms, the rotor blades scoop air in and hurl it into the following stator row. The stator vanes then slow the flow down again, converting its kinetic energy to potential energy in the form of raised pressure. Thus, each stage creates a pressure increase and the compressor as a whole generates a pressure ratio greater than unity between its inlet and outlet.

The slowing of a flow to increase its pressure is a diffusion process (section 5.3). It is the static pressure which increases, while there may be a slight loss of stagnation pressure resulting from frictional losses along the vane surfaces (section B.2). Only the rotor can put energy into the flow but, in practice, diffusion takes place in both the rotor and the stator. The proportion of the stage total diffusion which occurs in the rotor is called the *degree of reaction, D,* and the designer's choice of this parameter establishes the relationship of the flow angles in the rotor and stator (section E.2). The diffusion occurs because the rotor and stator passages diverge: their curvature causes the cross-sectional area to increase towards the trailing edge. In contrast, turbine blade passages converge, as shown in figure E.1.

The changes in flow direction and velocity through a stage are shown in figure 6.2, a section taken through the blades at their mean radius. The numbering used through the stage should not be confused with the numbering through the engine as a whole. The flow angles shown are typical of normal practice, while the blade angles are slightly different from the flow angles, as determined by cascade testing (section E.4). The absolute velocity approaching the rotor, V_0, is increased by the rotor to V_3 at entry to the following stator, which slows it to V_4 by diffusion; this is then V_0 for the next stage. Within the rotor, the blade motion causes a high relative value of approach velocity, V_1 which is reduced by diffusion in the rotor to a lower relative value, V_2 at rotor outlet. Adding the blade velocity U to the velocity V_2 (relative to the blades) yields the absolute rotor outlet velocity V_3.

The changes in flow velocity and direction in the rotor result in a force applied on the blades by the flow (section E.3); this force acts partly in a tangential

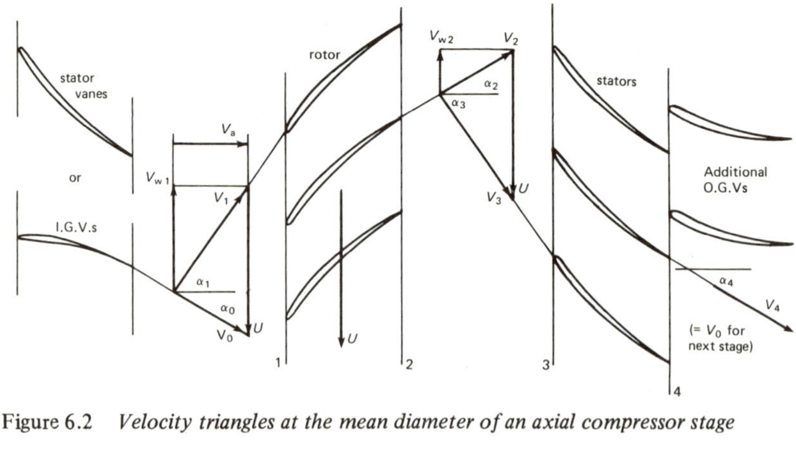

Figure 6.2 *Velocity triangles at the mean diameter of an axial compressor stage*

direction and is reacted by the torque applied to the compressor shaft. The force acting on the flow causes a change in momentum, while the *torque* causes a change in angular momentum. Thus

$$\tau = \dot{m}R(V_{w3} - V_{w}0)\ \mathrm{kg\ m^2/s^2\ or\ N\ m}$$

where V_W is the whirl velocity and R the radius.

The *power* input is torque times rotational speed

$$P = \tau\omega = \dot{m}U(V_{w3} - V_{w0})\ \text{(based on stator geometry)}$$

$$= \dot{m}U[(U - V_{w2}) - (U - V_{w1})]\ \text{(convert to rotor geometry)}$$

$$= \dot{m}U(V_{w1} - V_{w2})$$

so

$$P = \dot{m}UV_a(\tan\alpha_1 - \tan\alpha_2) = \dot{m}C_p\Delta T_{os} \qquad (6.1)$$

The *work done* is another common term, and is given by $W = P/\dot{m}$, or power per unit mass flow. The work done is actually less than that in equation 6.1 because of the radial variation across the annulus of axial velocity (figure 6.3). The high value of axial velocity, V_a, at mid-blade height causes a local negative angle of incidence (section E.4) and reduces the work done. This should be compensated for by the highly positive incidence at the blade root and tip, where V_a is low, but the flow in the boundary layer at the annulus walls is too disturbed and lacking in energy for the work deficit to be recovered.

This deficit becomes worse with the build up of boundary layer towards the later stages, and may be corrected for by the *work done factor*, Ω. At the first stage, $\Omega = 0.98$ is taken as a normal value. An empirical relationship, which expresses the mean value of Ω for the first n stages of a compressor and which was derived from test results by Howell and Bonham,[47] is: $\Omega_m = (5n + 5)/(6n + 4)$. This may not apply exactly to compressors of different annulus geometry or hub/tip ratio (section E.6), but indicates the general effect as the value of Ω for successive individual stages falls rapidly from 0.98 to about 0.8. In design work, Ω is often replaced now by a blockage factor (section E.5).

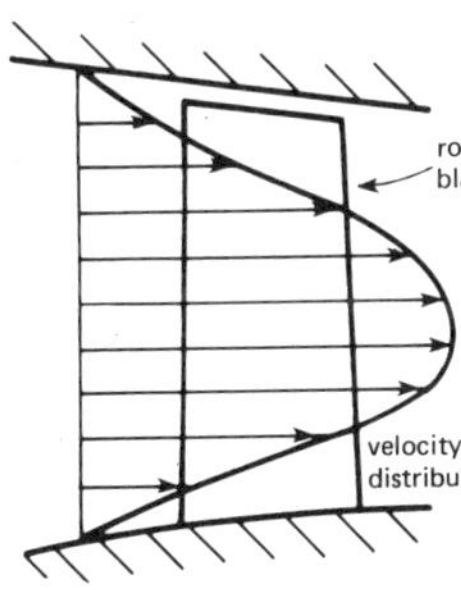

Figure 6.3 *Radial variation of axial velocity at a late stage in the compressor*

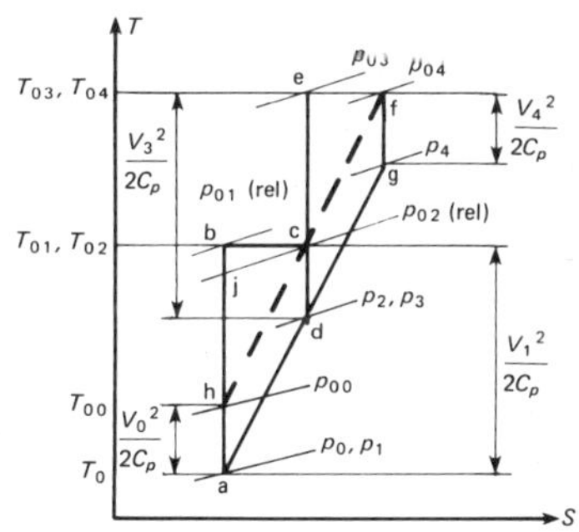

Figure 6.4 *Temperature–entropy diagram for one stage*

The work done by any stage is therefore

$$P = \dot{m}C_p\Delta T_{0\text{s}} = \dot{m}\Omega U V_\text{a}(\tan\alpha_1 - \tan\alpha_2) \tag{6.2}$$

This shows that the work input to the stage can be expressed simply in terms of the rotor geometry, the blade tangential velocity and the flow velocity. The work input can be made dimensionless in the form of the *stage temperature rise coefficient*, which is used as the blade loading design parameter

$$\psi = \frac{P}{\frac{1}{2}\dot{m}U^2} = \frac{C_p\Delta T_{0\text{s}}}{\frac{1}{2}U^2} = \frac{2\Omega V_\text{a}}{U}(\tan\alpha_1 - \tan\alpha_2) \tag{6.3}$$

although some organisations use the form $P/\dot{m}U^2$, omitting the value ½. A *stage pressure rise coefficient* may also be used, in the standard form

$$\frac{\Delta p_{0\text{s}}}{\frac{1}{2}\rho U^2} = \frac{2\eta_\text{s}\Omega V_\text{a}}{U}(\tan\alpha_1 - \tan\alpha_2) \tag{6.4}$$

Note that $\Delta T_{0\text{s}}$ and $\Delta p_{0\text{s}}$ represent the change in stagnation conditions across the stage, and η_s is the stage isentropic efficiency (section C.2). The link between equation 6.3 and 6.4 is $\Delta p/\rho = C_p\eta_\text{s}\Delta T_\text{s}$, which may easily be derived using equations B.2, B.6 and C.3. If the stage reaction is D, the stage efficiency is related to the blade efficiency η_b by

$$\eta_\text{s} = D\eta_{\text{b(rotor)}} + (1 - D)\eta_{\text{b(stator)}} \tag{6.5}$$

The *blade efficiency* is assessed in terms of the loss of total pressure, $\bar{w}$, of the flow through the blade row (section E.5). This is represented in figure 6.4, which shows the changes in condition and velocity of the working fluid at the various points through the stage (numbered as in figure 6.2). The temperature–entropy diagram shown is a small, detailed part of figure C.2*a*, which applies to the compressor as a whole. The absolute condition of the flow approaching the rotor blades is indicated by T_{00} and T_0, their difference being $V_0{}^2/2C_p$ (from equation B.13). T_0, being a static temperature, equals the value of T_1 at entry to the rotor because static temperatures are scalar quantities, unaffected by direction of motion. The high value of V_1 relative to V_0 puts T_{01} at a higher level than T_{00}, T_{01} being the temperature measured by a probe attached to the moving blade. No work is done on the fluid relative to the blade passage, so $T_{02} = T_{01}$ while the total pressure falls by $\bar{w}$, from p_{01} to p_{02}. Diffusion in the blade passage makes V_2 lower than V_1, so T_2 is therefore higher than T_1.

The change in condition from point 1 to point 2, within the blade row, is represented by the shape abcd in figure 6.4. This shape can also be drawn for the changes through a stator row. This is represented by defg, which joins the first diagram at d where $T_2 = T_3$, just as $T_0 = T_1$ at a. This figure has been drawn with the value $D = 0.6$, which implies that $T_2 - T_1 = 0.6(T_4 - T_1)$, from equation E.2. The line hf is the contribution of one stage to the line in figure C.2*a* from T_{01} to T_{02} (in engine nomenclature), while ag represents the change in static condition through the stage. The *blading efficiency* is portrayed by jh/bh in figure 6.4, and is expressed in pressure terms for cascade tests or temperature terms for compressor tests.

The pressure ratio generated by the stage is related to the work input, equation 6.2, because it is related to the temperature rise by

$$r_{p0s} = \frac{p_{04}}{p_{00}} = \left(1 + \frac{\eta_s \Delta T_{0s}}{T_{00}}\right)^{\gamma/(\gamma-1)} \tag{6.6}$$

This comes from equations B.12 and C.1, with the numbering applied for the stage. T_{00} is the stagnation temperature approaching the stage, but its value rises with each successive stage through the compressor. It has the effect that, if the same amount of work is put in at each stage, the rising temperature causes the pressure ratio generated by successive stages to fall. The first stage of a subsonic compressor may generate a ratio of about 1.2, while a supersonic fan may reach a ratio of 1.8.

The use of a polytropic efficiency (section C.3) gives the stage pressure ratio in the form

$$r_{p0s} = \left(1 + \frac{\Delta T_{0s}}{T_{00}}\right)^{\eta_{\infty c}\gamma/(\gamma-1)} \tag{6.7}$$

The over-all pressure ratio generated by the compressor is the product of the individual stage pressure ratios, and is given by the equations

$$r_{p12} = \frac{p_{02}}{p_{01}} = \left(1 + \frac{\eta_{12}(T_{02} - T_{01})}{T_{01}}\right)^{\gamma/(\gamma-1)} \tag{6.8}$$

or

$$r_{p12} = \frac{p_{02}}{p_{01}} = \left(1 + \frac{n\Delta T_{0s}}{T_{01}}\right)^{\eta_{\infty c}\gamma/(\gamma-1)} \tag{6.9}$$

where n is the number of stages and $n\Delta T_{0s}$ is the same as $T_{02} - T_{01}$, ΔT_{0s} being taken as the same for all stages.

6.3 Compressor Materials and Manufacture

Compressor blade materials include fibrous composites, aluminium, titanium, steel and nickel alloys. These are listed in increasing order of weight and their ability to withstand high temperatures. The last three are the most common, and a compressor rotor may use each in sequence to match the rate of temperature rise through the machine. Glass or carbon-fibre blades have high strength-to-weight ratio, but the soft matrix cannot withstand the impacts from continuous rain or hail ingestion, or from heavier foreign objects (section 6.4.2). Aluminium blades operate very successfully, although their softness renders them easily damaged or eroded by runway grit or unfiltered atmospheric smoke particles.[48] Titanium is widely used in the front stages of aircraft engines for its toughness, strength and lightness, but it is prone to rapid fatigue failure from notches or damage at vibrational anti-nodes (section F.1). Steel is safer but heavier, while nickel alloys are needed for temperatures above about 700 K.

The rotor material is generally steel for shafts and discs, but aircraft engines may use titanium at the front stages and nickel alloy at the rear. All components are usually polished to remove highly stressed sharp edges and generously filleted to avoid sharp internal corners. They are designed to withstand cyclic fatigue stressing, but some titanium materials are sensitive to creep as well as fatigue, the time spent under high stress and the number of stress cycles both being important.

Stator vanes may be of the same materials, but steel is the most common. Their stressing is different, involving no centrifugal loading, and long vanes are usually brazed or welded to inner shrouds, which helps prevent vibration failures. Casings may be of cast magnesium, aluminium, steel or iron, or fabricated from titanium or steel. The choice of material is influenced largely by the suitability of its weight and cost for flight, ground-level portability or stationary installation. The loading patterns are discussed in section 14.2.

Blade and vane surfaces are manufactured by specialist techniques, as they involve airfoil sections and may be twisted or curved or tapered along their length. The simplest way is to determine the profiles required at the hub and tip, and join them by straight, 'ruled' lines. If this 'single ruled' design is inadequate, a profile at mid-blade height may be chosen and the surface 'double ruled', from the middle to each end separately. Once the geometry of the family of ruled lines is established, they may be machined in turn by a milling machine, reset carefully for each line, to generate the shape required in a master block from which the forging die may be copy-machined. This permits the accurate forging of blades to their finished size, requiring only fettling and polishing.

The advanced method, for companies with the required plant, is to use computers to determine the blade shape required for aerodynamic and stress criteria. The computer may then instruct a numerically controlled milling machine to make the dies, which no longer need be of simple, ruled form.

6.4 Operational Problems

Some of the problems which affect the design, development and use of axial flow compressors are discussed below.

6.4.1 Performance Deficit

The compressor may fail to generate the required pressure ratio at any part of the operating speed range, or it may lose efficiency (section C.1). At the design stage, the best available data from cascade tests (section E.4) and experience (section E.5) is used. The performance may be improved in development, either at the design point or by compromising design point performance to improve operation at other speeds. The main method involves twisting sets of blades to different angles, to suit the flow at all radii. Early efforts in this field are described by Constant.[49]

In service, *blade profiles* change for several reasons. Dirt deposits may build up in industrial atmospheres on the blade concave surface or some distance behind the leading edge on the convex surface. Salt deposits may form in the same way in marine environments. Ice may form on the front stage blades or vanes in wet atmospheres near freezing point. Sand in deserts or atmospheric dust and smoke may blunt the leading edge, leaving a very coarse surface; or erode the concave side of the blade even to the extent of eroding it right through within a few hours. All shape changes result in departures from the carefully chosen profile and cause a loss of performance.

Performance lost as a result of deposits can be restored by periodic cleaning. This can be achieved by washing; spraying water, or water with alcohol or detergent into the engine intake while it is at idle speed or being rotated by the starter. Stickier deposits may require grit blasting, by walnut shells or granulated coke. Service air offtakes and other orifices may need to be blanked off before using these procedures, to avoid blocking heat exchangers and other small passages. The performance benefit is lower s.f.c., with correspondingly reduced exhaust gas temperature.

Eroded blades can also be saved provided that the material lost has not reduced the blade area (height times chord) excessively. A significant improvement may be obtained by filing and polishing the blunted leading edge back to its original small radius, and blending this smoothly with the remainder of the blade. In dirty or gritty atmospheres, filtration efficiencies exceeding 99.99 per cent have been found necessary for reasonable life expectation.[4C] Filter types are discussed briefly in section 15.4.

The other significant cause of performance deterioration with time is the increase of blade *tip clearance*. This may result from erosion, by the atmospheric grit which is centrifuged to flow predominantly along the outer annulus wall, or by blade location problems. Location errors may result from the whirling and general flexibility of the rotor and engine structure (section 14.2), particularly in long compressor assemblies, or by inadequate design precautions against the different radial growth of blades and casings throughout the working envelope (section 14.2.2). Tip clearance effects are most severe when particularly short blades are used (section E.6). Aero-engine efficiency in some cases falls by 6 to 8 per cent (increase in s.f.c.) after 2000 hours. This is mainly recovered at overhaul by replacing the *abradable linings* in the casings.

6.4.2 Foreign Object Damage (FOD)

This is the damage resulting from the ingestion of stones; small machine parts;

hand tools or tool kits; hats, ear muffs or headphones; birds; puddles or slush or snow sprayed up by an aircraft wheel; hailstones, or people. It is commonly an aircraft problem, resulting from the inlet suction or from chance collisions in flight, but may affect any engine with an unprotected inlet.

The inlet suction can create a vortex in the entering airflow, similar to that of water entering a plughole. The forward end of the vortex may weaken and die in the atmosphere, but it is more likely to turn down to terminate at the ground (figure 6.5), where it wanders about until its very low pressure centre finds and lifts a loose object. Effectively, the static pressure (section B.2) reduces with reducing radius towards the centre of the vortex, while the whirl velocity increases. Local variations in atmospheric wind plays a large part in vortex formation but a strong enough wind can prevent the vortex from reaching the ground. Glenny [50] quantifies the wind conditions causing vortex formation relative to the intake centre line height-to-diameter ratio, H/D: one simple conclusion is that the vortex will not reach the ground if the cross-wind velocity V_c is related to the intake axial velocity V_a by the expression

$$\frac{V_c}{V_a}\left(\frac{H}{D}-0.5\right) > 0.1$$

If there is no wind, a vortex may be induced by the upstream effects from the compressor rotation, particularly if there are no inlet guide vanes. A better way of preventing FOD than assessing the wind condition is to ensure that all loose items are removed from the intake region (section 4.3), but the vortex will still increase the compressor noise (section 13.2).

Small ingested items may simply cause a dent or tear, from which a fatigue crack may develop if not found by routine inspection (section 4.3). Some small marks can be removed by filing and polishing. Larger objects may tear a blade off at its root, with severe consequent damage to much of the remainder of the engine. The use of clappers in a shroud ring strengthens each blade by spreading

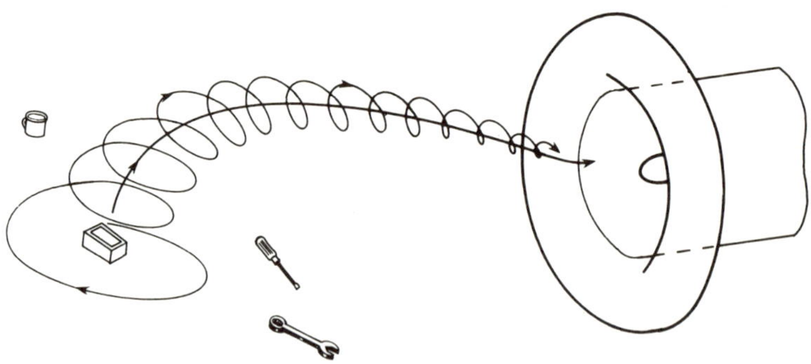

Figure 6.5 *An intake vortex turned downwards to terminate on the ground, where its low-pressure centre wanders about*

the impact load to its neighbours. Turbofan engines fare better than other types, as the debris usually passes down the fan duct and misses the core engine.

In a severe case, it is mandatory that an aircraft engine casing shall be strong enough to contain any broken pieces, to safeguard the engine (section 6.1). The testing to demonstrate this capability involves procedures such as shooting freshly killed chickens into a running engine. One requirement is that the engine shall continue to run and deliver not less than 50 per cent power after ingesting two 1½ lb birds, one on the nose cone and one at the shroud ring radius, while running at take-off conditions. The second requirement is to be able to ingest a 4 lb bird at take-off condition, without disintegrating or jumping off the wing.

6.4.3 Operational Malfunctions

Compressor malfunction generally means a reduction in flow and pressure ratio at a given speed. It affects centrifugal as well as axial compressors, whether working in compressor plant, in turbochargers, or as part of a gas turbine engine.

The malfunction is always associated with aerodynamic stalling of some of the blading. The change from unstalled to stalled operation causes a sudden change in operating point, as the compressor can no longer sustain the pressure ratio it has generated. The change may be of a variety of types, the most serious being *surge*, in which the working fluid surges backwards and forwards through the compressor (section 10.1). The other serious malfunction is *deep stall*, which is a severe case of *rotating stall* (section 10.4). It involves a stable and sustainable condition of greatly reduced mass flow and pressure ratio, which may result in turbine overheat (section 11.6).

Both surge and deep stall may also be known just as stall. Continuous operation in either condition can result in the breakage of many blades (a 'haircut') if undetected. They may be encountered at any speed from zero to maximum, and may cause a *hot start* (section 4.1) if the rate of overfuelling is excessive. The surge itself may vary in strength from a dull boom to a sharp bang (section 10.5): it may damage compressor blades by bending them into contact with the stator vanes, dislodge the intake ducting, or bring the combustion flame out through the front of the engine.

The *causes* may be internal, including malfunction of the compressor variables (section 11.4), incorrect system matching (sections 10.5 and 11.1), FOD or component failure. External factors include the surge of a compressor upstream or downstream in the same ducting, or operation in parallel with a stronger compressor which raises the pressure ratio in the common ducting. More common is *intake flow distortion*, the variation in pressure, and flow velocity and direction, round the compressor inlet. To be serious, it needs to involve significantly different intake conditions over at least a 60° arc relative to the rest of the intake. It may result from a cross-wind condition or intake vortex, or intake ducting of unusual shape or with significant bends, and has the effect also of increasing blade noise (section 13.2). It may also involve temperature variations if it results from exhaust gas reingestion, and may require the imposition of minimum permissible taxi speed in thrust reverse after landing, or a maximum permissible tail wind.

7

Combustion

Open-cycle, air-breathing engines usually add heat to the compressed air by internal combustion of fuel in the air flow. The combustion process was introduced in sections 1.1 and 1.3: it supplies an energy input from which a part provides useful work and the remainder goes to waste as heat in the exhaust gas. Some engines have a second combustion system which either reheats the flow for the later turbine stages (section 3.6.2), or burns more fuel in an afterburner behind the turbines to provide the high exhaust velocity required for the propulsion of supersonic aircraft (section C.6).

Combustion in a gas turbine engine is a continuous process, like an open fire or the fire in a boiler, except that it takes place at high pressure, in a smaller space and usually at a temperature above the melting point of its container. The incoming air is slowed down and a stable flow pattern is generated, to provide adequate residence time for the fuel mixing and chemical processes to take place before the outlet flow is accelerated into the turbine. The design of hardware is intimately linked with these aerodynamic and combustion processes, with the engine operational and performance requirements, and with its survival from exposure to the intense heat.

This interweaving of the various processes and the hardware geometry is reflected in the empirical nature of much of the design process. While the individual processes are well known, their combination into a working combustion system owes much to experimentation and experience. Combustion still defies a fully mathematical approach.

The interweaving also makes it hard to find an easy starting point for explanation of the system. This chapter therefore starts with the chemistry of the process, on which the other aspects depend. Although it is not discussed until much later (section 7.4), the hardware is presented initially in figure 7.1, with the flow conditions to show the direction that the discussion will take.

7.1 The Chemistry of Combustion

This topic is so vast, with relevant literature dating back into the last century,

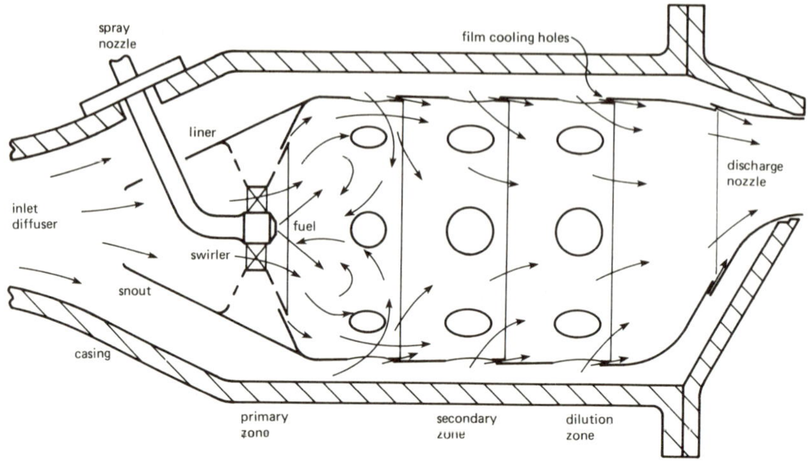

Figure 7.1 *Airflow pattern and hardware arrangement at a typical section through an axial-flow combustion system*

that only a few aspects can be summarised here. The range of fuels is also very wide, so this discussion will be tied mainly to liquid hydrocarbon fuels of the form $(CH_2)_n$. Using kerosine as an example, with an average molecule of $C_{12}H_{24}$, its combustion in air to form carbon dioxide and water is expressed by the simple equation

$$C_{12}H_{24} + 18O_2 + 67.7N_2 \rightarrow 12CO_2 + 12H_2O + 67.7N_2 + \text{heat} \quad (7.1)$$

The ratio by weight of this *stoichiometric* mixture, in which exactly the right amount of air is available to complete the combustion of the fuel, is determined from the molecular weights

$$\text{air:} \quad (18 \times 32) + (67.7 \times 28) = 576 + 1895.6 = 2471.6$$

$$\text{fuel:} \quad (12 \times 12) + (24 \times 1) = 144 + 24 = 168$$

The air/fuel ratio is thus 2471.6/168 = *14.712*, while the fuel/air ratio (f/a) is given by 1/14.712 = *0.06795*. The *equivalence* ratio ϕ is used to relate a particular fuel/air ratio to the stoichiometric value

$$\phi = (f/a)_{\text{actual}} \div (f/a)_{\text{stoichiometric}} \quad (7.2)$$

and is discussed further in section 7.2.

The *temperature* of the products of combustion of a stoichiometric mixture is about 2500 K in a typical engine, rising slightly with increased starting air temperature and pressure.[51A] The heat released by the completed reaction of equation 7.1 is normally taken as 42000 kJ/kg of fuel.[52] This value varies only from 40000 to 46000 kJ/kg for a wide range of liquid hydrocarbon fuels, and is called the *lower calorific value* (LCV). The higher calorific value is slightly greater, as it includes the latent heat of the steam as it condenses to water at

298 K. Liquid fuels which contain oxygen, such as the alcohols, have substantially reduced calorific value because the carbon and hydrogen atoms only comprise a portion of the weight: with hydrocarbon fuels, the oxygen is supplied by the air and does not need to be carried in the fuel tank.

In practice, the chemical reaction is by no means as simple as suggested by equation 7.1. There are far more reactants than the basic fuel molecule and oxygen, and the over-all reaction is made up by the combination of twenty or more simpler reactions.[51A]

The reactants present include the various constituent molecules of the fuel and a number of temporary unstable combinations of carbon, hydrogen and oxygen. *Hydrocarbon fuels* contain numerous different sizes and forms of molecules, from the light gases to the heavy oils and waxes, of which $C_{12}H_{24}$ describes only the mean composition of the molecules present in kerosine. The different molecules have a wide range of boiling points, specific gravities and other properties, with slight variations in LCV. The heavier fractions take longer to break down than the lighter fractions, while branched molecules take longer than straight-chain molecules.[32B] As the molecules are broken down by the heat of combustion (a pyrolysis or cracking process), a number of smaller molecules and radicals (incomplete, unstable molecules) form which interact with others as they combine and divide in a continuing sequence. Some of the reactants which are present during the combustion process have been shown by spectrographic analysis to be: C,CH_3, CH_5, CH_7 (alkyl radicals), C_2H_2, C_2H_4, O,O_3 (ozone), CO, CO_3, OH, HO_2, H_2O_2 (hydrogen peroxide), H,H_2CO and C_4H_8O. Some of these may also be seen in the *colour* of the flame: blue indicates ideal combustion of a fully mixed supply of fuel and air, white or yellow regions show the formation of carbon and green shows a locally rich mixture with considerable formation of radicals.

The many reactants take part in a wide variety of reactions, some giving out heat (exothermic) and some absorbing heat (endothermic). The reactions may go forwards or backwards, depending on the equilibrium concentration of the different reactants. High values of *temperature* and *pressure* increase the energy and frequency of collisions between reactant molecules, but some collisions are glancing blows yielding no reaction, while the chance of which reactions occur depends on which reactants collide. As combustion continues, the reactant concentrations reduce exponentially while the temperature rises: the rate of reaction is proportional to $e^{-K/RT}$ where K is the activation constant of the substance.

When this sequence of reactions ends, the effect is that the reaction in equation 7.1 is complete and the constituents of the outlet gas are the low energy, inert final products. Sometimes, however, the reaction does not run to completion. The chemical processes all require a period of time, the duration of which varies for the different reactions. This time may sometimes be too short, if the reactants are quenched by contact with a draught of cooler air or gas or by contact with the cooled container: this leaves carbon monoxide (CO) and unburned hydrocarbon (HC) molecules in the outlet gas. Alternatively, the time spent at high temperature may be excessive, permitting the formation of oxides of nitrogen (NO_x) and the dissociation of water and carbon dioxide. The effects of incomplete combustion are seen in the emission of various pollutants (section 7.3) and in the loss of combustion efficiency and performance (section 7.5).

7.2 The Combustion Process

This may be considered in terms of the types of flame, the effects of turbulence and local variations of equivalence (ϕ), the three main regions into which the process is divided, the stability of the flame and ignition.

The first mode of burning is by the laminar diffusion flame, which forms a 'sheet' between separate supplies of fuel and air. The sheet is like the envelope of the flame of a candle and is located between the fuel and air supplies, where they form a stoichiometric mixture ($\phi = 1$). It may also form round a burning fuel droplet, with the heat radiating inwards evaporating the droplet to continue the outward flow of vapour for combustion. Although the eye sees a smooth, continuous flame sheet, instantaneous Schlieren photographs of a Bunsen flame show considerable small scale turbulence,[53A] where the rate of flame propagation varies with local concentrations of the various reactants.

In contrast, the premixed laminar flame occurs where the fuel and air are already mixed in a combustible form. The velocity of the flame front through such a mixture is only about 5 to 7 m/s, which precludes combustion of such a mixture if it is flowing at higher speed through a simple, straight duct. The flame front velocity falls to a lower value still if the equivalence goes below unity (a lean mixture) or above unity (a rich mixture). In practice, these laminar flames only exist locally in a gas turbine within an over-all very turbulent flow field, which increases the effective flame speed. The turbulence mixes rich and lean, and hot and cold eddies continuously to maintain stable combustion, and is more significant to the process than the fuel spray pattern.[53B]

The flow pattern can be broken into three zones in which different stages of combustion occur. The fuel is mostly burned in a recirculating *primary* zone, as shown for a typical combustion chamber configuration in figure 7.1. This primary zone is in the form of a toroidal vortex, but other types are also used (section 7.4). The vortex causes the partly burned mixture to recirculate several times back to the point of fuel entry: this allows ample residence time for the reactions to proceed, as the spiral flow path lengths are quite long, and the process is continually regenerated by re-entry into the hottest region. The size of the fire-ball in this zone varies with the fuel burned as well as flow pattern, and dictates the size of the container (section 7.5). Combustion products with high radiation emissivity (appendix D) radiate heat away, reducing the intensity of the flame, so may need a hot ceramic-lined container to radiate heat back. Excess hydrogen, which burns at fuel/air ratios from 2 to 60, helps raise the combustion intensity, for instance when burning methane. A layer of burning carbon surrounding the flame also raises the intensity by radiating inwards as well as outwards. The equivalence is close to unity in this zone: the reaction is therefore unable to run to completion, because the various radicals are dispersed and do not all find an appropriate mate in the given time.

Additional air is introduced in a *secondary* zone, to provide an excess of oxygen in which all the remaining fuel molecules, carbon monoxide, hydrogen, and radicals may be fully oxidised. ϕ here is about 0.5, making the over-all value of ϕ in the first two zones less than unity. The temperature after combustion is therefore less than the stoichiometric level (section 7.1) but still too high for the turbine blades downstream. A *tertiary* zone is provided in which still more compressed air mixes in to dilute the combustion products and cool them to a safe temperature. The success of this mixing process and its effect on the turbine are assessed by the evenness of the temperature across the discharge outlet (section 7.5).

The rate of burning, in terms of velocity limitation and variation of equivalence ratio, sets the limit of *stability* for any given combustion system. The stability envelope is shown in general terms in figure 7.2: the flame blows out if operation crosses any part of the boundary as the rich and lean limits interact with the velocity limit. The horizontal scale is a mass flow parameter, whose limiting values will vary from system to system: the values shown were obtained from a spherical reactor.[53C] At the operating mass flow of the system, a typical requirement for combustion stability is to ensure that ϕ_{rich}/ϕ_{lean} exceeds 2.5.

As it is harder to ignite than to sustain combustion, the envelope in which light-up can be achieved is slightly smaller than the stability envelope. *Ignition* is achieved by the heating effect of an applied source of energy (section 7.4), either a high-voltage spark of energy level from 3 to 12 J, which is lethal to

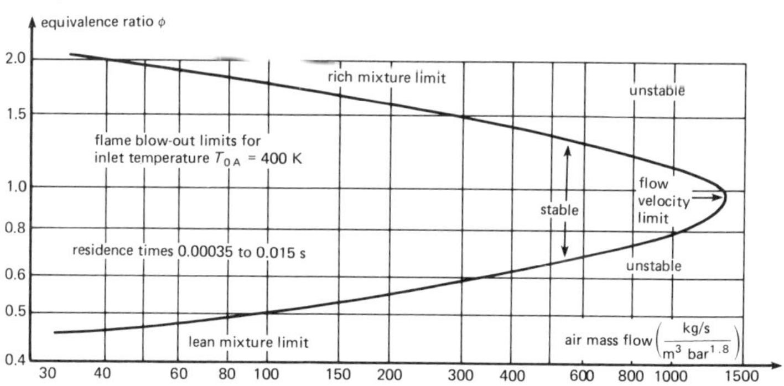

Figure 7.2 *The combustion stability limits of aviation kerosine burning in air in an ideal reactor. (Adapted from reference 53C by permission of the Council of the Institution of Mechanical Engineers)*

humans, or by a torch igniter which throws a flame into the primary zone from a small auxiliary burner. The heat raises the temperature of a local pocket of the mixture to above its spontaneous ignition temperature, as in the case of detonation. This pocket gradually enlarges to heat and ignite the whole primary zone. Ignition is made easier by raised initial mixture temperature and by significant fuel evaporation, which reduces the mean droplet size. The various molecules of hydrocarbon fuels have different ignition temperatures, the lowest applying to the long chain molecules.

Much of the foregoing applies to gaseous as well as liquid fuels. Solid fuels are quite different, as they could not be sprayed into an airflow pattern except in gasified form (section 7.6), but they will be a significant fuel in the future of gas turbine engines.

7.3 Efficiency and Pollution

The heat release from the combustion process derives from the eventual formation of CO_2 and H_2O, the former releasing about 65 per cent of the total. If combustion only proceeds as far as producing CO and H_2O, the amount of heat released is approximately halved. The *efficiency* of combustion (η_B) is therefore determined by the quantity of CO_2 and H_2O in the exhaust gas, relative to the maximum possible quantity. It can be expressed in terms of the heat or enthalpy increase, the temperature rise obtained or the quantity of fuel used for a required temperature rise. Thus, using the suffix A for the inlet air condition and B for the outlet gases after the burning of the fuel

$$\eta_B = \frac{(H_B - H_A)_{\text{actual}}}{(H_B - H_A)_{\text{ideal}}} = \frac{(T_{0B} - T_{0A})_{\text{actual}}}{(T_{0B} - T_{0A})_{\text{ideal}}} = \frac{\dot{m}_{\text{f ideal}}}{\dot{m}_{\text{f actual}}} \tag{7.3}$$

Before combustion is completed, the gas contains various molecules and radicals of hydrocarbon as well as carbon monoxide. These represent unused chemical energy and, if they remain present in the exhaust gas, constitute two of the major *pollutants*. The hydrocarbon forms present cannot be separately identified, so are generally denoted just by HC: they cause the fuel *smell* at many airports.

The reactions by which CO_2 and H_2O are formed are, typically: $CO + OH \rightarrow CO_2 + H$, and $H + HO \rightarrow H_2O$. There are also other reactions producing these products, and the hydroxyl radical (OH) is formed and consumed by many other reactions. CO and HC are used up faster in various reactions if the temperature and equivalence are increased, thereby improving efficiency, and their formation reduces sharply if the combustor inlet temperature (T_{0A}) is raised from below 400 K to above 500 K[51B]. However, CO_2 and H_2O begin to dissociate and continue to do so at an increasing rate as the temperature rises above 1650 K. This dissociation absorbs heat and therefore reduces efficiency at very high temperatures, although the tendency to dissociate into a greater number of molecules is reduced by the higher pressures of modern engines. The CO_2 and H_2O balance is therefore very sensitive to temperature, as it is also to oxygen availability: in an oxygen shortage, water is formed in preference to CO_2, with a resultant significant loss of efficiency and increase in CO emission.

The important pollutants in addition to HC and CO are the oxides of nitrogen (NO_x), sulphur dioxide (SO_2) and smoke. HC and CO have been covered already, except that HC can also form a type of smoke, as discussed later. Concern about all forms of pollution, visible or noxious or both, led to the U.S. Clean Air Act of 1970, which requires the Environmental Protection Agency (EPA) to control pollution. Their requirements effectively set the requirements for the rest of the world.

The *oxides of nitrogen* are mainly NO and NO_2. NO is colourless and toxic, sunlight converts it to the brown, toxic NO_2, so NO_x is generally regarded as being all NO_2. It forms in the hottest regions of the combustion process where ϕ exceeds unity and where the stoichiometric temperature is high. A significant correlation has been demonstrated between NO_x emissions from many engines and their combustor inlet temperature (T_{0A}) [54] : three typical points were 2.5 (g NO_x/kg fuel) at 400 K, 10 g at 600 K and 40 g at 790 K. As T_{0A} varies intrinsically with compressor pressure, NO_x emissions may also be plotted against pressure ratio,[51B] yielding a curve which may be expressed empirically by

$$NO_x \text{ (g/kg of fuel)} = 0.25 \times {p_{0A}}^{1.6} \qquad (7.4)$$

NO_x forms from the nitrogen in the fuel as well as from atmospheric nitrogen, making the fuel specification important.

Sulphur dioxide results from the combustion of the sulphur content of the fuel. As SO_2 is toxic and can be harmful to turbine blades (8.4), a fuel specification with minimum sulphur content is also desirable.

Smoke may be of carbon or hydrocarbon form. The pollutant HC content from a light fuel, such as kerosine or gasoline, is in the form of an invisible vapour in the engine exhaust. A heavier fuel such as diesel has reduced volatility, from which the unburned fractions may condense in a cold atmosphere during low speed operation, causing a noxious grey *vaporous smoke.* [6C]

The more common *black smoke* consists of carbon particles. Even a very opaque smoke represents negligible wastage of fuel or loss of η_B, [55] but it is harmful to the combustion hardware (7.4) and erodes the turbine blade coatings (8.4). It is measured as specified in reference 52, as also is NO_x. Carbon is left unreacted mainly in rich pockets of the primary zone, where ϕ exceeds about 1.5, and increasingly as p_{0A} rises above 10 bar. The high air pressure slows the evaporation of the fuel droplets, while the available oxygen is taken preferentially by the hydrogen atoms, which are the first to leave the fuel molecule. This process is accentuated with aromatic fuels, which start with relatively low hydrogen content, and with the large molecules of heavy fuels. Smoke may also form where the partially completed reaction is quenched after insufficient residence time (7.2), by contact with cooler air. Hardware design to minimise smoke generation includes the air blast fuel nozzle, which premixes air with the fuel before it enters the flame, and the annular chamber which has a shorter cooled perimeter than multiple cans (7.4). A wide fuel spray cone angle, [55] with high momentum to aid penetration [53B] is also beneficial. Metal-based additives may be used to help catalyse the reaction.

The conditions required to minimise pollution during normal operation are therefore: an inlet air temperature of about 500 K to keep HC, CO and NO_x emissions low; considerable turbulence to obtain full and rapid mixing; and equivalence slightly below unity to control smoke, HC and CO while keeping the flame temperature low enough to prevent NO_x emission. The high inlet temperature of modern, high pressure ratio engines has to be compensated by greater attention to fuel and air flow.

The problem is much more severe for engines which are run at variable speeds, particularly aircraft engines. Control of the emissions at cruise power may result is excessive NO_x emission at the maximum *take-off* temperature and excessive HC and CO at *idle* when T_{0A} is low. The over-all air/fuel ratio might well be 60 at take-off but as lean as 150 at idle, where the efficiency falls drastically at low speeds, as shown in figure 7.3. This curve looks similar when plotted against either of two parameters: $p_{0A}T_A/V_r$,[32C] where V_r is the reference mean velocity through the region of maximum cross section, or $(p_A{}^{1.75}e^{T_A/300})/\dot{m}$.[56] The latter parameter θ applies with scales of different values for different engines, but can be made to apply to all sizes of engines in the form $(p_A{}^{1.75}AD^{0.75}e^{T_A/300})/\dot{m}$, where A is the reference cross-sectional area and D is the diameter of the liner for axial flow systems (section 7.4). Low speed or high-altitude operation inevitably results in this loss of efficiency, to 0.9 or less, if the same fuel supply system is used as at high pressure. EPA requirements to achieve 0.985 efficiency at idle call for variations such as those discussed in section 7.4.

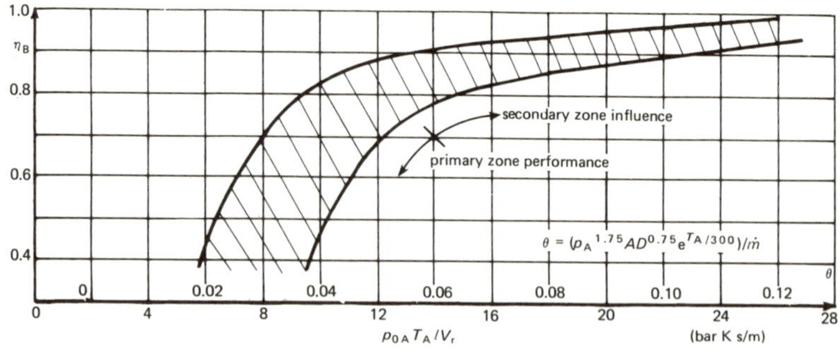

Figure 7.3 *Approximate envelope of combustion efficiency against two mass flow parameters*

The effects of two specific engine cycles on combustion emissions should be mentioned. The regenerative cycle (section 3.6.3) deliberately raises the value of $T_{0\,A}$ (and T_A) to improve engine cycle efficiency. This must also increase low speed combustion efficiency but at the risk of increased NO_x emissions at higher speeds. The other case is the reheat or afterburner system: being the second combustion system handling the same flow, the oxygen supply is reduced (vitiated). The stoichiometric temperature rise (section 7.1) is reduced in proportion to prior oxygen usage, because of the greater proportions of CO_2 and H_2O to be heated. The lower peak temperature results in low NO_x emissions, while the high exhaust temperature ensures very low emission of HC and CO, at least at low altitude.

7.4 Aerodynamics, Fuel Supply and Hardware

This section covers the engineering which harnesses the chemical and combustion processes. The aerodynamic processes create the three combustion zones as dictated by the hardware configuration, and the fuel supply is controlled to enter the primary zone as a finely atomised spray or vapour.

Several different configurations are used, and may be distinguished by their airflow patterns. The general *axial flow* type shown in figure 7.1 concentrates the primary zone into a small volume by the use of a recirculating toroidal vortex. This cross-sectional view could be found as indicated in any of the engine cross sections of figure 7.4: (a) the tubular system with separate cans, (b) the shorter annular system or (c) the tubo-annular or cannular arrangement. Of similar nature is the *reverse flow* type (figure 1.3) which may also be arranged radially (figure 1.1*a*): in the latter form, it may be used alone or as one of a few combustors around the engine. It may also be used remotely as a separate unit in the case of large industrial engines.

The other main configuration, though little used with gas turbines at present, is the *cyclone combustor*. The views in figure 7.5*a* and *b* show one form, while

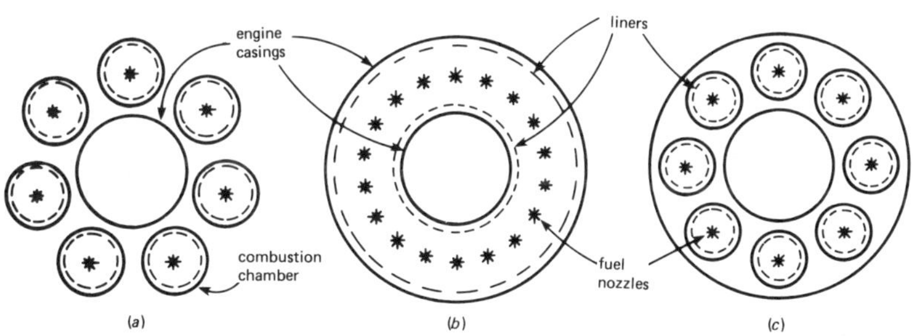

Figure 7.4 *Types of axial flow combustion systems: (a) tubular; (b) annular; (c) tubo-annular or cannular*

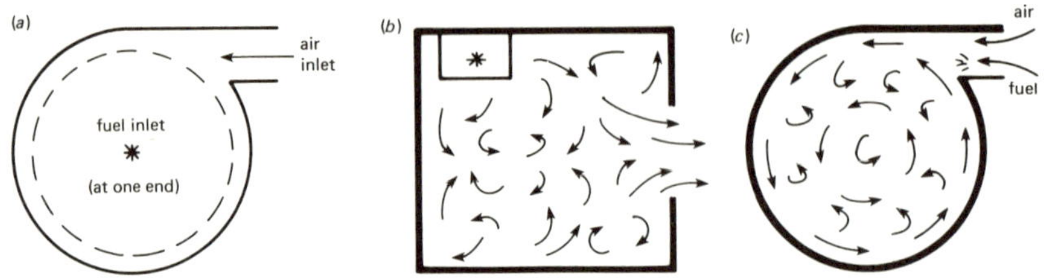

Figure 7.5 *Simple types of cyclone combustor*

figure 7.5*c* shows a variety with premixed fuel and air burning in a ceramic-lined casing. Air enters tangentially and spins towards the exit with a variety of internal turbulence patterns. Fuel droplets are suspended at a radius more or less proportional to droplet size: as the droplet evaporates, its reduced size reduces its centrifugal force and lets the air carry it towards the centre. The flame in such burners is very stable and may make it easier to develop than the axial type for land-based use. It is also particularly suitable for burning poorer quality solid fuels (section 7.6). [53D,57]

To return to the more conventional, axial type, the outlet flow velocity from the compressor may be as high as 150 m/s and must be slowed considerably to permit stable combustion. Diffusers of a variety of types may be used, as shown in figure 7.6: they may involve (a) the curved wall diffuser, which is long and has the separation risk of any diffusion process (section 5.3); (b) the dump diffuser, in which variability of the separation region is avoided; and (c) the newer *vortex controlled diffuser* (VCD).[58] The VCD peels off about 3 per cent of the flow from the low energy boundary layer in the intake duct to form a toroidal vortex and a Coanda bubble, which permit diffusion in a very short length. Over 80 per cent of the theoretical static pressure rise may be recovered, while the bleed

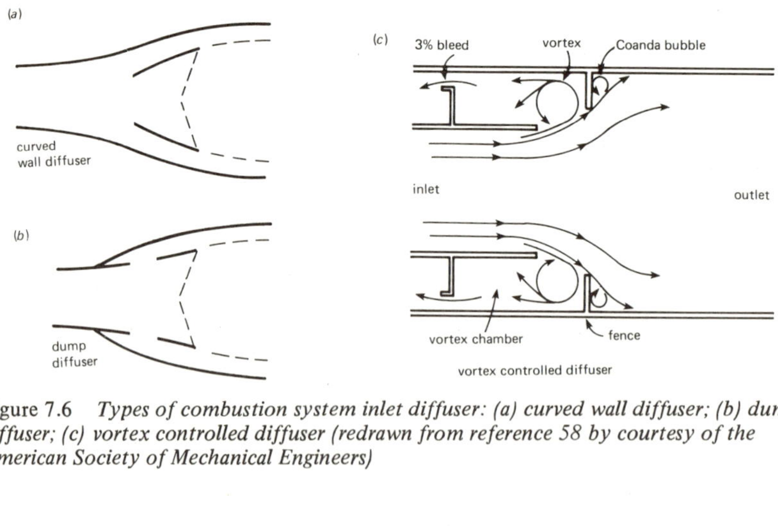

Figure 7.6 *Types of combustion system inlet diffuser: (a) curved wall diffuser; (b) dump diffuser; (c) vortex controlled diffuser (redrawn from reference 58 by courtesy of the American Society of Mechanical Engineers)*

flow may be used for internal engine cooling purposes. The effect of the diffuser is to reduce the flow velocity to about 25 m/s.

Combustion takes place in a perforated heat-resisting *liner* fitted inside the main casing, which is strong enough to contain the maximum pressure. The liner is typically made of a nickel alloy, which can operate up to about 1100 K, or of cobalt alloy, which may run above 1200 K. The perforations are in the form of many small holes, through which air flows to cool the liner and insulate it from the heat, and a few larger holes through which the main flow enters the three combustion zones.[6D] The sizes of all the holes are determined partly during development to tune the flow and heat patterns, and they cause a pressure drop between the outside and inside of the liner (section 7.5). The cooling holes may vary in size from those in a porous material up to about 3 mm: they may be aligned radially in the liner walls or axially in strips between the consecutive conical louvres which form the liner wall (figure 7.1). The temperature difference between the liner and the cooler surrounding structure creates different rates of thermal expansion. The liner is therefore supported on frettage-proof sliding surfaces and has the minimum of double-skin construction, as severe distortion occurs when a hot inner skin is restrained by a cooler outer skin. The liner is exposed to greater heat radiation if carbon particles are present, because they increase the luminosity of the flame.

The primary zone vortex is stabilised in position by flow round the sides of a plate or through the vanes of a swirler. A low pressure centre forms, which is filled by air entering round the liner and being induced forward. The fuel is sprayed directly into the primary zone, so that it recirculates with the flow. The gas escaping from the primary zone is fed with more oxygen from the mixing secondary air, and flows on to mix with the dilution air before escaping. The quality of dilution, in terms of the evenness of the temperature across the outlet, is described by the pattern factor (section 7.5).

The liquid fuel supply is fed through one of a variety of types of nozzle or burner. The main type is the *pressure atomising nozzle* (figure 7.7), which runs at a pressure drop of up to about 40 bar. The fuel enters tangentially into the spin chamber and the vortex thus formed centrifuges the fuel into a cone as it emerges from the orifice. Alternatively, the fuel is deflected into a cone by a

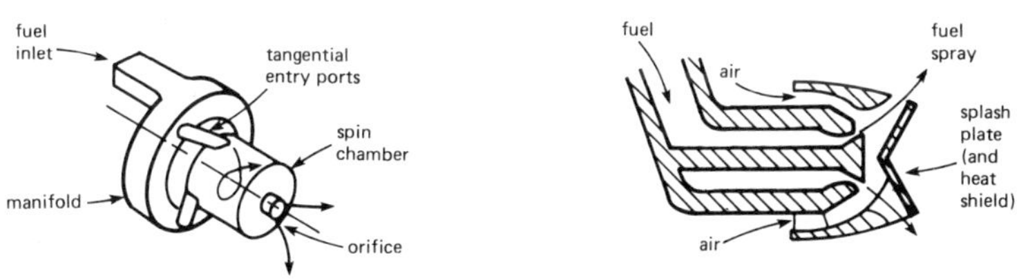

Figure 7.7 *High-pressure fuel atomising nozzles: (a) vortex type showing only the internal passage shape; (b) splash plate type*

splash plate, without spin. The conical fuel film thins with increasing radius and surface tension breaks the film into droplets of about 50 to 100 μm diameter. The included cone angle of the spray is typically about 80 to 100°.

The *air-blast nozzle* [59] is gaining favour because it forms a finer spray and reduces pollution by easing the evaporation of the droplets. It works like a surface carburettor (figure 7.8) with quite a low fuel pressure. The airflow on both sides of the circular lip atomises the fuel off the lip, but needs high air velocities to work well. The *vaporising nozzle* is a tube heated by immersion in the flame, which heats and evaporates the fuel before it emerges, as in a blow lamp. It is not common and can cause severe development problems.[2] The last type is the Turbomeca *flinger* system for small engines, which centrifuges fuel from orifices in the main shaft and flings it into the surrounding, radial flow annular combustor. It provides good atomisation, a natural increase of flow with shaft speed and needs only a low pressure supply.

The fuel flow is controlled (chapter 12) to match the engine requirements at all speeds, which calls for a very wide range of flow rates. The pressure atomising and air-blast nozzles cannot work adequately over such a range: if they can flow the fuel at the maximum rate required, they tend to dribble at low flows. Each can be improved by the use of a concentric, much smaller, primary orifice which atomises a small flow well and is used alone at low speeds. The combination of primary and main orifices creates the *duplex nozzle*, which is a very common configuration and may successfully atomise fuel flows varying over a flow range of 50 to 1. A duplex nozzle with single feed is shown in figure 7.9.

Even these may not make adequate allowance for the low over-all equivalence at low speeds, to control pollution adequately (section 7.3.) A number of techniques are possible[51B] to increase the local value of ϕ. One is to stop the flow to some nozzles round the engine, letting the others burn richer by increasing their supply. Another is to burn in the presence of a dispersed catalyst as in some vehicle exhaust systems, to provide stability at very low equivalence. Increased compressor air bleed may also help.

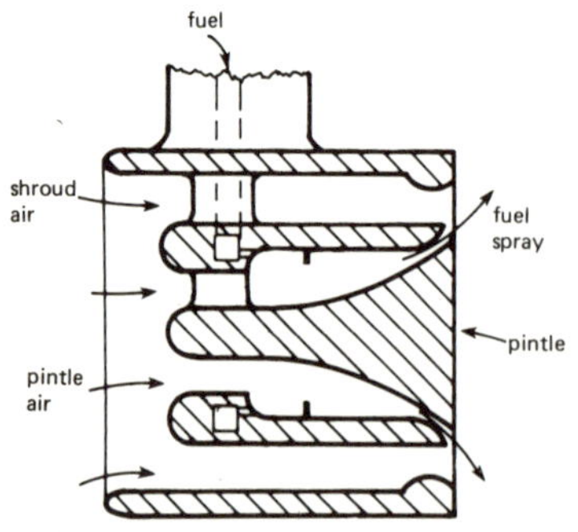

Figure 7.8 *The air-blast nozzle*

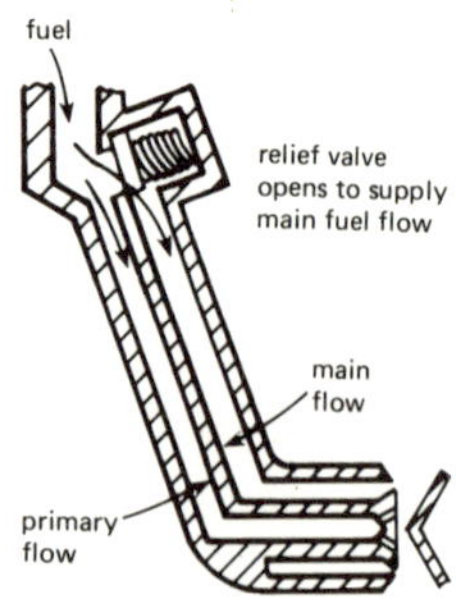

Figure 7.9 *A single-feed duplex nozzle (simplified)*

A major type of premix combustor is the aircraft engine *afterburner*. Fuel is sprayed in just downstream of the turbine. Some distance further back, figure 7.10, rings and spokes of V-shaped gutters form a large area of turbulence for flame stabilisation while presenting minimum cross-sectional blockage to the flow. Ignition is achieved automatically from contact with platinum-based catalyst inserts. A ventilated liner is used again to keep the outer casing cool, and an expanding nozzle is opened when the flame is lit to match the area to the considerably raised temperature and slightly lower pressure (section 14.1). This keeps the flow parameter $\dot{m}\sqrt{T_{0e}}/A_e p_{0e}$ constant (section B.5).

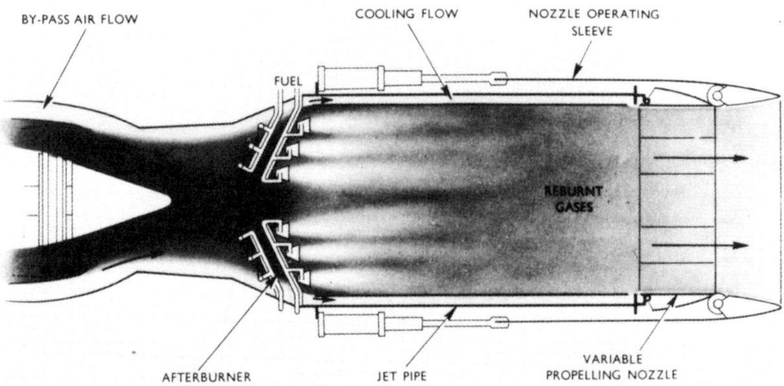

Figure 7.10 *A typical aircraft engine afterburner showing the V gutters which stabilise the flame, the cooled liner and the variable area nozzle. (By courtesy of Rolls-Royce Ltd)*

Ignition in the main combustor is usually triggered by an electric spark (section 7.2.) These are often applied at two locations round an aircraft engine, and the fuel from the other nozzles is lit in sequence by the proximity of the adjacent flame. In tubular or cannular systems, interconnectors or cross-fire tubes link the liners at the primary zone to promote lighting around. Annular systems can be harder to light, because the flame does not necessarily reach into the neighbouring, stable primary zone. Failure to light round can result in a *hung start* (section 4.1), where some parts of the combustor outlet periphery run hot but there is not enough energy input to accelerate the engine.

The *hardware* has only been mentioned briefly here. Many examples of current and future aircraft engine combustors are discussed in reference 51B, while early types are covered very fully in reference 32C. Heavy industrial combustors may use removable liner sections hung on a lattice frame, or even brick-lined systems. Vapour fuels are supplied through large nozzles of several types, and some engines have nozzles with different orifices for supplying either liquid or gaseous fuel.

7.5 Performance and Operation

The combustion process causes a *pressure loss*, mentioned in section 3.2.2, which results from two causes. Part is the *cold loss* which could be measured by flowing air without fuel through all the slots, holes, orifices, etc. To this is added the *hot loss*, shown in figure 7.11 to be proportional to the temperature rise. The effect of heat is to increase the velocity in the turbulence patterns, increasing the frictional loss and introducing a fundamental loss of pressure to offset the gain in velocity head. In a straight flow, velocity head could be gained at the expense of static pressure only, but in this very turbulent flow, stagnation pressure is also lost. The pressure losses are essential to the generation of the air flow pattern, and are assessed in terms of a pressure loss factor (PLF), which may be written as

$$\text{PLF} = \frac{p_{0\text{A}} - p_{0\text{B}}}{\frac{1}{2}\rho_{\text{A}} V_{\text{A}}^{2}} = k_1(\text{cold}) + k_2 \left(\frac{T_{0\text{B}} - T_{0\text{A}}}{T_{0\text{A}}}\right)(\text{hot}) \quad (7.5)$$

The values of k_1 and k_2 are determined by measurement of the performance of the system. As in section 7.3, the suffixes A and B are used rather than station numbers 2 or 3, etc., which would presuppose the number of shafts or compressors or other stations identified upstream.

Standard curves are available from a wide variety of sources to show the *temperature rise* to be expected from burning kerosine at various fuel/air ratios in air at various inlet temperatures. A few such curves are shown in figure 7.12: a lot of space is required to present them in a usable form for cycle calculations. As the top of the curve corresponds to an over-all stoichiometric mixture and is well outside the usual range of engine operation, the relatively straight lower

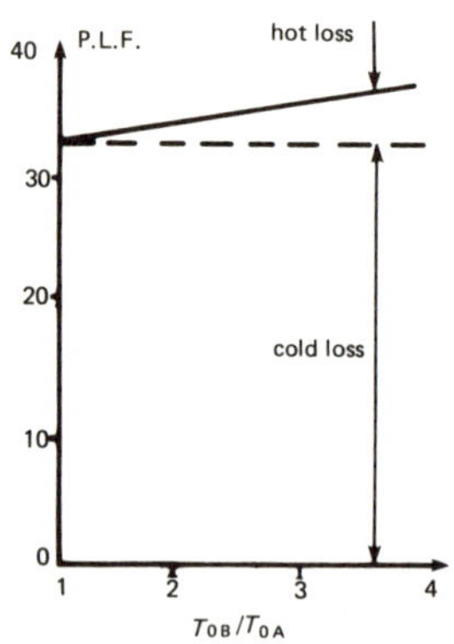

Figure 7.11 *Combustion pressure loss factor*

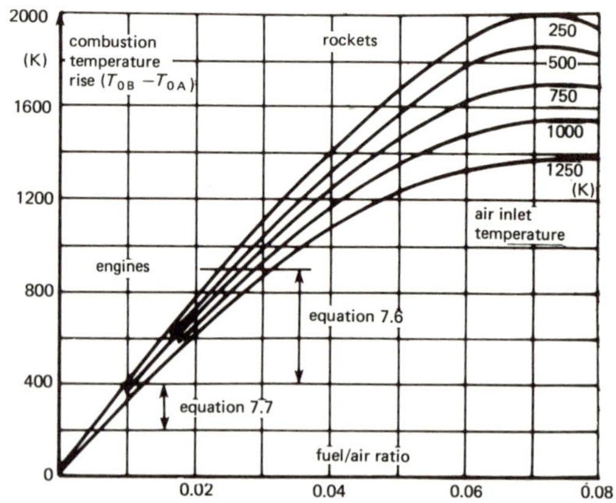

Figure 7.12 *Typical combustion temperature rise curves*

portion only is of interest: it can be described by two simple equations as follows

$$f/a = 1.1\,\frac{(T_{0\,B} - T_{0\,A} - 50)\,(1 + T_{0\,A}/3250)}{42\,000\,\eta_B}$$

$$\text{for } 400\,^\circ\text{C} < (T_{0\,B} - T_{0\,A}) < 900\,^\circ\text{C} \qquad (7.6)$$

$$f/a = 0.99\,\frac{(T_{0\,B} - T_{0\,A} - 10)\,(1 + T_{0\,A}/3250)}{42\,000\,\eta_B}$$

$$\text{for } 200\,^\circ\text{C} < (T_{0\,B} - T_{0\,A}) < 400\,^\circ\text{C} \qquad (7.7)$$

The values of f/a calculated from these equations are accurate within ±1 per cent over the temperature rise range shown, and may be used with discretion for comparative purposes to higher levels. They apply to fuel with an LCV of 42 000 kJ/kg[52]: if a slightly different value is applicable, the value of f/a for a given temperature rise should be corrected by inserting the LCV in place of 42 000 in the equations above. The effect of varying $p_{0\,A}$ has negligible effect on the accuracy of these calculations within the normal range of $T_{0\,A}$, $T_{0\,B}$ and f/a applicable to gas tubine engines. Rearrangement of equations 7.6 and 7.7 shows that $(T_{0\,B} - T_{0\,A})$ falls as $T_{0\,A}$ rises at constant f/a, mainly because the specific heat of the gas increases with temperature. At very high temperatures, the temperature rise is reduced also by dissociation of H_2O and CO_2 (section 7.3).

The mixing of the hot gas with the cool dilution air is dependent on the level of pressure loss through the liner. The outlet temperature is made more even by greater turbulence and duct length. In practice, the mixing is not complete and yields temperature contours such as those shown in figure 7.13 at the outlet behind two adjacent fuel nozzles or cans. It is desirable to locate any hot streaks at different radii, to spread the zone of greatest deterioration on the turbine blades and prolong their life. The over-all quality of mixing is assessed by the outlet traverse *pattern factor* (PF), where

$$\text{PF} = (T_{0\,B\,max} - T_{0\,B\,mean})/(T_{0\,B\,mean} - T_{0\,A}) \qquad (7.9)$$

A satisfactory value of PF is about 0.2. A zero value is virtually unobtainable because of the cooling air film along the liner walls: it would in any case be undesirable, as this film helps cool the blade platform and shroud, which cannot

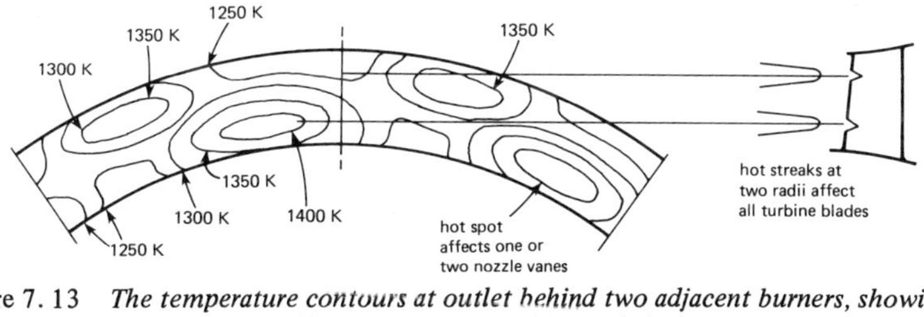

Figure 7.13 *The temperature contours at outlet behind two adjacent burners, showing the effect on the turbine blading of hot spots at two radii round the engine circumference*

be internally cooled as effectively as the blade airfoil (section 8.5).

The over-all performance of a combustor can be assessed by its rate of heat release and efficiency. The rate of *heat release* relative to volume and pressure is shown for typical systems in figure 7.14.[6E] The significant increase in heat release for shorter systems is related to the scale of the turbulence in liners with a small transverse dimension: more heat is released by using many small primary zones fed by individual nozzles, as is the case with radially thin annular combustors. Reduced length makes for a shorter and lighter engine, but can introduce operational problems (section 11.6.5).

The *efficiency* (section 7.3) is affected by engine speed and altitude, and can be discussed in terms of $p_{0A}T_A/V_r$ (figure 7.3). At low engine speeds, very low levels of p_{0A} and T_A are generated, while there may still be quite a significant draught through the system, a relatively high V_r. At *altitude*, reducing pressure and temperature again reduce efficiency as V_r remains high. The low pressure causes the burning gases to occupy a relatively greater part of the chamber volume: in the extreme condition, the combustion process no longer conforms to the three zones imposed by the liner hole geometry and the flame blows out. As p_{0A}, T_{0A} and V_r change at different rates with changes in engine speed and flight altitude, blow-out may occur at medium engine speed, but not at low or high speed at the same altitude.[33C] This can be seen in figure 7.15, which shows variations of operational altitude limit with changes in engine speed. It should be noted that the curve shape and altitude levels vary from engine to engine and that the low pressure conditions at altitude and during starting determine the required combustor *size*.

The θ (volume) parameter ($\theta_v = p_A{}^x e^{T_A/300} v/\dot{m}$) is useful for quantifying these operational limits. With pressure measured in bars and the volume in m^3, a simple calculation of θ_v determines whether a given combustion chamber will be satisfactory in two respects. If θ_v exceeds 200 when $x = 1.3$, then the engine should be able to relight at altitude. If θ_v exceeds 150 to 175 when $x = 1.75$, then combustion efficiency will be reasonably high. The corresponding values in lb/in.^{2}, kelvins, cubic feet and lb/s units are 1×10^5 and 2.5 to 3×10^5.

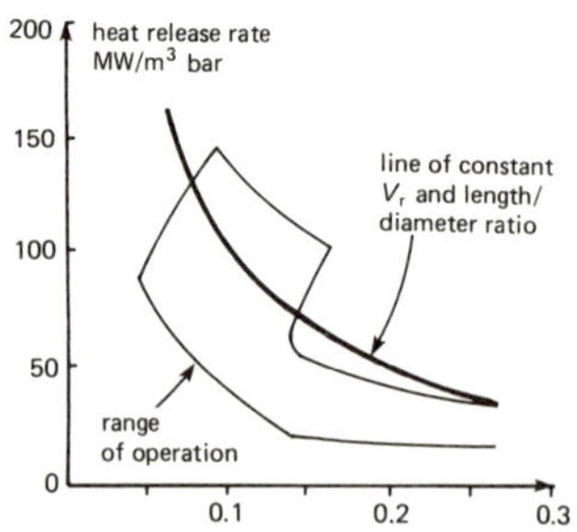

Figure 7.14 *Typical envelope of heat release rate. (Redrawn from reference 6E by permission of the Council of the Institution of Mechanical Engineers)*

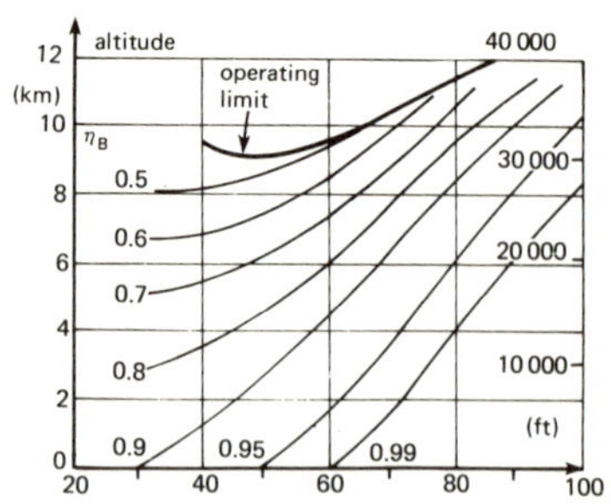

Figure 7.15 *An example for one engine of the loss of combustion stability at an intermediate speed within the running range, when operating at high altitude*

7.6 The Use of Solid and Low-grade Fuels

The use of *coal* in gas turbine engines is under active investigation.[8] It may be burned in solid form, pulverised into granules or powder with an energy input that is well under 1% of its calorific value. One burner for this is the *fluidised bed* of sand supported by a forced updraft of air. The solid fuel is dropped into the bed, which glows bright orange, and the combustion continues in a very stable fashion. A typical heat release rate of 3 MW/m^2 bar is achieved, with a bed depth of 200 to 300 mm. Coal up to 50 mm diameter is easily handled, burning as it floats in the sand.

The *cyclone combustor* (7.4) is also well suited to solid fuel.[57] Using coal crushed to 6 mm mesh, it could release heat at the rate of about 4 MW/m^3 bar (400 000 Btu/ft^3 h atm.), [53D] using the usual 5–20% excess air. An alternative figure may be 10 to 15 MW/m^3 bar at 3 mm mesh, but it depends on the cyclone action. The best results are obtained from multiple inlets, which generate a smaller scale of turbulence than a single inlet, and a size limit of 1 m^3 if $L/D = 1$. This arises from a scale effect, as the throughput varies with volume while the outlet flow varies with area. Clearly, as the volume is increased, the outlet diameter increases at a greater rate than the chamber diameter until the enclosing effect of the outlet end wall is lost.

The problems with coal are its high content of corrosive ash, sulphur and other undesirable elements (section 8.4) and that, even in powdered form, it is unsuitable for the modern small combustors used for liquid and gaseous fuels. For these, the way to use coal is in *gasified form*, as carbon monoxide rather than carbon, but there are still problems. About 10 to 20 per cent of its original energy content is lost during manufacture in its sensible heat at nearly 1200 K, which may be recoverable by generating steam in a combined cycle (section 2.5), but its sticky ash and grit and its reducing properties make filtration, heat removal and even containing it difficult.[8]

Several other papers in reference 53 describe the success and other experience of using other fuels. *Blastfurnace gas* may not be particularly corrosive, but its entrained grit particles may cause severe turbine blade and casing erosion. *Residual fuel* may have sodium and sulphur or, even worse, vanadium content, which causes severe corrosion at higher temperatures (section 8.4). Heavier fuels radiate more heat to the walls, requiring ceramic lining which must be adequately segmented to permit expansion and avoid thermal shock: a thicker lining is worse, while ceramic applied as paint to a metal container can only afford to be 0.1 mm thick.

The trend to cheaper aviation fuels also introduces problems. Their corrosive material content must still be carefully controlled, but their stability tends to be lower, with greater aromatic content and with the risk of 'cracking' in the plumbing upstream of the nozzle orifice.

8

The Axial Flow Turbine

The turbine is used to extract energy from a flowing stream of gas and has the general characteristics introduced in section 1.3. In the axial flow turbine, the gas flows through an annular passage surrounding the rotating power output shaft, in a direction substantially parallel to the axis of the shaft. As the gas flows through the turbine, the blading extracts energy by the action described in section 8.2. The loss of energy causes the pressure and temperature of the gas to fall from turbine inlet to outlet, and necessitates the use of an increasing passage area (normal to the flow) to restrain the flow to a constant velocity. This increasing area is the flare (F), shown with other typical turbine features in figure 8.1.

Apart from extracting energy from the gas flow, turbine design is concerned mainly with protecting all components from the heat of the inlet gas, which affects material strength and corrosion (sections 8.3 and 8.4). As the turbine life is drastically increased by small reductions in temperature (section 4.4),[4D] it is becoming increasingly common to artificially cool the blades to permit the use of higher temperatures for increased cycle efficiency (section 8.5). Turbines are also prone to vibration, leading to breakage and frettage problems: this is discussed in conjunction with compressor vibrations in appendix F.

8.1 Description

A turbine stage uses a ring of nozzles, or nozzle guide vanes (NGVs) (N), to direct the flow on to the moving rotor blades or buckets (B) attached to the periphery of a disc (D). The next stage repeats the process, and its disc may be attached to the first to form a two-stage turbine, as shown in figure 8.1, or may rotate independently, to drive its own shaft.

The static structure consists of the casing, nozzles, seals and supports for the bearing assemblies. The casing is protected from the heat by seal segments (H)

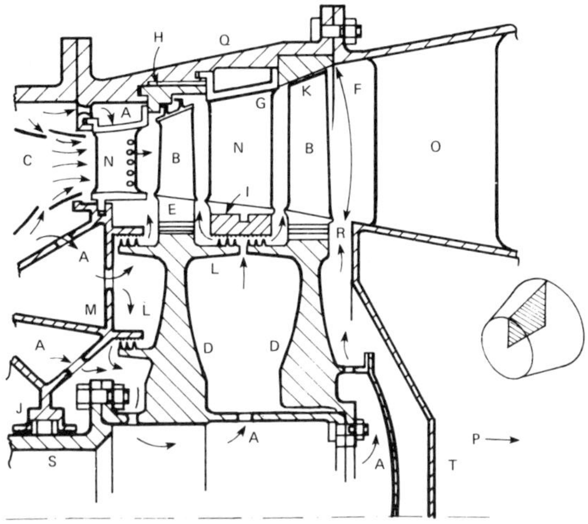

Figure 8.1 *A section through a two-stage turbine, showing: A cooling air flows; B turbine blades; C combustion outlet; D turbine discs; E extended blade root; F annulus flare; G blade shrouding; H seal segment; I NGV inner shroud; J journal bearing; K tip clearance; L labyrinth seal; M nozzle diaphragm: N nozzle guide vane; O outlet straighteners; P power turbine (downstream); Q turbine casing; R fir-tree root; S shaft; T tail cone (front piece)*

and an intervening air flow. In more sophisticated engines, additional cooling air may also be supplied to the casings during steady running, to minimise blade tip clearances and improve efficiency (section 14.2.2). The casing may be split in two halves across a diameter, and joined by two rows of bolts, or it may be a single piece. The former makes assembly and inspection easier, but may suffer distortion from an uneven stress pattern. The latter reduces distortion problems but requires sequential assembly of static and rotating components from one end.

The NGVs may be assembled singly, or in small welded groups, or as a complete ring into the casing. They are often cooled (section 8.5), being subjected to the highest temperatures in the engine. Their location within the structure must permit freedom of expansion while maintaining their position. This applies particularly at the first stage, where the diaphragm (M) may be located by an inner combustion system casing which expands axially at a different rate from the outer casing.

The rotor blade airfoils are commonly twisted between root and tip to cater for variations in the gas flow at different radii. The shape at any radius is defined partly by the degree of *reaction* (sections 8.2 and E.2). The blade twist is discussed with other typical data in section E.7 and the different shapes at root, mean and tip radius are shown in sequence in figure E.3. The blades end at a shroud (G) in higher duty turbines, to reduce tip leakage and vibration, but

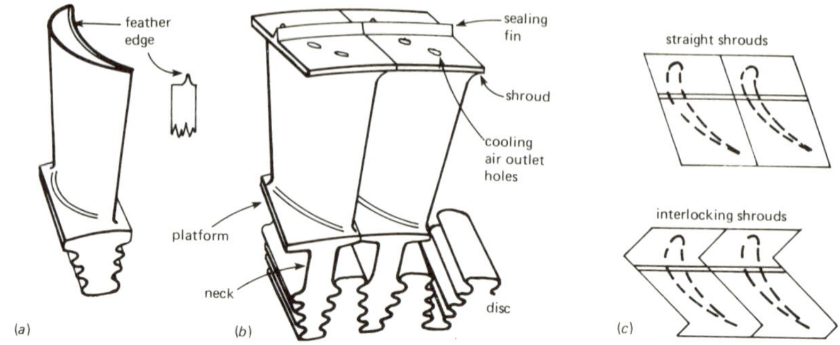

Figure 8.2 *Turbine blade features: (a) basic blade with featheredge which wears to minimum clearance; (b) blades with extended necks for cooling, tip shrouds and internal cooling passages; (c) shrouds with straight and interlocking edges*

earlier turbines had unshrouded blades; examples are drawn in figure 8.2. Like the NGVs, turbine blades may also be cooled by air flow through internal passages (section 8.5), particularly at the first stage. They are usually attached by fir trees to the disc, and locked in position by screws, pins, locking tabs or plates which are deformed in position. Whereas the fir tree used to be machined just below the airfoil root, higher temperature turbines now have an extended neck (E) below the inner annulus platform. This relies on the poor conductivity of the blade material to insulate the disc from the blade airfoil, which could well be hot enough to glow a bright red (up to 1250 K).

The disc stresses are such that a heavy hub is required to help the rim withstand the centrifugal blade loading and to achieve equal material stress at all radii. The stress levels are high (section 14.2) and the disc grows radially with increases in rotational speed; this complicates the design of the attachment to an adjacent disc, or to a shaft which does not expand as much.

The rotor is located axially relative to the casing by a thrust bearing and its position must be set very accurately to ensure minimum flow leakage between the blade tips and the seal segments (section 14.2). The turbine's capacity to extract a lot of work with very few stages means that there is a considerable pressure drop across each row of blading. It is therefore necessary to seal between the static and rotating assemblies between each blade row, and this is done by labyrinth seals (L) (section 14.2.1), which permit axial movement and adjustment. These seals permit small leakage airflows (A) which are used to cool the discs and shafts: adjustment of the radial clearance during manufacture determines the rate at which these flows occur.

8.2 Operation and Performance

The geometry creates accelerating flows within both the stator and rotor blading, but the motion of the rotor blades enables them to remove energy, retarding the flow and keeping the axial velocity substantially constant.

This can be seen from figure 8.3, a section through the blading at mean radius. The sonic velocity V_0 from the nozzles is reduced by blade motion to the lower relative velocity V_1 at rotor entry, and the high relative velocity V_2 at rotor exit leaves a moderate absolute velocity V_3 at entry to the next stage. The purpose of the nozzles is to accelerate the flow, exchanging static temperature for velocity (section B.2), and guide it into the rotor. The rotor blades cause a very significant change in the flow direction as well as the velocity change from V_1 to V_2; the results are a large force on the blading (section E.3) and the absorption of energy from the flow.

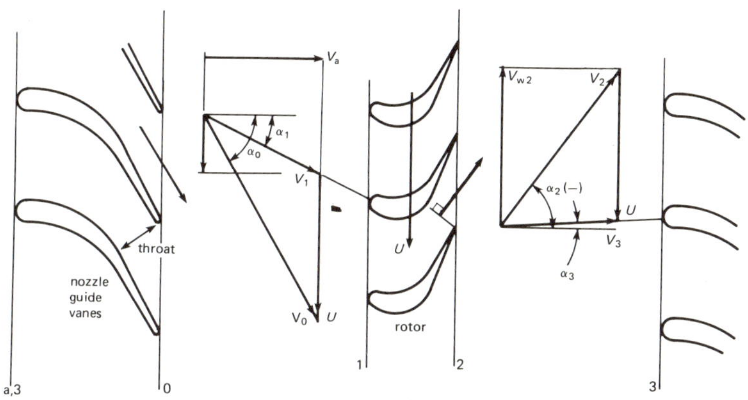

Figure 8.3 *Velocity triangles at mean diameter for a turbine stage*

The changes in velocity, along with temperature and pressure, may be represented on the temperature–entropy diagram, as shown in figure 8.4 This is built up in the same way as for the compressor, so that the small diagrams abcd and defg apply respectively to the conditions as the flow passes through the nozzles and the blades. Point d is the common link, where $p_0 = p_1$ and $T_0 = T_1$. The dotted line from c to h represents the condition change of the gas, while it is in the rotor, and the line from b to h represents the over-all effect on the gas as considered in simple cycle calculations. The isentropic efficiency (section C.1) is shown as bk/bj, or $(T_{0a} - T_{03})/(T_{0a} - T'_{03})$: this is in terms of the numbering through the stage, as against the numbering across the turbine as a whole used in section 3.3. The degree of reaction can also be seen from figure 8.4: $D = (T_1 - T_2)/(T_a - T_2)$ represents the proportion of the static temperature drop which occurs in the rotor, dm/(al + dm).

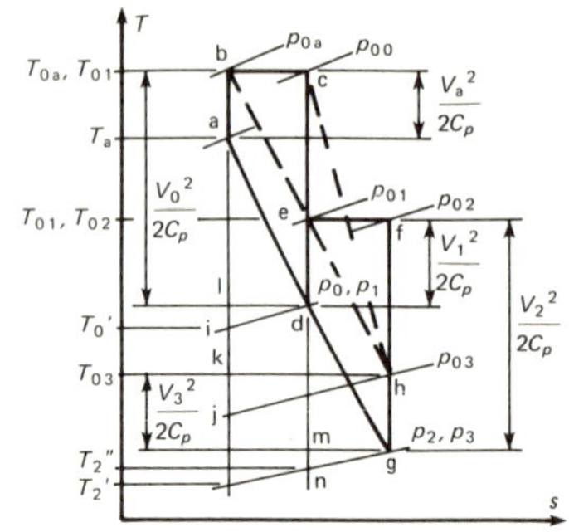

Figure 8.4 *Temperature–entropy diagram for a turbine stage*

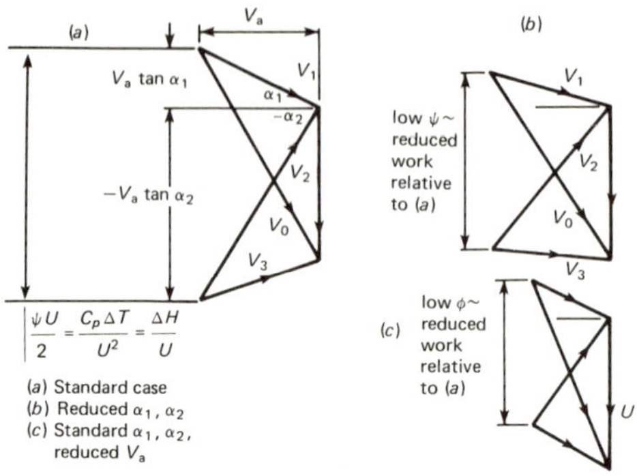

Figure 8.5 *Velocity triangles for: (a) standard case; (b) reduced blade loading ψ; (c) reduced flow coefficient φ*

Using α to designate gas flow angles, the power equation is similar to that for a compressor

$$P = \dot{m}C_p\Delta T_{0s} = \dot{m}UV_a\,(\tan\alpha_1 - \tan\alpha_2) \qquad (8.1)$$

Unlike the compressor, the turbine's value of α_2 is taken as negative (figure 8.3). Thus the deflection angle $(\alpha_1 - \alpha_2)$ is large and the numerical sum of $|\tan\alpha_1| + |\tan\alpha_2|$ is also large. In conjunction with the high value of V_a in the hot gas stream, this makes the turbine stage's work capacity six or seven times greater than that of a compressor stage (section 1.3).

A work-done factor, of the type used for compressor performance (section 6.2), is not normally used in turbines. The annulus wall boundary layer growth is limited, being continually re-energised by the secondary flow vortices (section E.5). A value of $\Omega = 0.98$, as for the first stage of a compressor, is used by some organisations, but the other way to handle this slight loss is to allow for it in the value of deviation angle, the difference between the angle of throat orientation and the outlet flow angle α_2 (section E.4).

The *blade loading* coefficient, or temperature drop coefficeint for the stage, is obtained by dividing equation 8.1 by $\frac{1}{2}\dot{m}U^2$. Then

$$\psi = \frac{C_p\Delta T_{0s}}{\frac{1}{2}U^2} = \frac{2V_a}{U}(\tan\alpha_1 - \tan\alpha_2) \qquad (8.2)$$

Some organisations omit the 2 for convenience.

Values of ψ, D (the reaction) and ϕ (the *flow coefficient*, V_a/U) are chosen by the designer to determine the blading geometry which will give the required performance. Relative to figure 8.5*a*, which combines the velocity triangles of figure 8.3, the effects of low values of ψ and ϕ are shown in figure 8.5*b* and *c* respectively. In both cases, low values reduce flow velocities and losses, with an increase in efficiency but a loss of work capacity. This is desirable in heavy industrial plant, where long-term operating costs count, but not in aircraft engines where the minimum weight, and hence the size or number of stages, is most important. Typical values of ψ are 3 for industrial use, 5 for aero use; ϕ may range from 0.6 to 1.0 and U is about 300 to 450 m/s.

The estimation of turbine performance in detail is commonly based on the work of Ainley and Mathieson,[60] which has been adjusted and extended by many other papers, a comprehensive one being reference 61. The basic method is to determine the values of pressure loss through the blade rows and of α_2 behind each blade row, for varying inlet conditions. The selection of different values of inlet conditions and blade speed U (or rotor speed N) then permits the operating characteristics to be determined. These are shown in figure 8.6, the flow *versus* pressure ratio curve at any speed being akin to that for flow through a fixed nozzle. The flow limit is set by choking, which occurs at a pressure ratio of about 2 for single-stage turbines, 2.7 for two-stage turbines and 4 for three-stage turbines. An alternative form of operating characteristic could also be drawn, as in figure 9.6.

Much of the data for this method comes from *cascade tests*, as described in appendix E.4 and treated much more fully in references 62 and 63A. These permit determination of the blade geometry and the leading and trailing edge angles which will generate the chosen flow angles. One parameter used is the pressure loss coefficient Y, which can be explained with reference to the $T-s$ diagram, as shown in figure 8.4. The value of Y in the nozzle is $(p_{0a} - p_{00})/(p_{00} - p_0)$, and in the blades is $(p_{01} - p_{02})/(p_{02} - p_2)$. Thus, Y represents the percentage of leaving energy degraded by friction, which drops the total pressure across the blade row. The equivalent loss coefficient in terms of temperature is X, which is defined as $(T_0 - T_0')/(V_0^2/2C_p)$ = il/ad for the nozzles and $(T_2 - T_2'')/(V_2^2/2C_p)$ = mn/fg for the blades. These pressure losses are equivalent to inefficiency: that is, a 10 per cent loss means an efficiency of 0.9. Typical *efficiency* levels are 0.85 to 0.9 but higher levels are obtainable with

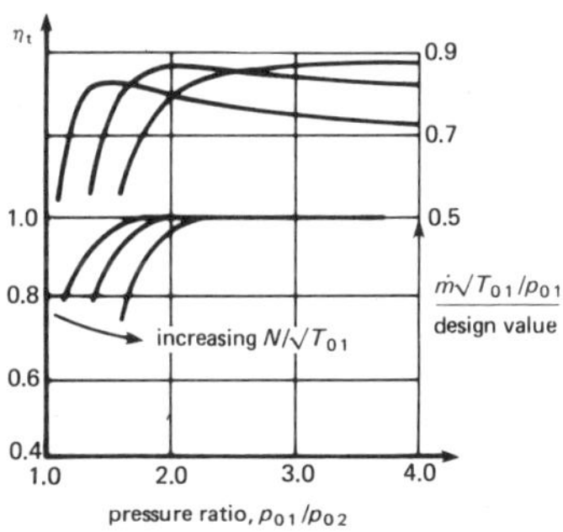

Figure 8.6 *Variation of mass flow and efficiency with pressure ratio*

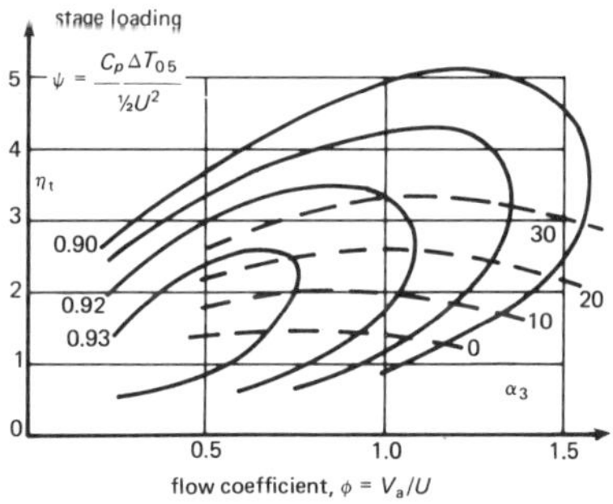

Figure 8.7 *Typical efficiency contours versus blade loading ψ and flow coefficient ϕ*

very low values of ψ and ϕ [6F] The standard efficiency contours relative to ψ and ϕ are shown in figure 8.7. The efficiency falls at low ϕ because of high deflection angles and at high ϕ because of high Mach number.

An implicit assumption in the foregoing is that the flow is steady and continuous. This is far from the case for the turbocharger fed from the exhaust of a diesel engine. Whereas the reaction at the blade roots may approach zero in steady flows (section E.7), a more appropriate minimum is 20 per cent in pulsating flow. [64]

8.3 Turbine Blade Materials

The materials available have always imposed a limit to the maximum usable turbine entry temperature (TET) in gas turbines and continue to do so, although blade cooling (section 8.5) permits a steep increase in TET when it is used. The properties required [6G] include tensile strength (relative to material density), resistance to high-frequency vibration fatigue stresses and low-frequency thermal fatigue stresses, and resistance to erosion and corrosion.

Steady high levels of tensile stress result in *creep*, with the material stretching progressively until rupture occurs at the end of the creep life. The approach of this condition can be detected by physical measurement of the blades, which are scrapped when stretched beyond allowable limits. One of the parameters used to assess creep strength is the *temperature capability*, the temperature at which a stress level of 9.0 tonf/in.^{2} ($\approx$ 139 N/mm^2) would cause rupture after 100 hours. [6H] A typical design creep life for commercial aircraft is 10 000 to 15 000 hours, but much longer lives are desirable in industrial applications.

Thermal fatigue results from sudden changes in gas temperature, during starting and engine transient operation (section 11.5). The material surface temperature becomes sufficiently different from that of the centre of the blade that, being restrained to the length imposed by the centre, it experiences a severe strain. Repetitions of this cyclic strain, compressive during start-up and acceleration and tensile during shut-down, results in the development of fatigue cracks, initially in line with the combustion outlet hot spots (section 7.5). The cracks often originate inside the cooling holes, where the predominant stress is tensile. The development of cracking can be slowed by reduced rates of engine transients and by reduced thickness of the walls of hollow blades, which prevent excessive strain. A deeper discussion of thermal fatigue and creep, and the combination of these two effects, is given in reference 65.

Rotor blades require high creep strength to counter radial accelerations of several thousand *g*, although this is compromised to achieve a satisfactory balance with the other properties. Nozzle guide vanes have low stress levels in bending, resulting from the gas loading, but operate at about 100 K higher temperatures than the following blades, as indicated by the length of al or ce in figure 8.4. The high temperatures are their most critical condition, resulting in severe thermal fatigue, erosion and corrosion, and even local melting.

The materials used are commonly nickel or cobalt alloys the former being favoured for blades, and either for NGVs. Nickel alloys are stronger but require very careful manufacture, including the use of vacuum melting. They retain better temperature capability in cast form than forged. Casting technology is advancing further, towards unidirectional solidification and even the casting of blades as a single crystal.[66] The main compositional changes to alloys for nickel blades, which have increased the temperature capability from below 800 K to over 1000 K in 20 years of progress, have been a reduction of chromium content from 20 per cent to about 5 per cent and the introduction of about 6 per cent aluminium, about 10 per cent tungsten and about 3 per cent tantalum: the content of carbon, cobalt, titanium, molybdenum, niobium and other elements has been variable. The effects of these constituents, the beneficial effects of boron and zirconium and the effects of manufacturing progress are discussed in reference 6H.

Nickel alloys for nozzle guide vanes retain about 20 per cent chromium content, to suit the corrosive operating conditions. Cobalt alloys are more expensive, but can be made and repaired with less expensive techniques and are better at resisting thermal fatigue and corrosion. Typical high-temperature, cobalt-based alloys contain about 20 per cent chromium, 10 per cent tungsten, 5 per cent tantalum and traces of titanium and zirconium.

Materials for the future include new chromium, niobium and molybdenum base alloys, the last two possibly being suited to use in the turbines for helium-cooled nuclear reactors. A new step increase in operating temperatures, to over 1750 K, may become possible with the introduction of ceramic materials, silicon carbide or silicon nitride. These have excellent properties except for brittleness, and are discussed in increasing detail in references 6G, 67 and 65A. Their stress temperature relationship is shown in figure 8.8.[6A]

8.4 Blade Corrosion: Protective Coatings

There are several corrosion mechanisms which affect nickel and cobalt alloys. 'Green rot' is a form of internal oxidation which deposits chromium carbide at the grain boundaries; when failure occurs, the fracture surface is coloured green by the presence of nickel oxide. Another involves the presence of vanadium in the fuel, resulting in the formation of vanadium pentoxide ash, which is deposited on the blades or erodes them. Poorer quality fuel oil and coal may also contain corrosive phosphates, or lead which is very corrosive in lead sulphate form. Sulphur is an undesirable constituent in the fuel, particularly in marine environments, where it causes the common and serious hot corrosion, also known as 'sulphidation' or 'black plague'.

Hot corrosion depends on the presence of sodium in addition to sulphur.

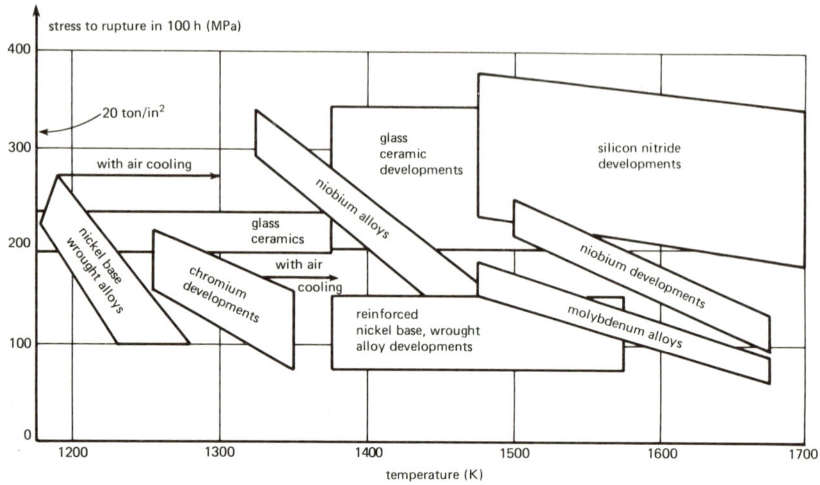

Figure 8.8 *High-temperature material developments: creep stress/temperature areas with material density corrected relative to Nimonic 115 (redrawn from reference 6A by permission of the Council of the Institution of Mechanical Engineers)*

Even minute quantities of sodium, 0.03 p.p.m. in air or 0.5 p.p.m. in fuel, cause the deposition of a semi-liquid slag of sodium sulphate on the blade surfaces.[4D] Under this slag, and above a threshold temperature which is about 100 K higher for cobalt-based blades than for nickel-based blades. the protective layer of combined nickel and chromium oxides is broken down to permit attack of the parent metal. This can proceed to the extent of corroding deep holes in the blade (figure 8.9), or blade breakage.[4C]

The chemistry is, briefly, as follows. Sulphur attacks the oxide layer by converting the chromium oxide to chromium sulphide, releasing the oxygen into the sulphate layer. This attacks the nickel oxide to form nickelate ions (NiO_2^{2-}), which migrate to the air/sodium sulphate interface and are then reprecipitated as nickel oxide (NiO) in a non-protective form. The chromium sulphide also changes back, in time, to chromium oxide. The sulphur and oxygen are thus both free again, and they continue this sequential attack until the oxide layer is penetrated. They then continue by attacking the parent chromium and nickel, in a similar auto-catalytic process. The breakdown of the protective oxide layer is basic to the corrosion processes involving the other elements mentioned above. Another example is chloride attack of the cooler blades, downstream of the blades suffering hot corrosion.

The blades may be damaged by other means than with a corrosive deposit. The operation of nickel/chromium/cobalt alloy blades in a *reducing atmosphere*, such as may occur with some fuels or in some chemical process applications,

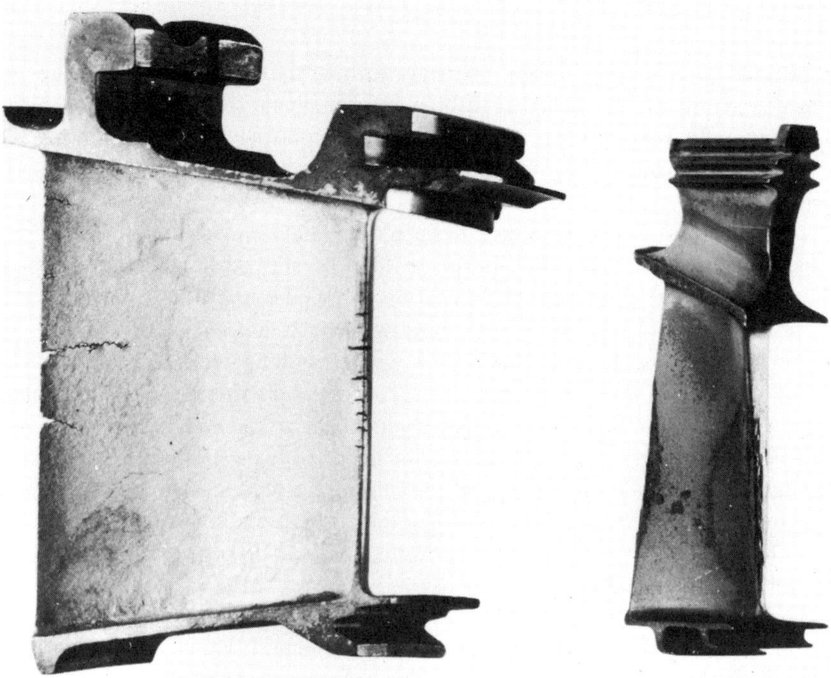

Figure 8.9 *Turbine blade and nozzle guide vane attacked by corrosion. The blade is still dirty, showing peeling deposit and products of hot corrosion and green rot. The cleaned NGV shows stippled contours where surface was attacked by hot corrosion. Leading and trailing edge cracks resulted from hot streaks (figure 7.13). Courtesy of Air New Zealand*

rapidly removes the oxide layer and attacks the alloy. For instance, carbon monoxide will remove the oxide and then combine rapidly with nickel to form nickel carbonyl, a volatile liquid which evaporates into the exhaust.

In an oxidising atmosphere, the materials can be made to resist hot corrosion and other mechanisms, or at least have a higher threshold temperature, by the addition of chromium to the alloy. This goes against the trend which resulted in increased creep strength, and calls for a careful compromise in the choice of alloy for a given duty. Fortunately, high-strength alloys gain considerable protection from the use of pack aluminised *coatings*, in which a carefully chosen mixture of metal powders is diffused into the blade at high temperature and pressure to alter its surface chemistry. The main elements used are aluminium, nickel, chromium and yttrium. The coating may last many thousands of hours until it is damaged, or eroded away by atmospheric dirt or carbon particles (smoke) from the combustion system (section 7.3).

Hot corrosion is discussed in greater detail in reference 23, which summarises the material in references 65 and 68.

8.5 Turbine Blade Cooling

Blade cooling technology for the first stage turbine started in Germany in the 1940s to permit the use of higher turbine entry temperatures than the available materials could withstand. Its use in production aircraft engines started in the 1950s, and it is now a well-established art which is extending into industrial turbine practice.

The objective of blade cooling is to keep the metal temperature at a safe level, to ensure a long creep life and low oxidation rates. Although liquid cooling is possible, using thermosyphon and heatpipe principles, the almost universal method is by a cool air or working fluid flow through internal passages in the blades. An example of a relatively simple system, with cooled nozzles and blades, is shown in figure 8.10. Air is introduced and distributed by a guide tube inside the nozzle, to impinge inside the leading edge at known local hot spots. It then flows round the guide tube against the inside of the airfoil wall to exhaust through holes at the trailing edge. Air for cooling the blades has to be accelerated tangentially so that the moving blades can take it in, using flow vectors similar to those in figure 8.3. The air then flows through passages in the blade in a single or triple pass, exhausting at the blade tip.

The effective gas temperature T_g is taken as the average level which an uncooled blade would reach if placed in the gas stream. It has an approximate value of: TET $- 0.15\, V_m{}^2/2C_p$.[63A] The difference between T_g and the mean temperature T_b of a cooled blade determines the rate at which heat flows from the gas into the blade. Thus, using the nomenclature listed on page xvi

$$\Delta H = h_g \times A_b \times (T_g - T_b) \tag{8.3}$$

If this is compared with the available heat in the flow, $H = \dot{m}C_p(T_g - T_b)$, then the parameter $\Delta H/H$ expresses the fraction of the available heat in the stream which is lost to the cooled blades. This parameter is akin to the loss coefficients X and Y which allow for the frictional losses through the blade row (section 8.2). It has been shown [63A] that $\Delta H/H$ can be expressed in terms of the Nusselt and Reynolds numbers, thus

$$\frac{\Delta H}{H} \approx 3.24 \frac{Nu}{Re} \frac{c}{s \cos \alpha_2} \tag{8.4}$$

The heat flows through the blade to the cooled areas, and then into the cooling air. The cooling air passages are generally made with a large perimeter-to-area ratio (in cross section) to achieve the maximum rate of heat transfer. The passage geometry can be related to the cooling achieved using the Z factor [69] where $Z = (S_c/c)^{1.2}/(A_c/c^2)$. All hole patterns with the same value of Z give the same cooling performance in terms of the relative blade temperature, $(T_b - T_c)/(T_g - T_c)$, where T_c is the cooling air temperature. This parameter can be plotted against Z, as shown in figure 8.11, the laminar and turbulent

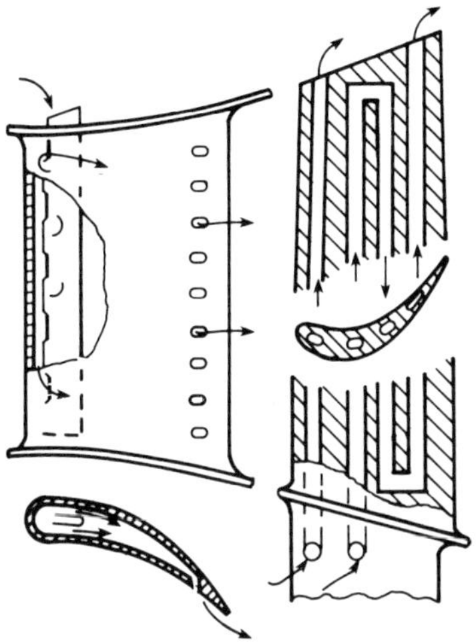

Figure 8.10 *Typical cooled NGV and blade; with impingement and film cooling of the NGV leading and trailing edges, and single and triple pass cooling passages in the blade*

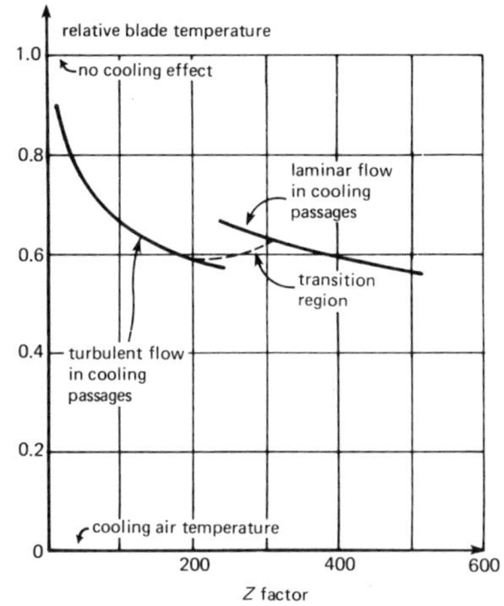

Figure 8.11 *The effect of cooling-hole geometry (Z factor) on heat transfer rate from the blade to the cooling air flow. (Redrawn from reference 69 by permission of the Royal Aeronautical Society)*

conditions applying to the flow in the cooling passages. The cooling performance also varies with the cooling air flow, which is chosen in combination with the value of Z.[38B] An alternative parameter is the cooling effectiveness, (TET – T_b)/(TET – T_c), for which a typical current value is 0.4 and projected values from 0.5 to 0.55 can be foreseen.[4B] A further, significant increase may be obtained by reducing T_c even to 288 K, by water cooling the compressed air.[8]

In practice, there are several departures from the mean conditions. The static temperature and pressure vary round the airfoil, approaching stagnation levels at the leading edge and falling to low levels on the convex side, where indicated in figure E.11 by the increase in velocity. These conditions cause variations in the rate of heat transfer to the blade around its profile, and the rate is strongly increased where the flow changes from laminar to turbulent, as in figure 8.12. Although it can be considered theoretically, reference 7 suggests that there is a particular departure from the theory, measurement showing that the heat transfer to the concave surface is up to twice as great as for the convex surface. This arises from vortex flows which continually feed hot gas into the boundary layer. The rate of heat transfer to the cooling air also varies, particularly at the long trailing edge which is remote from the nearest cooling hole, and also in local variations round the holes. The effect of all the variations, in heat input and removal, is shown in figure 8.13, the temperature contours of a typical advanced rotor blade section.

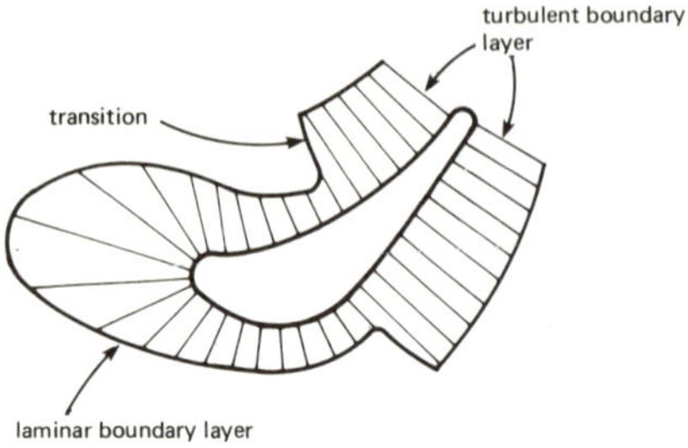

Figure 8.12 *Change of heat flow into blade with laminar and turbulent flow. (Redrawn from reference 69 by permission of the Royal Aeronautical Society*

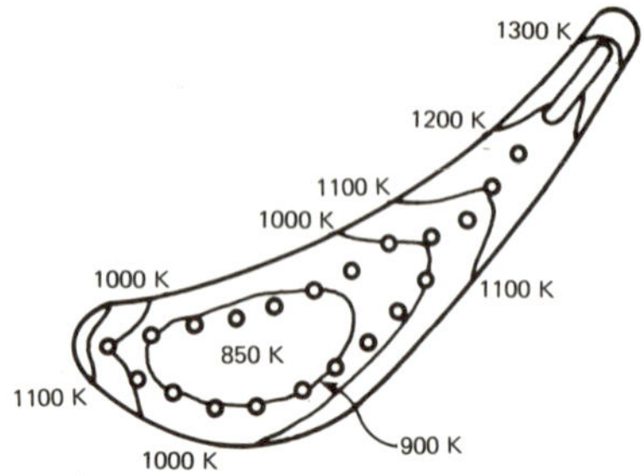

Figure 8.13 *Typical temperature contours across a cooled blade section. (Redrawn from reference 6F by permission of the Council of the Institution of Mechanical Engineers)*

The foregoing applies generally to blades or vanes with the cooling air separated from the hot gas, using a configuration like that of figure 8.10. As much as 5 per cent of the working fluid may be required from the h-p compressor delivery, missing the combustion system to cool the first-stage nozzles alone. In one advanced design, a further 2 per cent of the h-p flow was required to cool the first-stage rotor blades, reducing the metal temperature by 220 K from a TET of 1360 K to permit the blades to last 10 000 h. Uncooled, they would survive 0.2 h.[38]

An even higher level of cooling performance is permitted by effusion cooling, also known as film or transpiration cooling. This lets the cooling flow leak outwards through small holes in the airfoil surface, to form a film of cool air isolating the metal from the hot gas. Particular areas requiring this are the leading edge, and the concave side of the trailing edge. The blade walls may be relatively cooler, especially in split-shaft engines where the high h-p balde velocity minimises the stagnation effect. Cooling air supplies at different pressures may need to be matched to the static pressure variations around the airfoil to obtain the required flow rates. Film cooling can be applied to blades and vanes; a very sophisticated level of vane-cooling technology is shown in figure 8.14. The over-all cooling effect in the same engine is such that the mean rotor blade temperature is about 350 K below the prevailing gas temperature.[7]

Passing the air two or three times along the blade, and then exhausting it as a film yields the optimum cooling effect for a given air flow. The limit to blade cooling is set when the cooling air flow required is so wasteful as to offset the gains in engine performance resulting from the use of the higher TET. In fact, sometimes a different reference temperature is used, the *nozzle exit temperature* (NET); this is the temperature from which the turbine rotor's extraction of energy starts, which is already degraded from the turbine entry temperature level by the cooling of the nozzles.

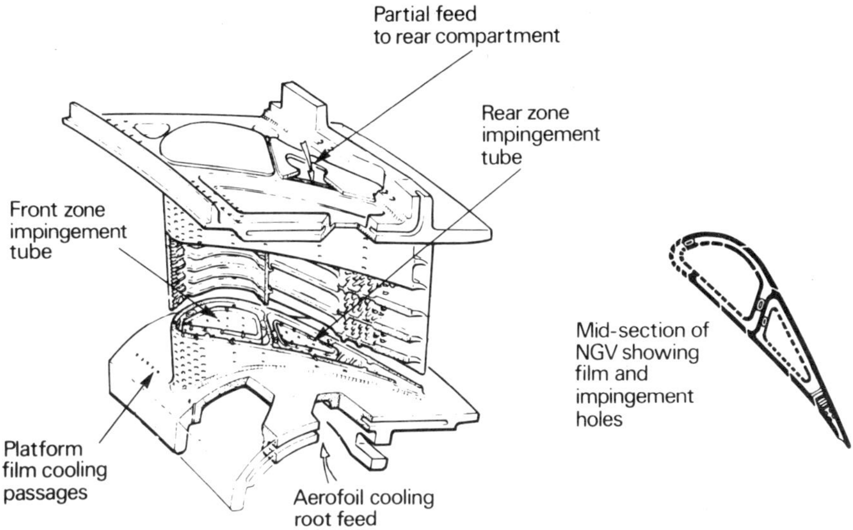

Figure 8.14 *Example of the advanced level of cooling technology used in a modern aircraft engine nozzle guide vane (RB211–22B) (by courtesy of Rolls-Royce Ltd)*

A number of the comments above have referred to the effects on performance. To collect these, the *penalties* of blade cooling include the loss of working mass flow, the reduction of inlet temperature to the later stages, pressure losses from the mixing of the cooling and main flows, and the loss of energy in pumping the cooling air.

8.6 Turbine Blading Manufacture

Nozzle guide vanes, involving an airfoil between an inner platform and an outer shroud, have complicated geometry and are almost always cast by a sophisticated process. Investment or lost wax casting is a common technique, giving the accuracy required for the location of the cooling passages. The airfoil, being untwisted, is relatively easily dressed smooth, and the end fixings are machined for location.

Turbine blades are commonly forged, although new casting techniques offer improved material properties in some applications (section 8.3). The forging processes may include extrusion, and the elliptical cooling passages may be formed at the same time. One method requires the cooling holes to be drilled initially in the blank and plugged with steel rods; the steel then deforms with the blank as the airfoil shape is formed. The forging pressures required are extremely high with nickel and cobalt alloys, and the lubricant between the workpieces and the dies may even be a soft steel. The residual steel can be removed by boiling in nitric acid, without detriment to the alloy.

The turbine blade airfoil is formed by dies, which are shaped by the methods discussed in section 6.3. The machining of the fir-tree root is often done by broaching, and electrochemical machining may be used in some parts to avoid the conventional cutting processes, which are difficult with these materials. Small holes, particularly the effusion cooling holes (section 8.5), may be made by spark erosion.

9

The Radial Turbine

The radial, or centripetal, turbine is used for very small flows, where it is cheaper and simpler to make than an axial flow turbine. It is used in the smallest gas turbine engines, as in some automotive or fire pump or other portable applications, and is used extensively in the smaller turbochargers. It is also suitable for turbo-expansion applications (section 2.9). Its size ranges up to about 0.15 m diameter, and its rotational speed is likely to be between 60 000 and 100 000 rev/min.

9.1 Operation

The appearance of the turbine may be seen in figures 1.5*b* and 9.1. It can be seen to resemble a centrifugal compressor and, indeed, the roles could be reversed, although with considerable loss of efficiency. The flow approaches the impeller tip with considerable whirl velocity: in a free vortex condition, the whirl velocity would increase as the flow radius decreased. The turbine rotor absorbs this whirl velocity, however, and causes the flow to exhaust approximately axially. The torque exerted on the rotor by the flow is exerted by the *Coriolis* effect (section 5.2).

9.2 Design Features

As with the axial flow turbine, the stage is taken in order of a nozzle followed by a rotor. The inlet to the nozzle usually involves a single tangential inlet duct into a scroll casing, as the gas supply usually comes from a single combustor or reciprocating engine exhaust. The scroll is a curved passage of reducing area (section 9.3) which terminates where it surrounds the impeller tip.

The *nozzle* may take several forms: a fixed set of vanes, a variable set of vanes or no vanes at all, which are all indicated in figure 9.1. The fixed vanes would

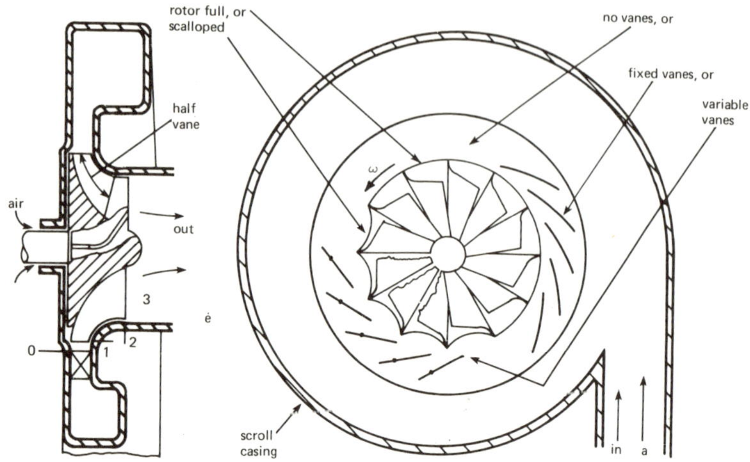

Figure 9.1 *The main features of a radial turbine*

typically be set for a flow angle α_0 of 70°, as would the variable vanes for normal operation. However, the variable vanes may be turned to about 80° to maintain rotor speed when the flow is low, or could even be turned to drive the rotor backwards in transport applications. The vaneless gap takes the form of a peripheral opening between the tips of the front and rear casings: this width is adjusted when matching the turbine to its duty. The lack of vanes permits the flow angle to adjust itself, and avoids putting in aerodynamic disturbances which might excite rotor vibrations.

The *rotor* is usually of cast, nickel-based alloy, for the same reasons of high-temperature oxidation, corrosion and erosion as in the axial flow turbine. A typical number of vanes is ten or twelve, many fewer than in a compressor because the accelerating flow is easier to handle. The vanes are usually radial at the tip, but are bent to form the exducer at the outlet. Half vanes may also be used, as shown in one part of figure 9.1: they bring a slight improvement in efficiency but add to the weight and complexity.

The main rotor stresses are those causing creep in the hub and at the exducer vane root.[61] These limit the life and are affected by the operating temperature. The thinness of the vanes precludes the use of internal air cooling passages, but an air supply which cools and helps seal the bearing system (section 14.4) may also flow outwards to cool the front of the hub. The inexactness of the means of calculating the rotor stresses is usually compensated for by the use of a moderately ductile alloy, which allows the overstressed regions to relieve themselves during the first run.[61] The stresses are reduced by cutting away the hub between the vane tips, which also reduces the rotational inertia and improves the acceleration characteristics for a very slight loss in efficiency.[70]

Usually the shroud is a part of the fixed casing, and there must necessarily be a running clearance. A gap of about 5 per cent of the passage width at the impeller tip has been found suitable;[61] this is about the same as for a centrifugal compressor. The outlet duct from the turbine simply takes the exhaust away to a convenient place, but it can be made slightly divergent to reduce the pressure at rotor outlet and improve its performance.

9.3 Design Point Performance

The velocity triangles are shown in figure 9.2. The relative velocity at entrance to the impeller tip (V_1) is aligned approximately with the vanes. The angle of the exducer vanes (α_2) causes exit velocity V_2 to sum with the blade velocity U_2 to leave the outlet flow V_3 approximately axial.

In the ideal case, it is assumed that the relative flow enters the rotor radially, that it maintains constant velocity as seen in side elevation ($V_1 = V_{a2} = V_3$) and that there is no swirl in the flow exhausting from the rotor ($\alpha_3 = 0$). The torque exerted is therefore the loss of angular momentum of the inlet flow, $\dot{m}V_{w1}R_1$ (figure 9.2). The power is given by torque times angular velocity, so

$$P = \tau\omega = \dot{m}V_{w1}U$$

where U is the tip velocity, equal to $R_1\omega$. As the inlet flow is radial relative to the rotor vane, V_{w1} is the same as U and the expression for power per unit mass flow, in W/kg s or J/kg, is simply

$$P/\dot{m} = U^2$$

This is a simpler expression than the equivalent for the centrifugal compressor, equation 5.2. As the power extracted is also the enthalpy change, $C_p(T_{0a} - T_{03})$ as shown in figure 9.3, the full expression becomes

$$W = P/\dot{m} = U^2 = C_p(T_{0a} - T_{03}) \tag{9.1}$$

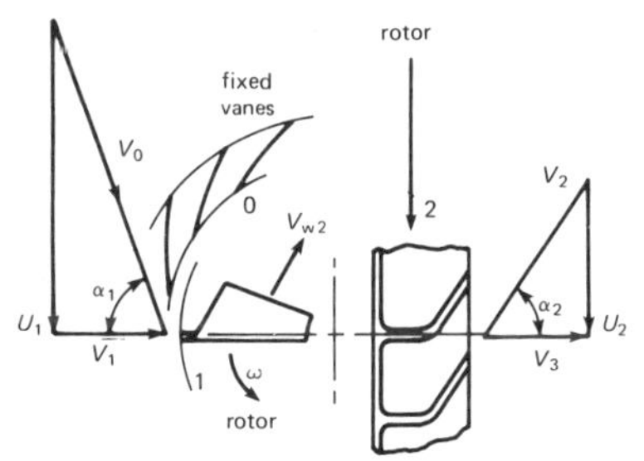

Figure 9.2 *Velocity triangles for a radial turbine*

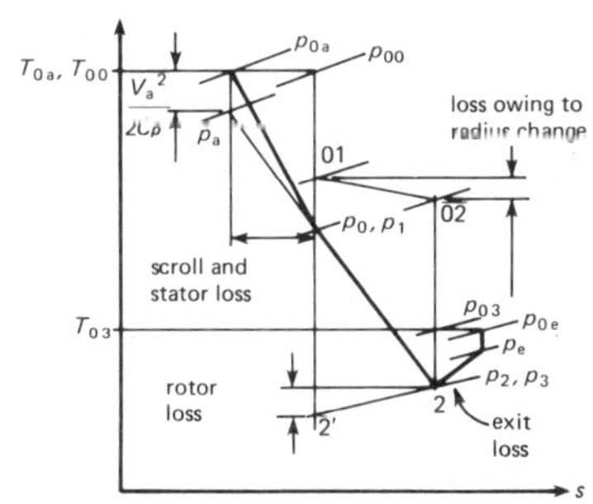

Figure 9.3 *T–s diagram for the case of shock-free entry to the rotor. (Redrawn from reference 71 by permission of the Council of the Institution of Mechanical Engineers)*

The performance is assessed in terms of non-dimensional (section B.5) speed, inlet and outlet pressure and temperature; the torque relative to inlet pressure and the outlet swirl angle. The pressure ratio is taken as either the ratio of stagnation (total) inlet to outlet pressure or as the *total-to-static* ratio. The latter is the ratio of total inlet to static outlet pressure and is sometimes also used with centrifugal compressors: it avoids the effect of an inherent change in total pressure with flow radius, as the blade speed is different at inlet and outlet so affects the velocity triangles differently. The total-to-static pressure ratio is therefore higher than the stagnation pressure ratio for a turbine and lower for a compressor. This accounts for the use of static conditions in figure 9.3, in contrast to figure 8.4. Typical values of performance parameters are a pressure ratio of up to 4 for one stage, efficiency of 0.8 to 0.9 and reaction of about 50 per cent.[61]

A sample calculation of the geometry and performance is given in reference 70, and may be summarised as follows. The isentropic temperature drop is greater than $(T_{0a} - T_{03})$, in proportion to the over-all efficiency (section C.2), and leads to the pressure ratio required, as in section 3.3, step 3. The known outlet pressure (approximately atmospheric) leads to the required inlet pressure. The pressure loss through the scroll and nozzles yields the conditions at rotor inlet, from which the density change and flow passage cross-sectional area change through the rotor may be calculated.

Hiett and Johnston [70] found that the efficiency was at a maximum when the tip width was about a tenth of the tip diameter. Using this ratio, the relationship between power requirement, mass flow and tip speed would permit determination of the rotor inlet area required, and therefore of its diameter and rotational speed. Typical data would be $U = 400$ m/s, $V_{a2} = 150$ m/s and the outlet diameter to tip diameter ratio would be about 0.75 to 0.8.

The scroll loss depends very much on its cross-sectional shape. A circular cross section (figure 9.4*b*) provides the minimum constraint on the flow direction at all points and therefore causes the minimum pressure loss, but it may be bulky. Losses rise as a smaller area is used, while the use of a square section or the ultra-small diameter, triangular design (figures 9.4*a* and *c*) have been shown to reduce the over-all efficiency by about 1 per cent. [6l] The scroll and nozzle

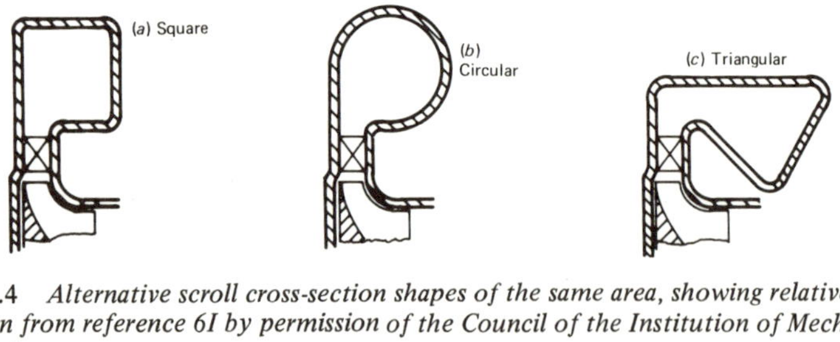

Figure 9.4 *Alternative scroll cross-section shapes of the same area, showing relative sizes. (Redrawn from reference 6I by permission of the Council of the Institution of Mechanical Engineers)*

loss is assessed as a pressure loss coefficient, $Y = (p_{0a} - p_{00})/(p_{00} - p_0)$ as used in section 8.2, for which typical values range from 0.05 to 0.15.

The rotor loss arises from friction along the walls and a significant velocity gradient across each passage, as explained by figure 5.6. A development of the blade-to-blade plane AA, midway between the hub and shroud (figure 9.5) shows the typical sudden change of direction of the passage. Flow velocities are high on the inside of the bend because of relative circulation, and may fall to negative values at the concave corner of the driving face.[71] This causes a recirculating flow, and occurs particularly when the mass flow is low relative to the rotor speed, causing a condition of negative incidence (section 9.4). The loss for the design point, zero incidence condition is seen from figure 9.3 to be $X = (T_2 - T_2')/(V_2^2/2C_p)$, as in the nomenclature of section 8.2.

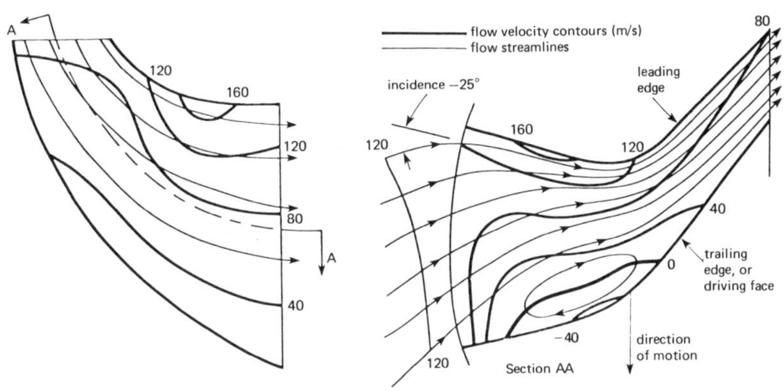

Figure 9.5 *Flow pattern and velocity gradient in one rotor passage. (Redrawn from reference 61 by permission of the Council of the Institution of Mechanical Engineers)*

9.4 Off-design Performance

Operation at off-design conditions implies that the load driven by the turbine shaft is not matched appropriately to the flow conditions. In running fast or slow, the relative velocity V_1 is misaligned with the radial vanes, α_1 being positive or negative, and the variation in torque may cause swirl in either direction in the exhaust. This occurs at the relatively small exit radius, however, so represents only a small variation in exhaust kinetic energy. For this reason, the radial turbine is able to operate at high efficiency over an extremely wide speed range.[72]

The reaction driving the rotor, as the flow is accelerated out through the angled exducer, enables the rotor to exert a torque over a wide range of inlet conditions. In fact, the *incidence* may vary from strongly positive (impinging on the driving face of the vane) through zero to strongly negative (opposing the rotation by impinging on the trailing face of the passage). One factor influencing the incidence is the irrotational nature of the flow in each passage, which tends to cause negative incidence in the same way as it causes *slip* in the centrifugal compressor (figure 5.6).

The incidence may be related to the *velocity ratio* U/V_T, where U is the tip velocity and V_T is the theoretical velocity which the gas would reach if expanded isentropically from the given inlet conditions to the static pressure at rotor outlet. Zero incidence typically occurs at a value of U/V_T between about 0.6 and 0.8. A lower value means that the rotor is running slower, perhaps because the resisting torque is high, and the incidence is then positive.

The typical performance over a wide range of speeds and pressure ratio is plotted in figure 9.6, derived from reference 72. The vertical scale of non-dimensional temperature drop could easily be converted by equation 9.1 to indicate power output, and its values are roughly appropriate to a machine of the proportions given in section 9.3. It can be seen that the loss in efficiency is not very severe as the speed varies from the ideal.

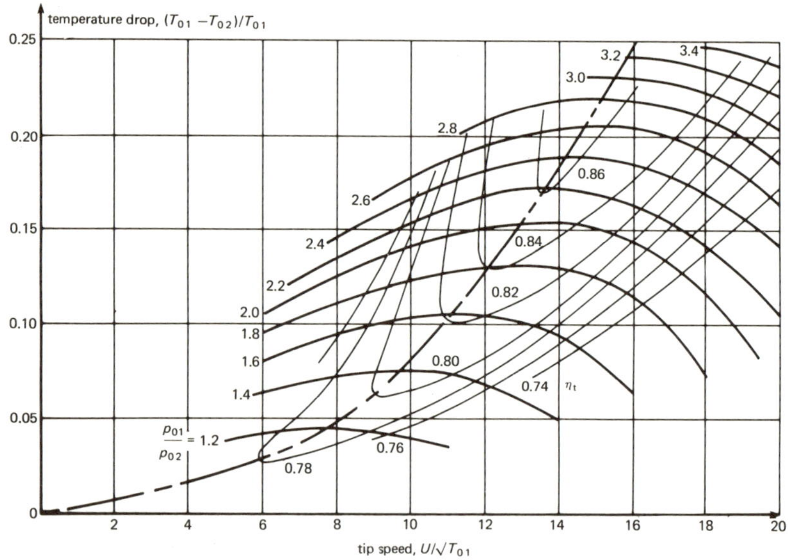

Figure 9.6 *Typical performance curves for a radial turbine. (Redrawn from reference 72A by courtesy of Newnes-Butterworths, Technical Publishers)*

The extremes of the speed range are stalling (lack of rotation rather than the aerodynamic nature of stall in a compressor), and runaway when unloaded. Hiett and Johnston show that the *torque ratio* (stalled torque/torque at peak efficiency) is about 2 for the radial turbine, compared with 2.7 for the axial turbine: the torque rises nearly linearly as the speed is reduced to zero. The *runaway* speed is about 1.6 times the speed at peak efficiency for the radial turbine, compared with about 2.5 for the axial turbine. The radial turbine can also be taken beyond stall, and made to run backwards by sufficient rotation of the nozzles. The flow pattern is then very disturbed, making the use of half vanes in the rotor beneficial. The excessive outlet swirl reduces the efficiency to about 0.5 and the torque to about half of the forward capability.

The general effect of running with incidence at rotor inlet is to introduce a *shock loss*, as shown in figure 9.7.[71] The over-all shock and rotor loss may be plotted against incidence (figure 9.8), in a form very similar to figure E.8, except that the radial turbine runs with quite low loss over the wide incidence range of about ±50°. This loss may be reduced by reducing the tip width, reducing the nozzle angle (α_0), by increasing the pressure ratio across the machine or by increasing the acceleration through the rotor.

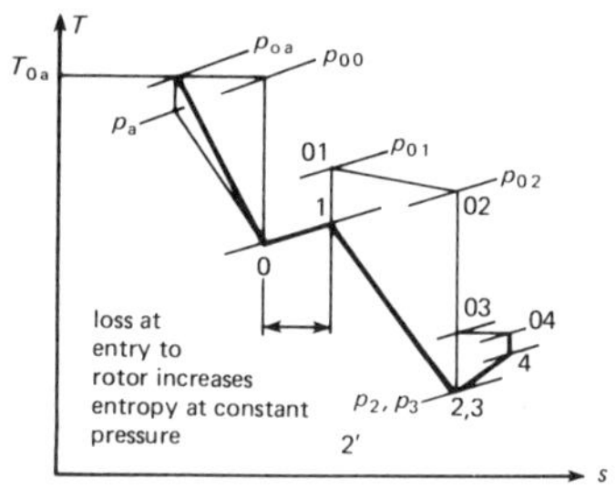

Figure 9.7 *T–s diagram for the case with shock loss at rotor entry (cf. figure 9.3). (Redrawn from reference 71 by permission of the Council of the Institution of Mechanical Engineers)*

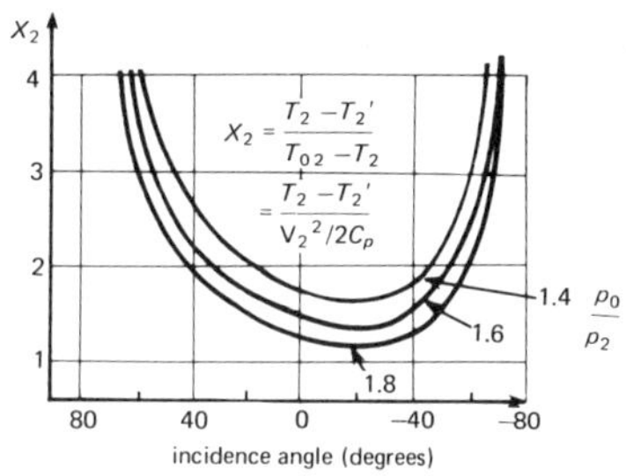

Figure 9.8 *Shock and rotor loss for a radial turbine, relative to incidence angle (Redrawn from reference 71 by permission of the Council of the Institution of Mechanical Engineers)*

The final, off-design performance condition to consider is the effect of a *pulsating* inlet flow, as may occur in a turbocharger driven by the exhaust gas of a reciprocating engine. Reference 64 indicates that the efficiency of axial flow turbines falls in such conditions if the reaction of the blading is low. It may therefore be inferred that, with its high reaction, the radial turbine copes well with pulsating flows. It has the further advantage that a pulse in the flow affects only a few vanes at a time where, in the axial turbine, the pulse tends to affect all blades at the same time because of its different mode of approach.

10

Compressor Operating Characteristics

The potentially tricky operation of dynamic compressors, with their propensity to surge, has already been mentioned in section 1.3 and chapters 5 and 6. It is a big enough subject to merit its own chapter, before the way in which the compressor is matched to the turbine and the over-all engine operating characteristics are discussed in chapter 11.

10.1 The Basic Compressor Characteristic

Both centrifugal and axial compressors have well-defined operating characteristics, which are normally plotted in dimensionless form. The characteristic is normally determined by rig testing: the test compressor is run at a series of different, fixed speeds while a valve controlling the outlet flow is steadily closed until the maximum pressure ratio is reached at each speed. Several steady-state performance points are recorded and plotted at each speed.

Figure 10.1*a* shows a typical constant speed characteristic for one stage of an axial compressor, in terms of stage pressure rise coefficient against flow coefficient (V_a/U). It is akin to the lift curve of figure E.8, but is plotted using different parameters and with the horizontal axis reversed. The two curves are also linked by equation 6.4

$$\text{stage pressure rise coefficient} = \frac{\Delta p_{0s}}{\frac{1}{2}\rho U^2} = 2\eta_s \Omega \frac{V_a}{U}(\tan \alpha_1 - \tan \alpha_2)$$

The vertical scale in figure E.8 is deflection, $\epsilon = \alpha_1 - \alpha_2$, which is related to ($\tan \alpha_1 - \tan \alpha_2$). The horizontal scale of incidence angle ($\alpha_1 - \beta_1$) is related to V_a/U, as demonstrated by figure 10.1*b*: the effect of reducing V_0 and V_a (and V_a/U) while α_0 and U remain constant is to increase α_1 and make the incidence more positive. Incidence values can therefore be shown on the V_a/U scale as in

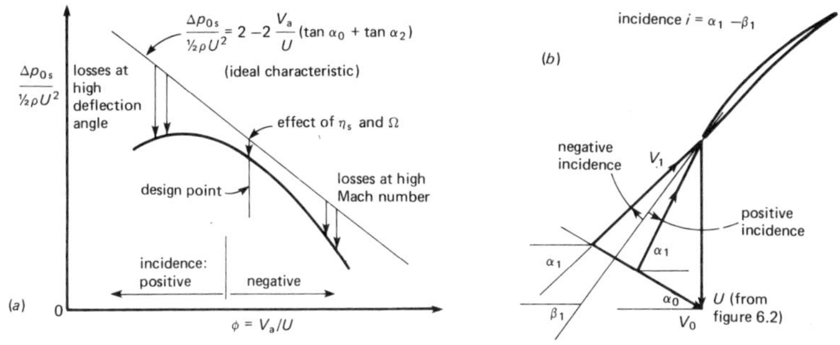

Figure 10.1 *Constant speed compressor stage characteristic showing positive and negative incidence and their effect on the stage pressure rise*

figure 10.1*a*, showing that the horizontal scale is reversed from that in figure E.8. The curve applies equally to the inducer of a centrifugal compressor. The straight line in figure 10.1*a* is a plot of $\Delta p_{0s}/\frac{1}{2}\rho U^2 = 2 - 2(V_a/U)(\tan\alpha_0 + \tan\alpha_2)$, which is an alternative derivation of equation 6.4 and in which η_s and Ω are omitted. It shows how the actual performance deviates from the theory as the incidence angle departs from the ideal.

The steady speed characteristic for a multi-stage axial compressor is shown in simplified fashion in figure 10.2, in terms of pressure ratio against mass flow. These parameters are comparable to those in figure 10.1*a* so the curves are of similar shape. The compressor operation is stable only to the right of the maximum pressure ratio because, if the flow were to drop slightly while operating at point P, the operating point would move up the curve to a higher pressure ratio. The increased pressure would increase the flow through the outlet valve, thus returning to point P.

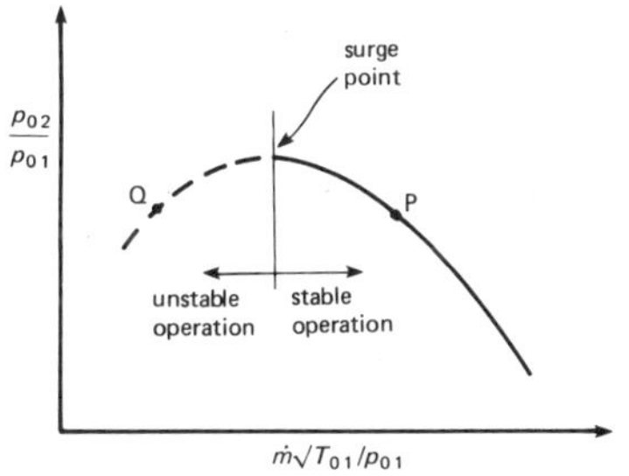

Figure 10.2 *Simplified constant speed characteristic demonstrating regions of stable and unstable operation*

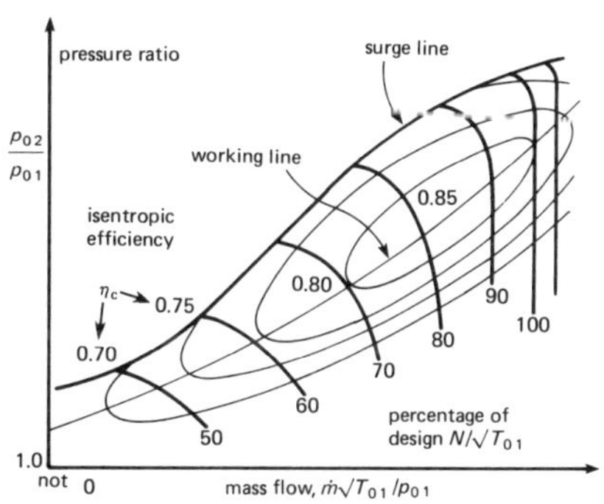

Figure 10.3 *Basic over-all compressor characteristic, covering the operational speed range*

On the other hand, when operating with the positive slope shown at point Q, a slight drop in flow would result in a slight loss in outlet pressure, a further loss of flow, greater pressure loss, etc. This is unstable, there being nothing to stop the operating point from accelerating off the bottom of the curve. This happens in practice, with the flow actually going negative and coming out of the compressor inlet, but it is only a momentary effect. When the back pressure has cleared itself, the positive flow is re-established but, if the outlet remains too restricted, the flow reverses again when the pressure exceeds the compressor's capability. Thus, the flow *surges* back and forth, and the limit of stability is called the *surge point.*

Surge results from flow breakdown or stall within the compressor (sections 10.2 and 10.3). Some operational causes and effects of surge are discussed also in section 6.4.3, while causes resulting from engine matching or compressor installation are discussed in chapter 11 and section 10.5. There is another significant compressor malfunction called *deep stall,* which is discussed later (section 10.4) and should not be confused with the term stall, which is sometimes used to describe surge.

When all the individual, steady speed characteristics are plotted together, the over-all compressor characteristic is obtained as shown in figure 10.3. The characteristics steepen at higher speeds and lower pressure ratio because of choking (section B.2) at the compressor inlet. The surge points are joined by the *surge line*, and only the region of stable operation is mapped. When the compressor is matched to the remainder of the engine, it is restricted further still to a narrow operating region through the speed range, as shown by the engine *working line*. A requirement of the engine design process is to ensure that the working line is below the surge line under all operating conditions (chapter 11). This also puts it near the peak of the efficiency contours, which complete figure 10.3.

10.2 The Centrifugal Compressor Characteristic

The constant speed characteristic shown in figure 10.3 applies in essence to the centrifugal compressor, but would be extensively broadened in flow range. An example of the characteristic for an advanced design is shown in figure 10.4.[41]

The impeller is very tolerant of different mass flows because its normal flow pattern is disturbed. Changing the flow will vary the degree of disturbance, as assessed by the proportion of jet flow to wake flow (section 5.2), without encountering the axial blades' transition from smooth flow to separated flow. The maximum flow limit is set by choking at the eye; the critical flow limit (equation B.25) causes the constant speed curves to turn down vertically. The minimum flow limit of the impeller is set by negative incidence stall of the inducer vanes: it may be a full stall of all vanes, causing outright surge of the impeller, or it may be the axial compressor type of rotating stall (section 10.4)

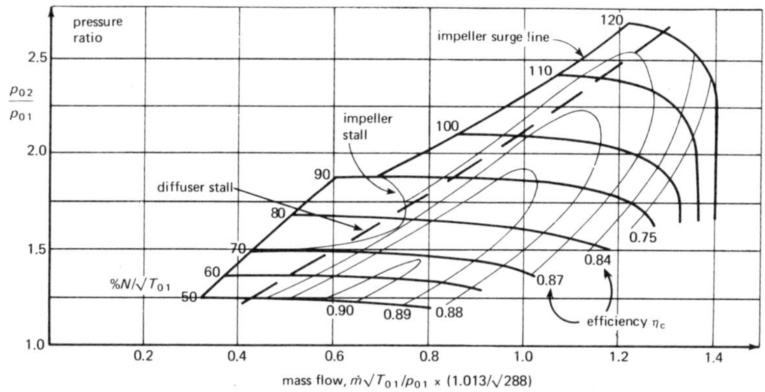

Figure 10.4 *Over-all characteristic of a centrifugal compressor showing flow limits and efficiency. (Adapted from reference 41 by permission of the American Society of Mechanical Engineers)*

at lower speeds. This wide operating range is shown by the continuous lines in figure 10.4, and applies to the use of an ideal, unstalled vaneless diffuser.

When a realistic, vaned or other compact diffuser is used, its own negative incidence stall (section 5.3) imposes a minimum flow level on the impeller's operating characteristic. This is shown as the dotted line, and represents the surge line for the compressor as a whole. It is therefore always the diffuser which sets the minimum flow surge limit, while the impeller limits the maximum flow. In this example, the mass flow range at design speed is 30 per cent, the increase from the surge flow to the choking flow. A typical working line may be located with 10 per cent *surge margin*, the mass flow being 10 per cent greater than at surge, or as defined in section 11.1.

10.3 The Axial Compressor Characteristic

The operating conditions within the region of stable operation vary considerably with speed and pressure ratio. In turn, this means that the limits of stable operation are set by a variety of different types of flow breakdown, as shown in figure 10.5. A few points within the characteristic are identified alphabetically, starting from the design point at A.

At design point the flow conditions are correct for all stages through the compressor, with incidence angles being ideal, the efficiency at its maximum and, usually, constant axial velocity. This condition corresponds to the majority of engine operation. At the same running speed, the pressure ratio could rise as far as B, or fall to C. At B, figure 10.1*a* shows that the axial velocity (V_a/U) is low. It decreases steadily through the compressor, until it is so low that the last stage stalls with excessive positive incidence, causing the compressor to surge.

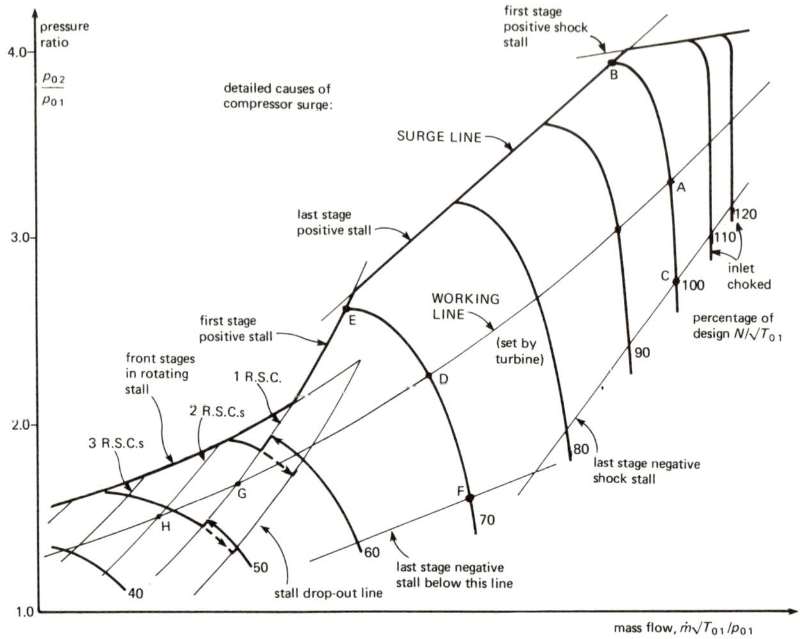

Figure 10.5 *Over-all characteristic of an axial compressor, showing the various operational limitations applicable at different speeds*

Conversely, at C the axial velocity increases progressively until the last stage encounters negative shock stall. This does not hamper operation in the manner of surge, but it may cause aerodynamic flutter and rapid fatigue failure of the blades (section F.3). Operation must therefore avoid both extreme conditions B and C.

At design speed and pressure ratio, the rate of reduction of the flow cross-sectional area matches the rate of increase in pressure and reduction in specific volume. At a pressure ratio of 5 the ratio of inlet to outlet area is also 5, excluding the effects of boundary layer blockage (section E.5). Consider now operation at lower speed and pressure ratio, as at point D. The flow ($\dot{m} = \rho A V$) an outlet still has a large specific volume, and ρ is low so AV is high. Thus a high velocity at exit is required to pass the flow through the available area. In practice the flow velocity at inlet reduces below the ideal, in partial compensation, and the mass flow is low. The value of V_a/U is now low at the front stages, high at the rear stages. Figure 10.6 shows in simple form how the value of V_a/U varies through the compressor at these operating points.

In raising the outlet pressure at lower speed from D to E, the initial positive slope at D becomes a reduced positive slope at E, whereas the slope at B was negative. Somewhere between B and E, the mechanism causing surge changes from stall at the last stage to stall at the front stage. At F, the rate of increase of axial velocity is accentuated and the last stage continues to stall with negative incidence. However, somewhere between C and F, the Mach number reduces to the point where choking and shock waves no longer occur.

The same sort of phenomena affect the operation of axial compressors with *supersonic* or *transonic* flow,[63B] but the proportions of the characteristic are quite different (figure 10.7). The flow relative to the rotating first stage blades reaches Mach 1 initially at the blade tips, when the rotational speed reaches a moderate level. Above that speed, there is very little increase in mass flow but the pressure ratio increases substantially.

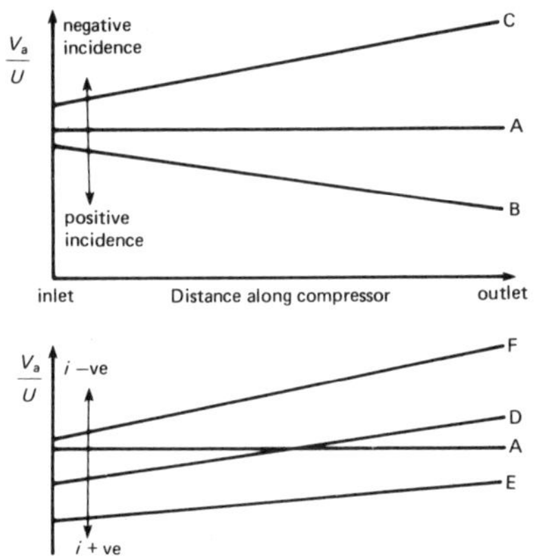

Figure 10.6 *Variation of V_a/U and incidence at successive stages through an axial compressor, at the points identified in figure 10.5*

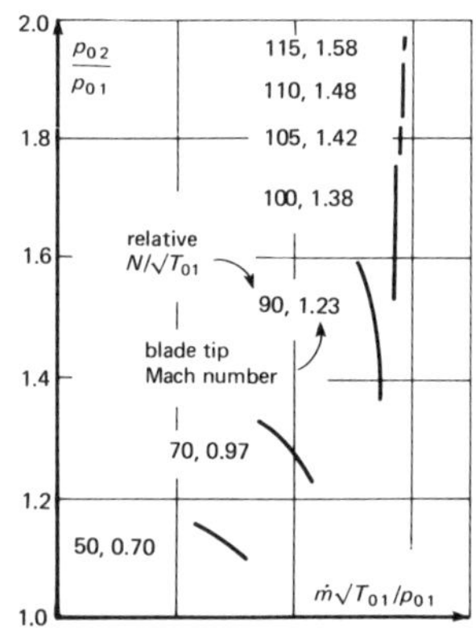

Figure 10.7 *Characteristic of a single stage, supersonic fan rotor, showing the mass flow limit at choking conditions (from reference 63B)*

10.4 Rotating Stall Mechanisms

In addition to the normal operation described, which could be called the primary characteristic, there are two further, stable operating modes which could be referred to as the secondary and tertiary characteristics. They involve two different rotating stall mechanisms (below) which, respectively, cause blade vibration problems (section F.3) and operational problems during starting (section 4.1) and at high altitude (section 11.6.5).

The *secondary characteristic* is the region of reduced pressure ratio shown in figure 10.5, in the characteristics drawn for 50 per cent and 60 per cent of design speed. With reduction in speed down the surge line from point E, the first stage stall progressively weakens until, at G, the stalled region involves only a few blades at a time and the compressor may continue to operate. Slowing down along the working line through H, a second region of stall develops, then a third and, in time, stalled regions may also develop in the later stages of the compressor.

They are known as *rotating stall cells* (RSC) because they rotate with the compressor at, typically, 20 to 70 per cent of the rotor speed.[73] The ability to operate stably in this mode raises the surge line relative to the first stage stall limitation. A hysteresis effect is shown by the loop in the 60 per cent speed line, as rotating stall continues before it drops out to a lower pressure ratio or higher speed than where it was initiated.

The traditional explanation for the *initiation* of an RSC is as follows. As the operating speed reduces from D to G, uneven turbulence pushes a blade into positive incidence stall. This is shown as blade X in figure 10.8: the stalled region effectively reduces the area of the passage on its convex side, so the flow approaching that passage is deflected to encounter blade Y with a greater angle of incidence. This blade therefore stalls, and the region of stall is transferred successively to further blades in a direction counter to the blade motion. Meanwhile, reduced positive incidence permits blade X to unstall, followed by Y, etc.

The rotating stall pattern round the compressor annulus may consist of several cells near the tip (part span) or one full span cell of very variable width (figure 10.9). Their rotational speed is shown as $f\omega$. Figure 10.10 shows the *flow pattern* in one cell of any type, which the rotor blades enter at $(1-f)U$ from one side and the stators enter at fU from the other side. Assuming zero mean axial flow through the cell, Day and Cumpsty[74] showed that the blade entering the cell experiences a complete flow reversal (at K). It then accelerates the air to the point where the flow is re-established (at L). This acceleration is greater than the rotor normally imparts to the flow and is achieved very inefficiently; the torque resisting rotor motion therefore requires a substantial *power input* which is largely dissipated as heat.

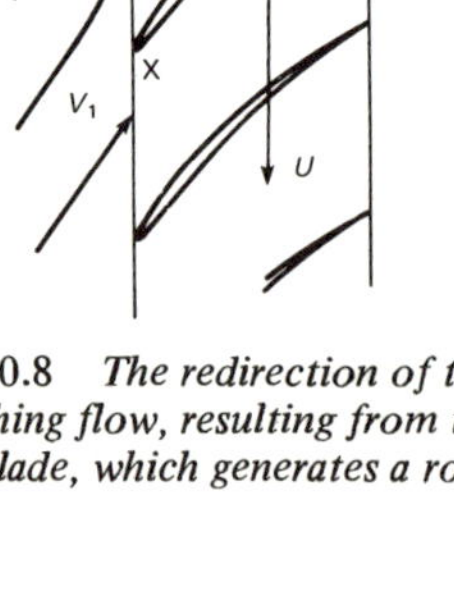

Figure 10.8 *The redirection of the approaching flow, resulting from the stalling of one blade, which generates a rotating stall cell*

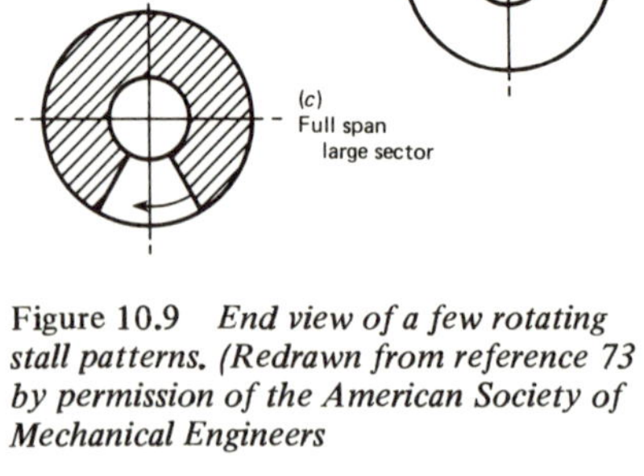

Figure 10.9 *End view of a few rotating stall patterns. (Redrawn from reference 73 by permission of the American Society of Mechanical Engineers*

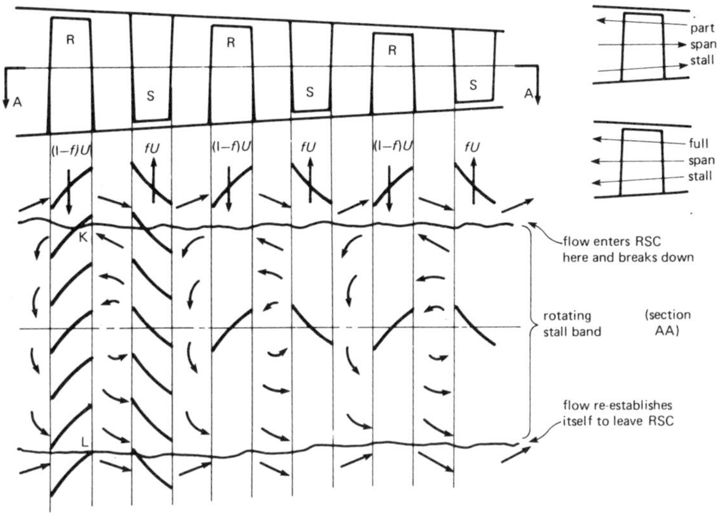

Figure 10.10 *The flow directions inside and ourside a rotating stall cell. (Redrawn from reference 74 by permission of the American Society of Mechanical Engineers)*

The *tertiary characteristic* may now be simply defined as full span stall, while the secondary characteristic involves part span stall. Figure 10.11*a* [73] takes figure 10.2 to a more advanced level on a plot of non-dimensional, total inlet to static outlet pressure rise (ψ_{ts}) against $\phi(V_a/U)$. This could represent the complete characteristic at 60 per cent speed as compared with that in figure 10.5: the continuous lines all represent stable modes of operation. The tertiary characteristic is also known as *deep stall*, and the circumferential width of the stall cell varies from a small arc (figure 10.9*b*) at its right end to a large arc (figure 10.9*c*) at its left end. The secondary characteristic drops out at higher

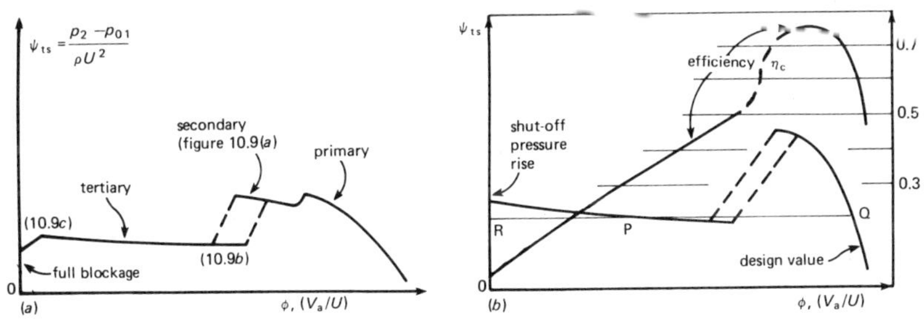

Figure 10.11 *Actual constant speed characteristics for an axial compressor showing: (a) the three stable operating conditions with the locations of the rotating stall patterns of figure 10.9 at low speed and; (b) the two operating conditions at higher speed. (Redrawn from reference 73 by permission of the American Society of Mechanical Engineers)*

speeds, for instance at 80 per cent, as shown in figure 10.11*b*. The efficiency curves show typical low values for the tertiary characteristic, which account for the excessive heating. The tertiary characteristics show little increase of pressure rise with speed: they could be plotted on figure 10.5 and would be found to cross the surge line, which applies only to the primary and secondary characteristics.

The line RPQ in figure 10.11*b* shows how the compressor mean flow condition may be divided into the region of normal flow at Q and the average zero flow of the stall cell at R. They are at the same level because of the constraint of the common, approximately constant, static pressure round the outlet. The pattern within the stall cell varies substantially with the design value of ϕ[74], as do many other parameters; the flow pattern shown in figure 10.10 applies with the intermediate value of 0.55, and the characteristic shapes and hysteresis band widths of figure 10.11 apply to $\phi_{\text{design}} = 0.71$.

Some *typical values* have been found to categorise some aspects of rotating stall.[73] The proportion of the annulus cross section occupied by the stall cell may be described by the blockage, λ (not the same blockage as in section E.5), where

$$\lambda = \frac{\phi_{\text{unstalled}} - \phi_{\text{mean}}}{\phi_{\text{unstalled}}} = \frac{\text{PQ}}{\text{QR}} \tag{10.1}$$

as shown in figure 10.11*b*. Full-span stall tends to occur if λ exceeds about 0.3 and ϕ_{design} exceeds about 0.5: below these values the stall is of part-span type. Multi-stage compressors typically generate a value of $\psi_{\text{ts}} = 0.11$ times the number of stages, in full-span stall, and 0.17 times the number of stages when in part-span stall, almost regardless of speed. The unusual use of the parameter ψ_{ts} allows for realistic measurements. It would be meaningless if the total outlet pressure of this interrupted flow pattern were to be measured (section B.2).

10.5 The Influence of Installation on Surge and Rotating Stall

Although surge mechanisms vary with the compressor hardware (sections 10.2 and 10.3), the surge is caused by incorrect engine matching (chapter 11), external factors (section 6.4.3) or the design of the ducting of a compressor installation.

The surge may be regarded as an axial resonance in the flow, excited by the compressor when its operating conditions are sufficiently adverse. The presence of a large plenum volume between the compressor outlet and the downstream controlling orifice makes the level of surge more severe because the pressure takes longer to decay as the compressor flow reverses. Greitzer [75] showed that the limits of stability may be determined by a single parameter

$$B = \frac{U}{2\omega L_c} = \frac{U}{2c}\sqrt{\left(\frac{v_p}{A_c L_c}\right)} \tag{10.2}$$

since

$$\omega = c\surd(A_c/v_p L_c) \tag{10.3}$$

where ω is the Helmholtz resonance frequency of the column of air in the compressor duct (area A_c and length L_c) as it resonates axially relative to the enclosed plenum volume v_p behind it, and c is sonic velocity. The main variables affecting B are thus the mean blade speed U and v_p.

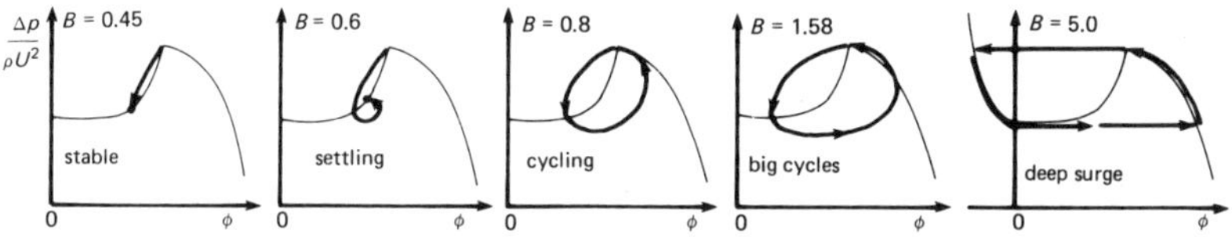

Figure 10.12 *The effect of compressor outlet ducting, plenum geometry and blade speed on the severity of instability or surge. (Redrawn from reference 75 by permission of the American Society of Mechanical Engineers)*

Unstable operation permitted temporary departure from the steady-state operating characteristic of figure 10.11*a*. As B rose to 0.45, the test compressor would enter deep stall (figure 10.12). At $B = 0.6$, the flow would oscillate before settling into deep stall. At $B = 0.8$, the operation entered a continuous loop of surge, with the flow and pressure cycling continuously. This loop was predicted to enlarge with rising values of B until, for example, at $B = 5$ a deep surge condition occurs with the pressure climbing to its peak, the flow falling back to a negative level and the pressure falling to let the flow re-establish itself. In a lighter level of surge, it was found that part span stall could develop briefly at a certain stage during each cycle.

11

Engine Matching and Transient Characteristics

The performance of an engine at any constant speed may be determined from a set of simple dimensionless curves (section 4.2). The level of pressure ratio (and temperature ratio) could lie anywhere within the stable operating range of the compressor (section 10.1), but in practice is set at the highest practicable level to operate at high efficiency and obtain the best performance relative to weight and cost.

The way in which the turbine is matched to the compressor, to establish a *working line* which permits stable operation at steady speeds and during transient operation, is discussed in this chapter. The working line defines the pressure ratio generated at any running speed.

11.1 Matching of a Compressor, Turbine and Nozzle or Load

The working line of the compressor in a given engine is determined by plotting turbine performance data on the compressor characteristic. The starting point is the knowledge of the individual compressor, combustion and turbine dimensionless characteristics, and the engine inlet temperature and pressure. The following basic assumptions are then made.

(1) Compressor shaft speed = turbine shaft speed (unless gearing is used, in which case the gear ratio is inserted).

(2) The compressor air flow plus the fuel flow but less any air bleed flow equals the turbine flow (that is, the requirement of flow continuity is satisfied).

(3) The power flows also balance.

An iterative process is then used to determine the position of a simple (for example, 3.5 or 4.0) over-all temperature ratio value on one of the lines of constant speed on the compressor characteristic. The steps in this calculation are listed in table 11.1. When the appropriate compression ratio has been found, at

Table 11.1 Determination of One Turbine Operating Point for a Single-shaft Gas Turbine Engine

Given:	Compressor operating characteristic (section 10.1): $N/\sqrt{T_{01}}$ and η_{12} at any p_{02}/p_{01} and $\dot{m}\sqrt{T_{01}}/p_{01}$. Combustion pressure loss relative to temperature rise (section 7.5), and efficiency (section 7.3). Turbine operating characteristic: $N/\sqrt{T_{03}}$ and η_{34} at any p_{03}/p_{04} and $\dot{m}_3\sqrt{T_{03}}/p_{03}$. Inlet conditions: p_{01} and T_{01}.
Compressor:	Select a point on a constant speed characteristic, $N/\sqrt{T_{01}}$: p_{02}/p_{01}, $\dot{m}\sqrt{T_{01}}/p_{01}$, η_{12}. Calculate the temperature rise and power input (section 3.3, step 1): $P_{12} = \dot{m}_1 C_{p12}(T_{02} - T_{01})$.
Combustion:	Calculate the actual air flow into the combustion system (section 3.3, step 3): $\dot{m}_2$. Choose a simple value of over-all temperature ratio: T_{03}/T_{01}. Calculate the values of outlet pressure and fuel flow (section 3.3, step 2): p_{03}, $\dot{m}_f$.
Turbine:	Calculate the dimensionless mass flow and speed: $\dot{m}_3\sqrt{T_{03}}/p_{03}$, $N/\sqrt{T_{03}}$. Determine the corresponding pressure ratio and efficiency: p_{03}/p_{04}, η_{34}. Calculate the appropriate temperature drop and power output (section 3.3, point 4): $P_{34} = \dot{m}_3 C_{p34}(T_{03} - T_{04})$.
Check:	Subtract the accessory power requirement (section 3.3, point 3) and any external load or losses: $P_{34} - P_{ext} = P_{12}$. Compare the new value of p_{12} with the compressor input requirement, and, if different:
Iterate:	Choose a different compressor operating point at the same $N/\sqrt{T_{01}}$: new p_{02}/p_{01}, etc. Repeat the calculation and iteration using the same value of T_{03}/T_{01}, until P_{12} values match. Plot this point on the appropriate speed line.
Nozzle:	Calculate the turbine exit and nozzle conditions: p_{0e}/p_a and $\dot{m}_e\sqrt{T_{0e}}/p_{0e}$. Determine if the nozzle is choked (section 3.3, step 6) and calculate (equation B.23 or B.25): $\dot{m}_e\sqrt{T_{0e}}/A_e p_{0e}$. Calculate effective nozzle area A_e and indicate its value at the plotted T_{03}/T_{01} point.

which the power flow is in balance, the process is used again until several values of over-all temperature ratio have been located as points on each constant speed characteristic. These may then be joined by contour lines of constant T_{03}/T_{01}, as shown in figure 11.1.

Each iterative process is then continued to determine the controlling condition which would set the working line level. For instance, if a nozzle were to be fitted downstream of the turbine, a different effective nozzle area would be required at each calculated point and it could be determined as shown at the bottom of table 11.1. Comparing the calculated value of $\dot{m}_e\sqrt{T_{0e}}/p_{0e}$, with the value of $\dot{m}_e\sqrt{T_{0e}}/A_e p_{0e}$ which matches the pressure ratio, yields the value of A_e. This will not be a simple number, so intrapolation will be required to find a range of simple area values on each constant speed curve. These may now be joined by the lines shown in figure 11.2, which cross the lines of constant T_{03}/T_{01}.

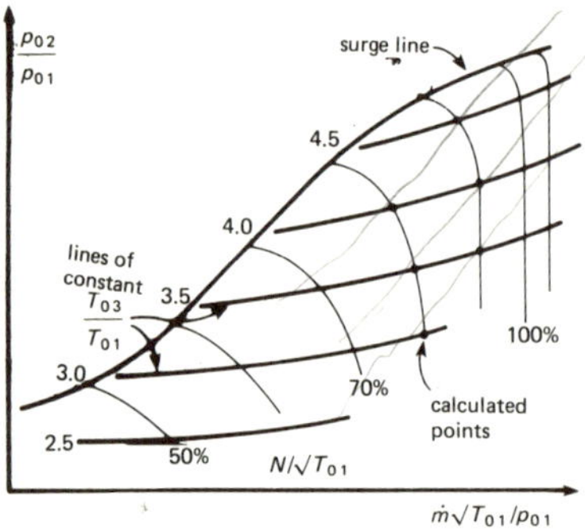

Figure 11.1 *Lines of constant over-all temperature ratio plotted on the compressor characteristic*

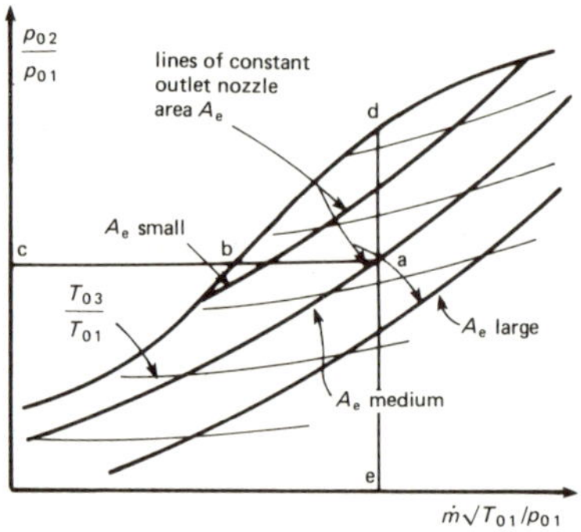

Figure 11.2 *Lines of constant nozzle area plotted over figure 11.1, showing working lines and surge margin*

The choice of any value of exhaust nozzle area now establishes the working line level of the compressor. As might be expected, a small nozzle area restricts the outlet and raises the internal pressures, while a large nozzle drops the pressures. The obvious choice is the medium area, which locates the working line below the surge line at all conditions and enables the compressor to run at high efficiency. In an aircraft, the low speed working line is lowered by increasing ram pressure ratio, causing the feathered effect shown (section 4.2).

Alternatively, a single shaft engine may be coupled direct to a *load* (figure 1.1), leaving its exhaust at low pressure and not restricted by a nozzle. The working line is then set by the power demand of the load at any speed, requiring load control. The unacceptable extremes may be represented by a generator delivering constant power over a wide speed range, or a propeller of fixed pitch absorbing power proportional to the cube of the speed. The load control system which finds the happy medium will therefore necessarily form part of the engine control system (section 12.3).

The distance between the working line and the surge line is called the *surge margin*, which is defined in several different ways. For centrifugal compressors, whose constant speed characteristics are nearly horizontal, it is convenient to use the mass flow axis to assess the surge margin. Thus, at a given pressure ratio, the surge margin at a particular operating point is defined as the percentage increase of flow above that at which surge occurs: in figure 11.2 it is shown as ab/bc. Axial compressors have more vertical constant speed characteristics, so the surge margin is assessed in terms of pressure ratio. For instance, it could be depicted as ad/ae in figure 11.2, although different companies use slightly different methods of assessment.

Numerical examples of matching calculations are included in reference 31.

11.2 Matching with a Free Power Turbine

As an alternative to a nozzle, a second turbine may instead be installed downstream of the gas generator, or core engine, turbine. After the iteration to plot points of T_{03}/T_{01} on the compressor characteristic, the values of $\dot{m}_4\sqrt{T_{04}}/p_{04}$ at outlet from the core turbine could be calculated, and would be found to fall as p_{02}/p_{01} rises at constant speed. There is therefore a unique flow condition at the power turbine inlet for each gas generator operating point, just as there was a unique value of nozzle area.

The power turbine's flow capacity therefore establishes the core engine's compressor working line. The flow continuity between the turbines is shown in figure 11.3, which also relates the core turbine outlet flow to its pressure ratio: a particular flow condition necessarily determines the pressure ratios for both turbines, such as those shown asterisked. Linking these points with the compressor and the combustion systems, while continuing to satisfy the three criteria in section 11.1, calls for a turbine iterative process at each core engine

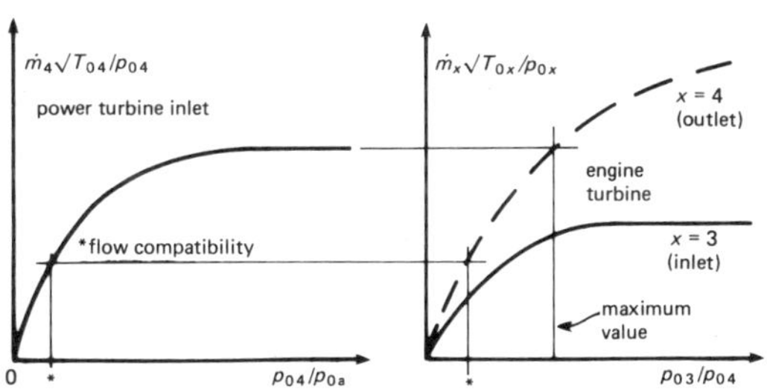

Figure 11.3 *The effect of flow continuity on the pressure ratio of two turbines in series*

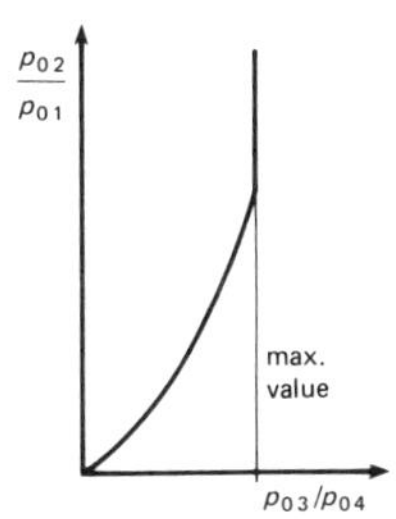

Figure 11.4 *Simple determination of the working line*

calculation, a laborious task. This may be short-circuited by considering a further assumption

(4) The over-all compression ratio equals the over-all expansion ratio, including the combustion pressure loss, that is

$$\frac{p_{02}}{p_{01}} = \frac{p_{02}}{p_{03}} \times \frac{p_{03}}{p_{04}} \times \frac{p_{04}}{p_a} \tag{11.1}$$

This enables the compression ratio to be plotted against the core turbine expansion ratio, as in figure 11.4 and establishes the core engine operating point at each speed.

11.3 The Matching of Multi-shaft Engines

The use of the power turbine to drive another compressor adds further to the complexity of calculating the matching, although such a split-shaft engine will operate in a well-defined manner. Several configurations are used (section 2.1) and these have significantly different effects. The 1-p compressor's outlet flow may all enter the h-p compressor (figure 1.7*d* and *e*), or it may divide in bypass or fan engine form (figure 2.1*b* or *c*). There may also be additional shafts.

In engines with undivided flow, the three criteria of section 11.1 are joined by

(5) The flow rates are compatible through all the compressors and turbines.

This imposes a unique relationship between the $N/\sqrt{T_{01}}$ values for the shafts (section 4.2), which further depends on the characteristic of the final nozzle or an additional, power turbine. In a split-shaft engine, the operation of the h-p rotor may be treated as for a single-shaft engine, in which the l-p turbine nozzle guide vanes act as the controlling area. This effect can be seen in figure 11.3, where the flow limit of the downstream turbine establishes the maximum possible pressure ratio for the previous turbine.

Bypass and fan-type aero engines are constrained by the same factors. Flow continuity determines the operation, whether the bypass flow rejoins the core flow to exhaust through a common nozzle, or whether the fan flow is controlled by its separate nozzle. However the two types have different effects on the l-p compressor (fan) working line. The fan nozzle is usually of fixed area and the fan blades set at a fixed angle: a common result is that the working line runs at a shallow angle across the compressor map, and would cross the surge line at medium speeds if precautions were not taken (section 11.4), as shown in figure 11.5.

This figure shows also the steeper angle of the bypass engine's l-p compressor working line. The change of slope results from an effectively variable nozzle controlling the bypass flow, which arises as follows. At low engine speeds, the core engine exhaust temperature is low, and it requires a relatively low proportion of the available final nozzle area. At high speeds, the much hotter primary exhaust needs a greater proportion of the available area. The remainder of the nozzle area which is available for the bypass flow is therefore large at low speeds and small at high speeds: the working line therefore crosses the lines of constant area. The relationship between the shaft speeds therefore also differs from the fan type.

11.4 Variable Geometry for Axial Compressors

Comparison of the characteristics of centrifugal and axial compressors shows that the former has a much wider operating range than the latter in terms of variation of mass flow at a given speed (chapter 10). The axial compressor may also generate a higher pressure ratio than most centrifugal compressors. The slope of the working line, and the risk of its crossing the surge line, is therefore much more critical in the axial compressor. With six or more stages the slope is likely to be too shallow, calling for a variety of means of changing the configuration to provide *working line control*. The devices used are discussed in this section, and their means of control in chapter 12.

11.4.1 Variable Final Nozzle Area

The effect of this has already been explained in section 11.3. For an engine with undivided flow, the final nozzle may be made mechanically variable, and programmed to maintain any desired working line level and slope. Engines with afterburning require nozzle enlargement when the afterburner is used (section 14.1), to permit the hotter exhaust to escape without raising the working line.

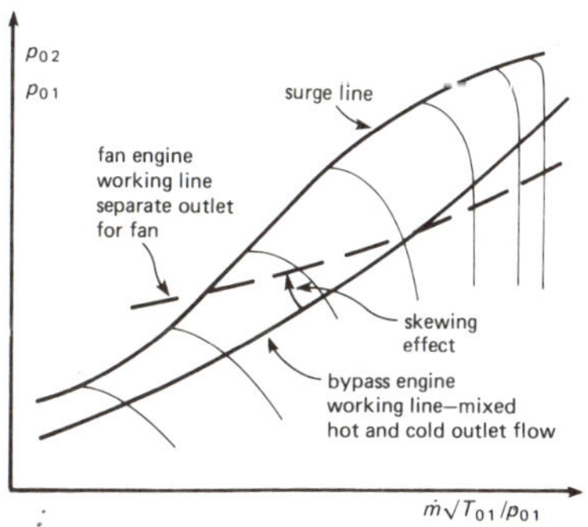

Figure 11.5 *Working lines of the l-p compressor twin-spool engines, in bypass and turbofan configurations*

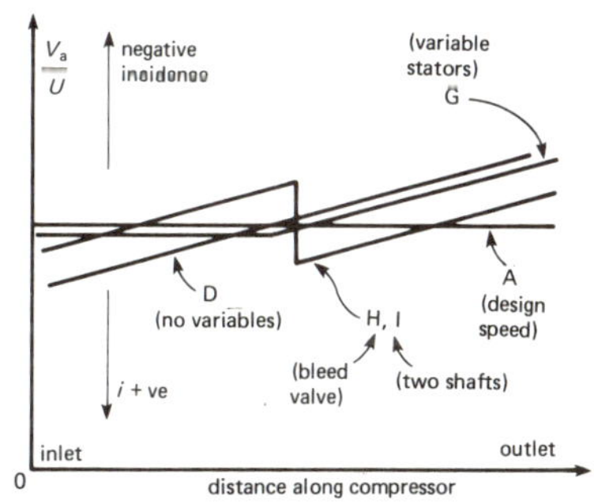

Figure 11.6 *The varation of V_a/U along the compressor(s) at low speeds, with different configurations. (cf. figure 10.6)*

11.4.2 Variable Compressor IGV and Stator Angles

VIGVs swivel to increase the flow angle α_0 (figure 6.2) as the speed reduces, to correct for the corresponding reduction of V_a/U (section 10.3) and maintain the correct flow angle of incidence into the first stage of blades. The changed flow angle also implies a reduced flow area; thus, the inlet flow area is reduced relative to the turbine nozzle area, and the working line is lowered.

Some engines use variable angle stator vanes in the first few stages, to correct the incidence for all stages which would encounter low V_a/U. With reference to line D in figure 10.6, the use of variable IGVs and stator vanes changes it to line G, as shown in figure 11.6. The corresponding effect is shown on the compressor map in figure 11.7.

11.4.3 Compressor Air Bleed Valve

Long compressors may require air to be bled away from the middle stages at low speeds, sometimes in addition to VIGVs. There may be one stage of bleed in a compressor of seven or more stages, or two or more stages of bleed with twelve or more stages. A typical bleed valve arrangement is shown in figure 6.1: it may open suddenly as the engine speed falls to a particular level, or it may open progressively as the speed falls beyond the set level.

The effect of the bleed valve is to provide a step change in V_a/U along the compressor, as shown by line H in figure 11.6. It permits an increased V_a through the early stages, and provides a flow discontinuity to relieve the swallowing capacity of the later stages and the turbine. The loss of air drops the working line as shown in figure 11.7, while the incidence angle is maintained at a satisfactorily low level at all stages, whether positive or negative in sense. If VIGVs are also fitted, the bleed valve could open at a lower speed than the VIGV kink point, permitting more flexibility for surge margin control.

Air which is bled off in this fashion is expensive in terms of energy content. In bypass or fan-type aircraft engines, it may enter the l-p duct through a peripheral slot round the h-p compressor, and be put to good use. In other cases, it may be possible to use it for aircraft or plant services, such as boundary layer control or recharging starting air bottles.

11.4.4 Extra Air Inlet Doors

Some fan-type engines (section 11.3) increase the value of V_a for the fan blading at low speeds by increasing the area available for air to enter. This is done by a series of doors in the inlet cowl, which open to form smooth ducts for the extra flow. The need for these was partly dependent on the earlier profiles used for the intake lip. While these help a fan-type engine, they would worsen the conditions for a bypass or undivided flow engine, where a bleed valve may be the appropriate choice. They also have a bad effect on compressor noise (section 13.5).

11.4.5 Variable Turbine NGV Area

The turbine flow capacity is set by its inlet NGV area, while its expansion ratio and specific work are determined by the ratio of outlet area to inlet area. The h-p NGV area may therefore also be used to control the working line: it is adjustable within small limits during the building of some engines by bending the trailing edge to adjust the throat opening (figure 8.3).

Automotive engines and other types sometimes use variable NGVs (figure 9.1) which swivel like VIGVs to change the flow angle and adjust the width of the throat at the trailing edge. These may be used to raise the speed of the core engine shaft at low power turbine load, to improve the rate of response to a subsequent demand for more power (section 9.2). In the power turbine, VNGVs may maintain the core engine back pressure at all loads, or reverse the power turbine's direction of rotation.

For completeness, the matching of a *turbocharger* should also be mentioned as it also involves control of the NGV area. The presence of the reciprocating engine using the air flow between the compressor and turbine results in the very steep working line J in figure 11.7. The compressor (usually centrifugal) in this case requires pressure relief at high speed, which is accomplished with a loss of engine torque at medium speeds by increasing the turbine inlet area (line K) or with a waste gate or blow-off valve (line L), preferably located before the reciprocating engine [76].

11.4.6 Two or More Shafts

Division of a single-shaft engine into a twin-spool type permits a reduction in the number of compressor stages and greatly eases the problems of compressor control. It works very similarly to the bleed valve in the terms of line I in figure 11.6, but does so by adjusting to a lower value of U for the first few stages rather than by increasing V_a. The l-p and h-p compressors thus run at their own, equilibrium speeds (section 11.3), and the over-all pressure ratio is the product of their individual ratios. Although line I is similar to line H in figure 11.6, line H in figure 11.7 is not appropriate to the split-shaft engine, as each compressor has its own separate characteristic.

11.5 Acceleration and Deceleration

The means of accelerating or decelerating an engine from a constant-speed

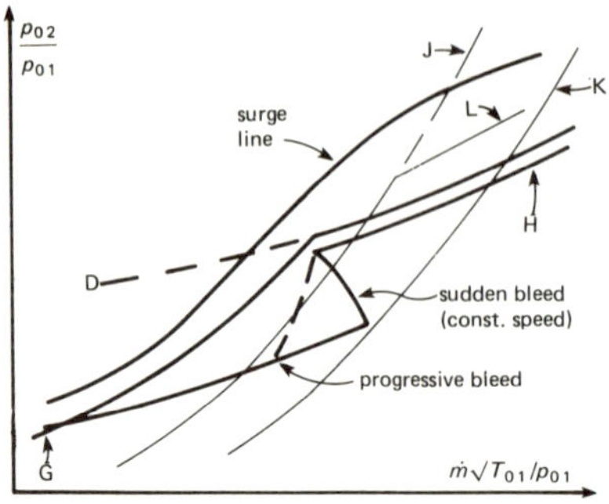

Figure 11.7 *The effect of VIGVs (G), middle stage bleed (H), and turbo-charger features (J,K,L) on compressor working line*

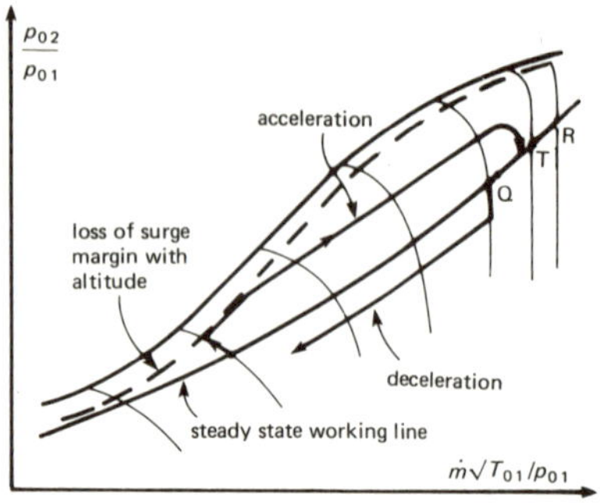

Figure 11.8 *The change of surge margin during acceleration and deceleration and at altitude*

condition is to increase or decrease the rate at which fuel is injected (section 12.3). On acceleration, the increased gas temperature raises the turbine work capacity at a given pressure ratio (section 1.1): its increased torque then accelerates the shaft.

Overfuelling also has an effect on the compressor working line, as may be seen from figure 11.1. Before the shaft speed has time to change, the raised value of T_{03}/T_{01} clearly pushes the operating point towards surge. The working line is therefore raised during acceleration, while deceleration lowers the working line (figure 11.8).

The permissible degree of over or underfuelling is limited by several factors, involving the compressor, combustion and the turbine. The *compressor limits* involve the risk of surge, or inadequate surge margin to cater for intake flow distortion (section 6.4.3) or altitude effects (section 11.6.1) during acceleration, or blade flutter during deceleration. The *combustion limits* are set by the risk of flame extinction at the rich or weak limits of stable operation (figure 7.2), of which the weak limit is usually more significant than the risk of blade flutter. Finally, the *turbine limit* is that of thermal shock with sudden rates of temperature change; this will lead to premature thermal fatigue (section 8.3), affecting engine life and overhaul costs rather than the immediate operation. The rates of acceleration and deceleration are therefore set as a compromise between these limitations and the need for rapid response, to suit the aircraft or plant requirements.

Multi-shaft engines behave in a similar fashion. All shafts accelerate or decelerate together, maintaining the steady state speed relationship unless the l-p shaft inertia is increased by connection to a large fan or heavy industrial load. Then the l-p shaft speed will lag, and the suction from the h-p compressor could lower the l-p compressor working line during acceleration. It would similarly rise during acceleration, thus behaving differently from the general pattern shown in figure 11.8. In aircraft engines with very high pressure ratio, the use of some nickel h-p compressor discs may change the relative inertia balance between the shafts. They also have the effect of slower heating during acceleration, which temporarily affects the blade tip clearance, compressor efficiency and surge margin and may call for positive means of tip clearance control (section 14.2.2). The calculation of transient operation is discussed in reference 77.

11.6 The Effects of Altitude on Engine Behaviour

Operation at altitude implies low levels of inlet air temperature, pressure and density, which are only partly offset by the ram effect of a moving engine. The low temperature causes the l-p shaft to run relatively faster than at ground level, for reasons explained in section 4.2. The low pressure causes some other effects which also affect engine behaviour, as follows.

11.6.1 Reynolds Number Effect ($\rho Vd/\mu$)

The ρV part of the Reynolds number is effectively the mass flow (ρAV), as the engine has fixed internal areas. The mass flow is drastically reduced by the loss of air density at altitude, with a resultant increase in the boundary layer thickness round all the blade and annulus surfaces. The blade wakes and profile drag therefore increase (section E.4) and the compressor efficiency and surge margin (section 11.1) are reduced. This effect is naturally more severe at lower engine speeds, causing a dip in the surge line (figure 11.8).

11.6.2 The Effect of Bypass Ratio

Pure jet and low bypass ratio aircraft engines usually operate at successively lower values of h-p $N/\sqrt{T_{01}}$ as operation proceeds from take-off through climb to high altitude cruise. With most engines, the surge margin increases as speed is reduced from take-off (T), and the cruise condition at Q is very stable (figure 11.8).

Very large fan engines of high bypass ratio (4 to 6) behave differently. The large inlet flow imposes a very high momentum drag (section 14.1), while the low fan outlet velocity contributes little thrust. The core engine shaft (h-p) is therefore run at higher $N/\sqrt{T_{01}}$ than at take-off to provide adequate primary jet thrust, although the change of temperature leaves the actual speed N lower than at take-off. The h-p compressor is therefore run in a region of reduced surge margin (at R), which may cause altitude surge problems.

11.6.3 Mis-scheduling of Compressor Variables

Engine designers attempt to make the VIGVs and bleed valves operate at unique dimensionless conditions, in terms of speed or pressure ratio, to suit the nature of the compressor. Often the control system has hysteresis, depending upon which activating signals are used, and the hydraulic or air ram or electric actuator may suffer from friction and require a significant time to complete its full travel (section 12.6). Air-operated systems suffer from reduced pressure at high altitude. The slow response during engine acceleration or deceleration may combine with the loss of surge margin to cause surge problems at altitude.

11.6.4 The Bodie Test

This manoeuvre, used during engine and aircraft certification trials, requires the engine power to be cut back suddenly (slam deceleration) from cruise speed and then, at any speed before the engine reaches idle and cools down, the power lever is pushed forward again to demand a hot re-slam acceleration.

This is usually the worst handling test an engine may experience. The seal clearance and tip clearance (section 14.2.2) may have increased, reducing efficiency; the acceleration starts with reduced surge margin and the compressor variables may not yet have caught up. In multi-shaft engines, a compressor may surge during deceleration (section 11.5), or acceleration, or the engine may not have surge problems: engines with only a few stages in each compressor are usually the most surge resistant in this test. Even they, however, may not be immune if the Bodie test induces deep stall, as discussed next.

11.6.5 Deep Stall, or Altitude Stagnation

Deep stall first became a serious altitude handling problem in the 1960s, affecting some engine types. Attempting the Bodie test at too high an altitude, with a hot re-slam acceleration from about 80 per cent shaft speed, could use up the compressor's reduced surge margin and trip it on to a stable tertiary characteristic (section 10.4), with drastically reduced mass flow and pressure ratio.

The engines affected tend to be of twin-spool, bypass or fan type. The symptoms are that the engine becomes unusually quiet, fails to accelerate (stagnates) and the turbine is rapidly overheated unless the power lever is pulled back. The stagnation occurs in the h-p compressor only, with some of its usual intake flow being diverted into the l-p air system: if the l-p compressor attempted to enter deep stall, ram pressure from the flight velocity and suction from the h-p compressor would soon restore the acceptable flow. The engine configuration also contributes in that the move in the 1960s to more compact combustion chambers created the small plenum volume which reduces the likelihood of outright surge (section 10.5).

In effect, the engine finds a new, stable matching point as an alternative to that of section 11.1. The high compressor outlet temperature and low mass flow increase the combustion efficiency (section 7.3), making the turbine entry hot, and increase the compressor power demand relative to the turbine capacity so that no excess torque is available for acceleration.

The fuel system also needs an adverse characteristic, such that the high shaft speed causes a high rate of fuel flow for which the reduction of the combustion system air pressure, to perhaps half the level normally encountered at high altitude, is unable to compensate (section 12.2). This factor then causes the turbine overheat while the reduced surge margin initially triggers the deep stall.

Stagnation may sometimes be avoided by very careful calibration of the overfuelling acceleration schedule (section 12.3). It may also be avoided by not conducting the Bodie test, but this may impose an unacceptable limitation on, for instance, a military operation. The use of three shafts, with the i-p and h-p compressors handling the same flow in series along the duct, may also inherently avoid it.

12

Gas Turbine Engine Control

The performance parameters of a running gas turbine are determined at any time by a unique set of curves, such as those in figure 4.2. Of all the parameters shown, it is the fuel flow which is the input and which controls the operation.

The need for accurate control of the fuel flow has already been discussed in connection with starting (section 4.1) and changes of speed (section 11.5). There are also several engine safety limits which must not be exceeded (section 12.1.1), and the operation of the compressor variables must be scheduled accurately according to speed or pressure ratio (section 11.4). When these and many less obvious engine requirements are considered, it is apparent that the incorporation of a control system will

(1) reduce the operator's work load, and
(2) safeguard the machinery.

The ideal control system will sense the operator's requirements from the position of his single control device and will set the engine speed and power accordingly, subject to any imposed safety limits. In aircraft, such ease of control is essential, particularly in difficult flying conditions. In industrial applications, control may go further still, to the point of automatic starting, running and stopping, as dictated by the needs of the plant.

Control system philosophy has changed considerably with time. The means of control to any set condition within the running range may involve controlling any one of several different parameters. The early aircraft systems (1940s) controlled the pressure in the lines to the spray nozzles (burners) by spilling fuel through a valve (section 12.2): such a system is exceptionally dangerous as, with any breakage or loosening of the nozzle components, the control unit will increase the flow in an attempt to restore the pressure and in turn may well be feeding a serious fire. The next generation of systems controlled the flow rate, again by spill control; however, once again this is not a dimensionless parameter with respect to power. Both systems could require regular manual correction as the power changed with engine inlet conditions.

In the 1960s, the trend was to speed control, particularly of the h-p shaft. Thus, a control lever setting would demand a particular speed. This was still not entirely dimensionless, as the $N/\sqrt{T_{01}}$ would rise with the reducing T_{01} during climb, steadily increasing the pressure ratio but losing thrust with reduced p_{01}. The trend in the 1970s was to control according to engine air pressure, so that the power lever selected thrust and the control system would provide it. The same form of progress would apply equally to industrial engines, but possibly with a different time scale: speed control or power control would be relevant to most applications.

So far, only aircraft and industrial applications have been mentioned. The land transport case will be similar to the aircraft case except for deletion of variation in altitude (section 12.2). Marine applications are more related to industrial gas generator usage. The discussion will therefore remain centred on just the two types.

This chapter will provide an introduction to control concepts and practice to the extent that they affect the gas turbine itself. Control system theory is omitted, being the subject of many other books and reference 78. Further information is given in references 9, 6J, 6K, 6L and 36, which present diagrams and explanations of control systems, hardware and plant requirements. Standard specification requirements for the fuel system functions and protection devices are given in ISO 3977.[52]

12.1 Safety and Operational Limits

12.1.1 Absolute and Working Limits

A few parameters must be controlled to maintain the safety of the engine and the plant. They are

(1) maximum rotational speeds – for rotor stress reasons
(2) maximum compression pressure – for casing stresses
(3) turbine entry temperature – material strength and creep.

The working limits imposed are usually the *type test* limits (section 4.3). The engine has a safety margin beyond these limits, which may be used occasionally for short periods: a typical absolute limit might be a shaft speed of 110 per cent. If the type test limits are exceeded for longer than a specified few seconds or minutes, or if the absolute limits are exceeded at all, then the engine is mandatorily shut down, dismantled and the affected parts scrapped. These limits affect the maximum engine power as discussed in section 12.1.3.

12.1.2 Parameters affecting Reliability and Life

There are other parameters which do not immediately affect safety but which help to preserve the machinery

(4) compressor variables control – for surge prevention (section 11.4)
(5) acceleration and deceleration rate – see section 11.5
(6) oil and fuel temperatures and pressures.

Surge can damage an engine, and often requires a subsequent *in situ* inspection for signs of damage. The risks associated with uncontrolled acceleration and deceleration rates, causing surge, flame-out or turbine damage, have already been discussed. The oil condition affects the life and safety of the gears and bearings (section 14.4). The fuel condition is discussed in sections 12.2 and 15.4.

12.1.3 Limits of the Performance Envelope

The maximum output power or thrust is set by the type test limits and varies according to day temperature. Figure 12.1 shows the limits imposed on aircraft engine thrust over the usual ambient temperature range: the effect on shaft output power would be similar.

The limit on cooler days is set by the internal static pressure, p_3 in the case of a two shaft engine. The thrust is constant because it is related to pressure (equations B.26 and B.27): as drawn, this engine would be described as *flat rated* to ISA plus 10 °C, an inlet temperature of 298 K. The effect of varying the inlet temperature is to change the temperatures and shaft speeds at constant pressure, thrust and non-dimensional speed. That is, at constant $N_2/\sqrt{T_{01}}$, N_2 must change slightly as T_{01} changes.

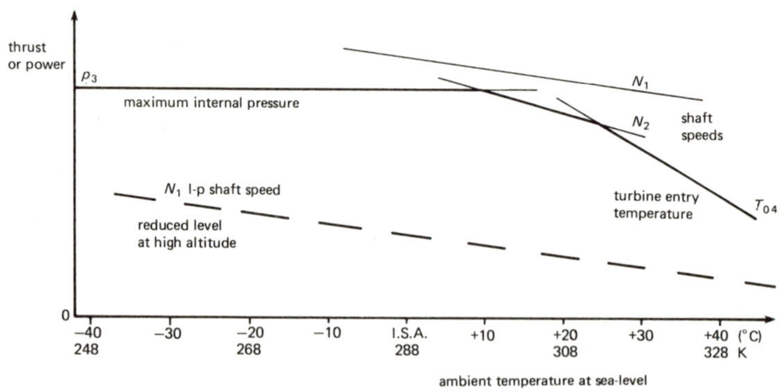

Figure 12.1 *Thrust or power limits imposed by different operating parameters with varying day temperatures*

On a warm day, the h-p shaft speed may well set the limit. The limit of flat rating is reached when N_2 is at its type test value. As T_{01} continues to increase at constant N_2, the value of $N_2/\sqrt{T_{01}}$ falls and so do the thrust and other non-dimensional parameters (figure 4.2). The l-p shaft speed usually imposes no limitation at ground level, but may take over control from the h-p shaft as their speed relationship changes with the reducing temperature during climb (section 4.2). In fact, the l-p shaft speed also takes over from p_3, which is reduced by altitude.

On hot days, the maximum value of turbine entry temperature, TET or T_{04}, takes control. While $N_2/\sqrt{T_{01}}$ falls with increased T_{01}, there is usually an increase in the absolute value of T_{04} even though T_{04}/T_{01} is also falling. This depends on the slope of the T_{04}/T_{01} line relative to speed: some engines may not encounter a TET limit, or may pass from a pressure to a temperature limit without encountering a speed limit.

12.1.4 The Imposition of Performance Limits

The restriction imposed by these predetermined working limits can only be applied by advising the control system when to stop further increases in fuel flow. All engines use instrumentation *sensors* to detect pressures, temperatures and speeds. Some of these signals are relayed to control room indicators, from which an alert operator may take action. They may also be fed to governing devices, which cause the control system to act when it reaches a pre-set limit.

The hardest parameter to measure is the TET, as its level is such as to melt most thermocouples if the draught does not erode or blow them away. Many engines have their TET measured indirectly by measuring the exhaust gas temperature (EGT), and using control adjustments to match the EGT uniquely to the calculated TET.[35] Even then, temperature measurement with thermocouples can be difficult (section B.2). It is sometimes replaced by optical pyrometry, in which the colour of the h-p turbine blading is observed and compared with a standard. Speed and pressure signals are usually easier to handle (sections 12.3 and 12.4).

12.2 Aircraft Engine Fuel Systems

Aircraft engine fuel systems usually form part of the collection of accessories and plumbing attached outside the main casings. Some of the fuel handling and controlling units are mounted on and/or driven by a gear box, while electrical or other controllers may be mounted elsewhere in the airframe. Low weight and compactness are essential attributes, as are reliability (section 12.6) and accuracy of the controls.

Two requirements which differ from those of most other gas turbine applications are a rapid response rate, and compensation for the considerable variation of inlet air pressure and temperature with altitude (section B.4.5). Rapid acceleration aids recovery from an impending premature landing or other in-flight emergency; a typical fast acceleration rate is about 5 seconds from flight-idle to 95 per cent of maximum power. (Note: some governors may affect the time to maximum power from 95 per cent (section 12.6)). Such short acceleration times are known to have prevented accidents. Ease of control, in terms of automatic response to movement of a single throttle lever, also plays its part in accident prevention.

Aviation *fuels* are of hydrocarbon type with densities usually in the kerosine to gasoline range, and this puts them between diesel and petrol as regards mean molecular weight (section 7.1). The engine is required to cope with the normal slight variations between batches of the same fuel and the different fuel supplies available in different countries. The properties of the various fuels are discussed in reference 36 and in oil company literature in considerable detail. A simple requirement is to have the minimum content of sodium, sulphur and vanadium to avoid turbine troubles (section 8.4).

The handling of the fuel for a typical twin-spool engine is shown in figure 12.2. A low-pressure centrifugal pump prevents cavitation and damage within the high-pressure fuel pump. It may be omitted in some aircraft if the fuel tank pumps are considered adequate. The fuel is then filtered to prevent internal damage to the subsequent fuel control units or blocking of the spray nozzles, which might shut the engine down, or distort the flame pattern with consequent damage to the combustion system or turbine.

A *flow meter* is usually included and may be of the turbine type (in which the speed of rotation of a small rotor in the flow is sensed) or of the type in which the flow deflects a vane against a spring. Its output is displayed in the cockpit and may also be recorded for future reference.

It is advantageous to *heat* the fuel prior to combustion, to speed the vaporisation process and to improve combustion efficiency. This is achieved by allowing the fuel to take up heat from the lubricating oil, thus serving to cool the oil; in supersonic aircraft, the fuel is also used to cool the wings. At low speeds, in cold weather or soon after starting, this method of heating the fuel may not prevent the risk of ice crystals (from entrained water in the fuel) from damaging or blocking the system. Hot engine air is therefore bled when there is a need to ensure that all ice is fully melted and dissolved.

The *high-pressure pump* may need to deliver fuel at as high as 80 to 100 bar, depending on the engine pressure ratio and the spray nozzle design (section 7.4). The types used include centrifugal pumps, gear pumps, and piston pumps with swash plate control.

The *flow control unit* (FCU) is very variable in nature. In this case, it is considered as controlling h-p shaft speed, acceleration and deceleration rates, compensating for variations of air pressure with altitude and setting all the operating limits (section 12.1.3). In practice some of these functions may be

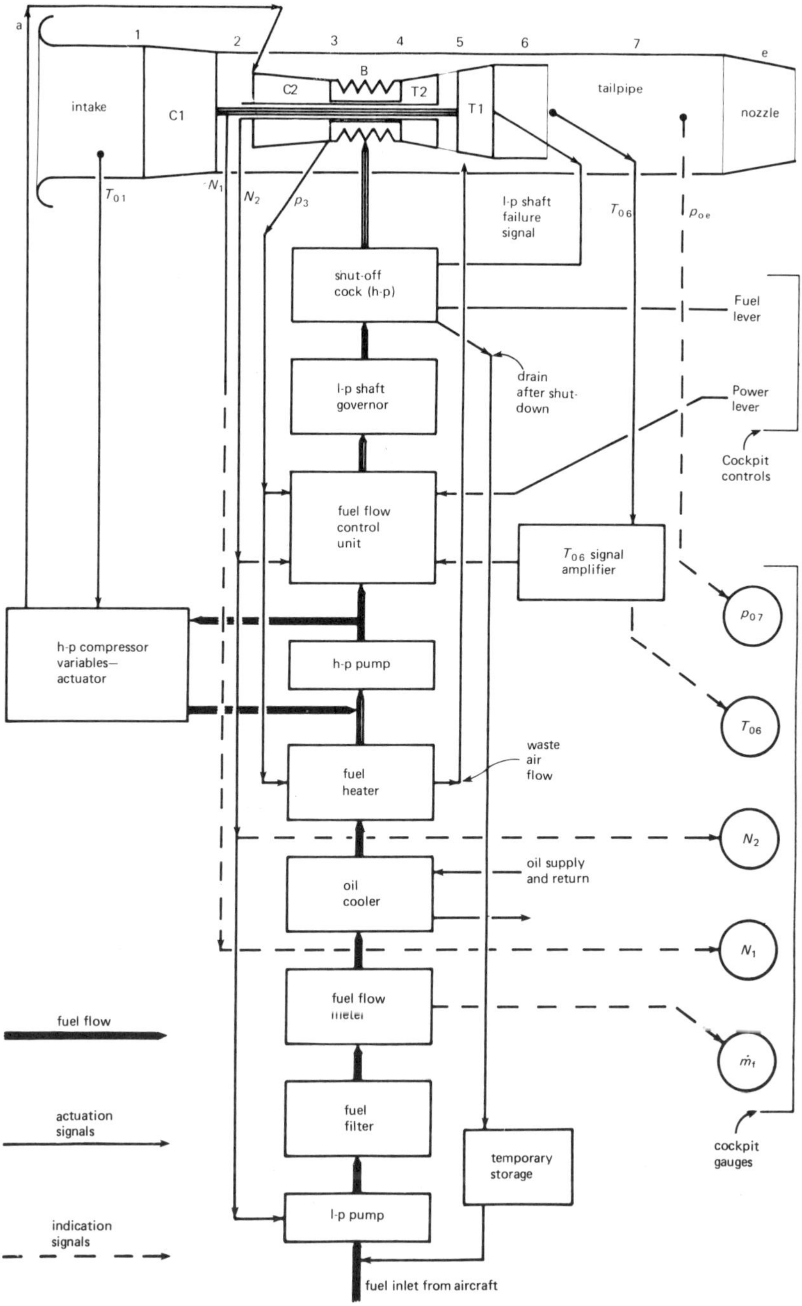

Figure 12.2 *Typical fuel handling and engine control system for a twin-spool bypass aircraft gas turbine engine*

handled by other components. The two basic methods of flow control are for the control unit to cause a swash plate pump to vary its delivery, to match all conditions and requirements, or for it to control a bleed or spill flow which reduces the supply to the combustion and returns some fuel to the high-pressure pump inlet. Some other aspects of an FCU are discussed in sections 12.3 and 12.4.

The *l-p shaft governor* in this system is shown as a separate unit, which has the final chance to adjust the fuel flow if the l-p shaft should reach its speed limit. It could operate by opening a spill valve, or by overriding the pump swash plate control by sending an appropriate signal to the pump control.

The last unit shown is the *shut-off cock*, which rapidly cuts off all fuel supply during normal engine stoppage, during an emergency manual shut-down or during an automatic shut-down in response to a serious failure such as, in this case, breakage of the l-p shaft (section 14.5). Another function is to dispose of the fuel in the pipes to the spray nozzles, which could dribble and cause fires in a hot engine after shut-down, or might 'crack' in the pipes and deposit carbon on the nozzles, which could in time seriously distort the spray pattern. A typical procedure is to use a three-way shut-off cock which, as it stops the main supply, opens a drain line from the nozzles to a *collecting tank*: the residual air pressure then blows the fuel into the tank and leaves the pipes empty and dry. The tank may also collect leakage from various seals in the fuel system to prevent external spillage on the airport tarmac: it is returned to the fuel supply during the next start by automatic means.

A unit which is shown but which is out of the main fuel supply line is a fuel-operated hydraulic ram to control the position of the h-p compressor variables according to a processed $N_2/\sqrt{T_{01}}$ signal. This is generated in the form of speed squared, which comes as a pressure signal from a governor device, relative to inlet temperature, $N_2{}^2/T_{01}$. An alternative means of control is by compressor pressure ratio, which can be handled by means such as those referred to in the following section. In this case actuation is by high-pressure fuel from the h-p pump or by compressed air in a ram or motor.

12.3 Hydromechanical Control Systems

As almost every engine design has its own unique control system, to deal with one system only in detail would be unduly narrow in scope. This section will therefore use a simplified system as a basis for discussion of some of the principles and hardware, as applicable to liquid fuels only.

The objective of the system is to satisfy the fuel demand curve of the engine, as shown in figure 4.2. A thin line is shown in this illustration which projects back to the origin at zero speed and zero flow. A fuel pump with a fixed displacement per revolution would work along such a line from the origin. It is therefore apparent that the pump flow is required to increase considerably faster than its rate of increase of speed, to satisfy the engine above 80 per cent speed. If the pump flow capacity is taken as 100 per cent at 100 per cent speed, then its required output per revolution reduces with speed, as shown in figure 12.3.

Consider also the delivery pressure required from the control system to the spray nozzles. The pressure drop through the nozzle will vary with the square of the flow velocity: if the value at 100 per cent speed is taken as 40 bar (section 7.4), then the static pressure p_3 must be added to determine the fuel supply pressure. This is also shown in figure 12.3 for a simple nozzle: these two curves now specify the control system output.

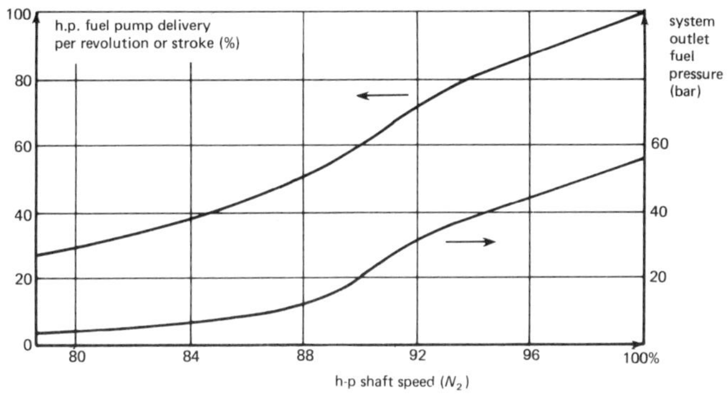

Figure 12.3 *The variation of fuel flow and pressure requirements with engine speed*

The design of an *h-p pump* which increases its delivery as it increases pressure may be difficult. The use of a relief valve behind the pump would allow flow to spill as its pressure increased: this contravenes the requirement. The opening of the valve could be servo-controlled to make it close progressively as the pressure increased: this is one solution. Another is to make the pump delivery proportional to speed at a constant high pressure, well above the peak required pressure of nearly 60 bar, and to make a control down-stream of the pump which will reduce the nozzle pressure as the speed falls.

Figure 12.4 shows a pump of this type, supplying fuel to the control unit, which will be discussed next. The pump is of swash plate, multi-piston type with its swash plate angle controlled by a servo-piston S. The position of this piston is balanced between the delivery pressure of the high-pressure fuel H, and the spring force aided by the low-pressure fuel L. If delivery pressure rises, the piston moves to the right and reduces the pump stroke and flow: the delivery pressure therefore rises slightly as flow is reduced, because of the increased compression load in the spring, so the delivery pressure is not absolutely constant but will suffice.

The *flow control unit* essentially acts as a valve of variable opening, through which the fuel pressure drop will also vary. By controlling the opening, according

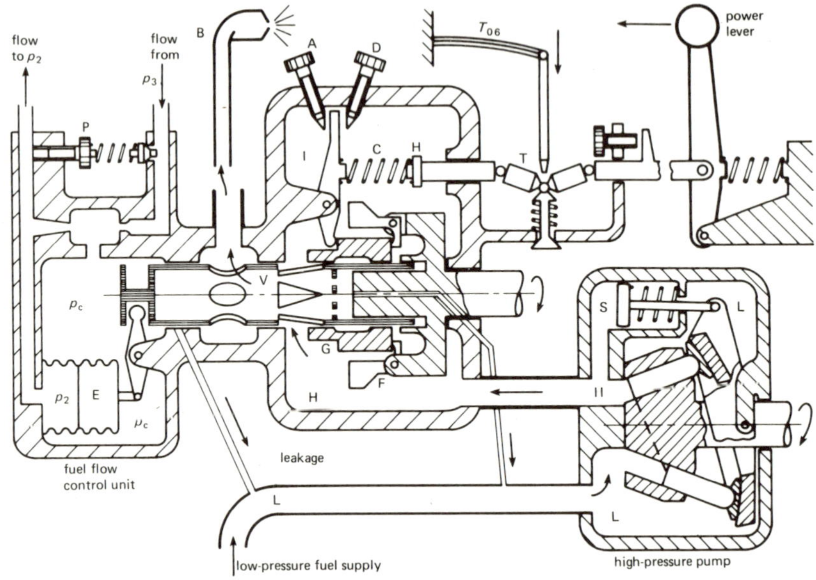

Figure 12.4 *A simple hydromechanical fuel control system, showing the high-pressure pump and the flow control unit for a twin-spool aircraft engine with the following features: A acceleration stop; B fuel spray nozzle; C control spring; D deceleration stop; E evacuated space; F flyweight governor; G governor sleeve; H high-pressure fuel; I input lever; L low-pressure fuel; N speed limiter; P pressure limiter; S servo-control piston; T temperature limiter;* $p_{2,3}$ *engine static pressures;* p_c *controlling pressure*

to the engine speed, the outlet pressure may be controlled to match the curve in figure 12.3, and the pump will vary its stroke automatically, so needs no separate control.

The control area in the unit shown in figure 12.4 is a series of carefully profiled, approximately triangular, variable metering orifices (VMO) V formed in the VMO sleeve. This sleeve is moved axially to expose more or less of its opening area to the gap formed between the casing and the governor sleeve G. Any hesitation in the movement of this sleeve reduces the precision of control: it is therefore rotated continually to break the frictional grip of the seals (section 12.6). In this design, it is shown to be rotated by viscous drag from the flyweight governor F, but it is more usual to rotate it slowly by gearing from the input shaft.

Consider operation at *constant speed*. The power lever will have been pushed forward to the required position or until it reaches the speed governing stop N, compressing the spring C against the input lever I and exerting a force through the governor sleeve to the flyweights. The speed is such that the flyweight centrifugal force balances the spring force; the flyweights, governor sleeve and input lever occupy the same position at any constant speed, creating a constant gap between the sleeve and the casing. The flow is set for each speed by the axial position of the VMO sleeve, which is moved relative to speed by a capsule that is sensitive to air pressure. An increase in the pressure fed from the engine compressor to the outside of the capsule causes the capsule to collapse and push the VMO sleeve to the right. The opening area is increased and the fuel flow is also increased with the air pressure, matching the dimensionless form $\dot{m}_f/p\sqrt{T}$.

Some *design faults* should be exposed here, before going more deeply into the operation. The first is that, as the VMO sleeve moves to the right, the fuel pressure inside it increases to feed more fuel to the combustor. This increased pressure may well push the sleeve back to the left against the air pressure, preventing high-speed operation. This is so significant that the sleeve cavity will be sealed by closing the dotted bulkhead, and the pressure on the right side of this bulkhead will be relieved through the input shaft to the low-pressure fuel system. This takes most of the load off the capsule arm that locates the VMO sleeve.

There are several other faults, which will be subsequently ignored. The pressure drop through the orifices may be too great at idle speed, and provision may be required to use two or more control valves in series between the pump and the combustor. The system shown only feeds a simple nozzle, which may not cope with the range of flows (section 7.4): a unit to control separate pilot and main flows would be much more complex. Finally, this type of unit cannot cope with a change in *fuel specification*. The fuel energy is related to its weight (that is, in J/kg), which allows for the number of molecules of carbon and hydrogen. A change in specific gravity would require a different volumetric flow for the same rate of energy release. The fuel-immersed governor can compensate for this, as an increased fuel density increases the buoyancy of the flyweights towards the axis of governor rotation and reduces their equilibrium radius. If the governor sleeve moved correspondingly, to reduce the opening and the volumetric flow rate, the rate of fuel mass and energy flow would remain constant. In this design, the governor sleeve moves the wrong way!

Means have to be found of changing the rate of increase of fuel flow at about 80 to 85 per cent speed, where the curve steepens and the engine enters the working part of its speed range. The means used in this example is the *air potentiometer*, which controls the pressure p_c in the capsule chamber. Air at static pressure p_3 is fed through an orifice and falls to the value p_c in the intermediate chamber. It then expands through a second orifice down to p_2, at which it returns to the h-p compressor inlet. Air at p_c is fed to the outside of the capsule and at p_2 to the inside. The capsule deflection is proportional to the pressure difference, and is increased relative to pressure by the incorporation of the evacuated section E.

The difference caused by feeding p_c to the capsule chamber, instead of p_3, is shown in figure 12.5. The non-dimensionalised deflection (deflection divided by chamber pressure) is plotted against p_3 for the cases where the chamber pressure is p_3 or p_c. While the first potentiometer orifice is choked at higher values of

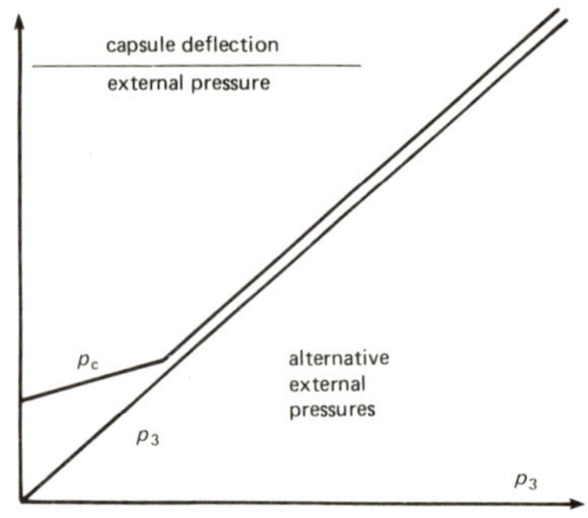

Figure 12.5 *Non-linear deflection of pressure capsule resulting from the use of the air potentiometer*

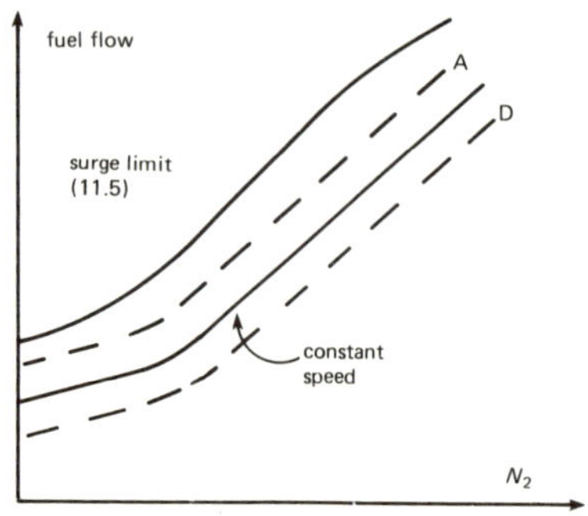

Figure 12.6 *Match of constant-speed operation to surge line, with flows for acceleration (A) and deceleration (D)*

p_3, p_c remains the constant fraction (1/1.893) of p_3 as fixed by equation B.17. The two lines therefore coincide, although they are separated in the figure for clarity. Below the choking condition, when the h-p compressor pressure ratio is about 70 per cent of its maximum in this case,the value of p_c becomes a larger fraction of p_3, lifting the capsule deflection at the lower speeds. This kink in the characteristic serves the required purpose and also allows the engine working line to follow under the dip in the surge line, as expressed in terms of fuel flow in figure 12.6.

This air system provides for easy *limitation of* p_3, of which the relief valve with the adjustable spring load is the simplest version. The air system makes control of the movement of the VMO sleeve so easy that, by providing additional controlled means of reducing p_c, by leaking air from the intermediate chamber to atmosphere, it may be used to control speed or acceleration. This may also be true if the air potentiometer system is replaced by a fluidic logic system.

The control unit shown handles *acceleration* very easily. Power lever movement to increase the load in spring C pushes the input lever I to the left and the governor sleeve G to the right. The flow-controlling area is increased, causing the fuel flow and TET to increase, and accelerating the engine (section 11.5). As the engine speed rises, so does its internal pressure, which progressively moves the VMO sleeve to the right. The orifice area continues to increase, leading the rising speed until the increasing centrifugal load on the governor flyweights matches the load in spring C and restores the input lever to its vertical position. Deceleration is achieved in the opposite fashion. The limit to the rates of acceleration and deceleration is set by the adjustable stops, A and D (figure 12.6).

The last control limit to apply is *temperature*. Using a bimetallic device to sense T_{06} rather than T_{04} (section 12.1.4), it moves a probe progressively as temperature rises. When the limit is reached, the probe contacts the throttle linkage at T and deflects it to a position which demands a lower speed. In practice, an electrical system is normally used, which has the same effect.

No mention has yet been made of *propeller pitch control*, which is integrated with the flow control system. In flight, the power lever typically requests a thrust, which is produced by the combination of propeller speed and pitch angle. The pitch angle is also varied to govern the l-p shaft speed. Some engines vary speed and pitch; while others vary pitch only to govern at constant speed, and vary power by changing fuel flow and turbine temperature. These control functions raise the system to a new level of complication.

However, this explanation should indicate at least what is involved in engine control. For further reading, one of the more advanced hydromechanical systems is described in reference 79.

12.4 Electronic Control Systems

The logic functions of the flow control unit may be handled instead by electrical/ electronic systems. This could include control at any speed and the governing to the various limits. The flow control unit may be replaced by a relatively simple valve, which opens and closes to control fuel flow to the same schedules as the hydromechanical systems. This is essential with gaseous fuels, which cannot be handled hydromechanically.

Integrated circuitry offers many advantages over the hydromechanical system, two of these being the steadily reducing cost and increasing reliability of electronic systems. A modular concept permits the use of standard basic scheduling units and governing units, sometimes in microprocessor form, to be assembled off the shelf for any engine. The same sized modules may apply to any size of engine, only the fuel valve and pump having to change size. Modules affecting plant control (section 12.5) may also be plugged in. The electronics may very easily be altered to adjust the fuel flow schedule, making the development much easier. Circuits may be duplicated or triplicated at little cost or bulk, with fault detection circuits ready to switch from a failed control without upsetting the engine operation. Many more functions can be monitored and controlled, but this may also be a disadvantage.

It does, however, have disadvantages which account for the continuing use of some hydromechanical control systems. The main disadvantage concerns its compatibility with the various control media involved in sensing and actuating. Pressures have to be transduced to appropriate electrical form from their essentially analogue nature. Speeds may be detected very easily in digital form, by inductive or capacitive pick-up off a toothed disc on the engine shaft. Temperatures may come in d.c. electrical analogue form from thermocouples or may require transducing from optical sensors. These signals must be compatible with the type of electronic control used, which may be [6L]

(1) d.c. voltage analogue – inexpensive but prone to drift
(2) a.c. voltage analogue – AM, more expensive and complicated
(3) pulse width modulated – requires extensive signal processing, or
(4) digital and frequency analogue – FM, more accurate, fail safe.

The controlling signals from any of these systems have next to be amplified or otherwise processed to enable them to control the actuators attached to the fuel pump, fuel valve, compressor variables, etc. These add to the cost and bulk of the system, and render the balance of advantages and disadvantages almost equal. The electronic system must also be protected against aircraft or plant electrical system failure, a fault that would have no effect on a hydromechanical system.

12.5 Industrial Engine Control Systems

These systems differ from aircraft types in dispensing with the altitude compensation and/or rapid acceleration features. They may instead incorporate automatic and plant compatible features, which favour the use of electronic control.

The associated fuel handling accessories are usually bulkier and heavier, and are often console-mounted separately from the engine and driven by base plant electric or hydraulic power. The units may be well spaced for ease of access, and connected by standard types of heavy pipes and fittings. The only connections needed from the engine are the instrumentation lines and the fuel feed to the combustion system.

The control and fuel handling functions are similar in nature to those already discussed. In this section, therefore, only the differences and also the features which match the engine to the plant, will be considered.

12.5.1 Automatic and Remote Starting

The starting process described in section 4.1 is often linked to the plant functions by a supervisory system, which ensures that the pre-start checks, the correct sequence of actions and the correct results are achieved in turn. The start may well be initiated remotely from a head office, hundreds of kilometres away and with no personnel on site: this particularly applies to gas pumping duty (section 2.4). The start sequence is then controlled automatically by relays, transducers and interlock circuitry.

The more difficult case for an industrial control system is that of starting a generating plant, in which the alternator must be synchronised with a grid system or network before connection. An example of some of the main activities for this case is given in table 12.1, based on a chart from reference 6K. For the gas pumping duty the engine would not need control refinement to permit synchronisation, but it would have a different set of plant functions to complete.

The *start-up time* depends very much on the engine size, and whether it is of heavy or lightweight design. A typical start-up time for an aircraft-originated gas generator is quoted as 2 minutes from cold to full power for a 25 MW plant.[6K]

Table 12.1 Sequence of Activities to Start a Gas Turbine Electrical Generating Set

The following activities are to be accomplished in turn before proceeding to the next.

Engine and Plant Starting

Select the required type of start control: remote or local.
Select type of operation required: emergency or peak
Start: timer, recorder, oil pumps, ignition, fuel heating.
Activate: voltage regulator, fuel supply, control system, fire system.
Initiate the engine start, delaying fuel input until speed is adequate (section 4.1).
Check: times to light up and to idle speed, rise of exhaust temperature.
Accelerate gas generator to working speed.
Check: that the alternator rotates and reaches governed speed.
Check: that the voltage regulation system is operative.
Adjust plant accessories: stop jacking oil pump, start oil cooler, etc.

Auto Synchronisation

Adjust power to equalise the alternator frequency to the grid frequency.
(Any error here means that, on connection to the grid, the rotor speed will change almost instantly and may strain or break the coupling shaft from the power turbine.)
Adjust power so that the alternator creeps to within 10° of the correct phase angle (also to reduce a sudden jerk).
Check: that the voltage is within 5 per cent of the grid voltage.
Close the circuit breaker to connect the alternator to the grid.

During the starting and operation of *multiple gas generators*, which feed gas into a common plenum to drive the single power turbine, care must be exercised to ensure that the back pressure from one engine does not upset the working lines of adjacent slower running or differently matched engines. An over-all control is exercised here which senses the engine outlet pressures and keeps them in balance by adjusting their fuel flows.

12.5.2 Operation with Different Fuels

There is a natural inclination to use locally available or cheaper fuels. In the case of liquid fuels, this may involve diesel or heavier oils, some of which may need to be pre-heated to reduce their viscosity for easy pumping and atomising. The quality of the actual fuel being used may need to be carefully monitored to avoid trouble (section 7.6).

The other fuel commonly used is natural gas, which is the obvious choice for engines that are working on gas pumping duty. Such a gas supply system involves larger plumbing than a liquid fuel system: both of these would be installed side by side on *dual-fuelled* engines. It is sometimes required to change over from gas to liquid fuel, or vice versa, without affecting engine operation. This is done by having the liquid and gas flow control valves linked so that the speed would not alter at the changeover. Here the fuel switching process involves selecting one shut-off cock 'closed' and the other 'open', at the same time.

The effect of having the plumbing full of compressed gas when the gas shut-off cock is closed is that its pressure at the nozzle falls slowly and the gas may continue to feed into the engine. Positive efforts may have to be made to clear gas rapidly from these pipes to avoid this occurrence.

12.5.3 Emergency Stopping

This may be a frequent occurrence with industrial engines, resulting from plant shut-down rather than an engine or plant breakage. This is particularly the case if an alternator trips off line: the unloaded power turbine could then rapidly overspeed and typically requires the fuel supply to stop within 0.1 s in order to ensure operational safety.[6K] Such a stop could have undesirable effects on the engine turbine if it was running at full power, involving the risk of seizure of labyrinth seals or of rubbing them to a larger clearance (section 14.2.2).

The means of shut-down may be by the shut-off cock already described (section 12.2), but there are other ways which may be kinder to the fuel control system. One is to redesign the h-p pump servo-control so that dumping of fuel from a control line rapidly collapses the pump to zero swash. Another is a manual override on the flow control valve, allowing it to be closed much more rapidly than normally.

12.6 Control System Problems

If there are no obvious breakages or signs of deterioration in the engine, the failure of a gas turbine engine to perform correctly can usually be traced to its accessories. Instances of mechanical breakage, leaks, excessive friction, dirt lodged in a small controlling or sensing orifice, corroded or wet electrical contacts, sensor faults, broken wires and other problems, all give rise to characteristic operating problems which are easily recognised by an experienced turbine engineer.

Breakdown in an industrial plant is a serious matter, as downtime costs may be extremely high. The cost of an engine failure while performing a peak lopping function may cause a substantial increase in the whole month's electricity costs (section 2.3). Assured reliability (section 14.5), with standby plant in case of failure and with more frequent maintenance and overhaul than normal, help to minimise the risk and cost of a breakdown.

In an aircraft, when problems occur causing an in-flight shut-down, an aborted take-off, flight delay or cancellation, the effect on passenger safety and convenience are also regarded as serious. Typical aero symptoms may be a failure to start or to reach maximum power, false fire warning or simply an inoperative cockpit indicator. Considering the number of components involved, a high degree of reliability from each one is essential (section 14.5). As with the engine itself, the accessories are monitored, maintained and overhauled to a planned servicing schedule.

Wear and friction cause hysteresis and delays in the control system response, or even seizure of a mechanism: yet the choice of material combinations which neither wear nor corrode is very limited. The hydromechanical flow control unit is prone to stiction problems because of the very low forces involved in the actual control movements, and this in turn causes starting and acceleration problems. The control sleeves are rotated by an internal transmission system, breaking the frictional grip which might delay the axial motion, but the friction of the driving mechanism may also cause problems.

Interaction of the transient characteristics with the aircraft operation is also a problem. An example previously mentioned was the overswing characteristic of a control system at a time when the temperature limit was in control, as shown in figure 12.7. The timing of the first underswing at A coincided with the attainment of V_1, when the pilot decides whether to take off or brake to a stop. With the engine thrust temporarily below the minimum required for take-off (section 4.4), the take-off had to be aborted. This problem was overcome by

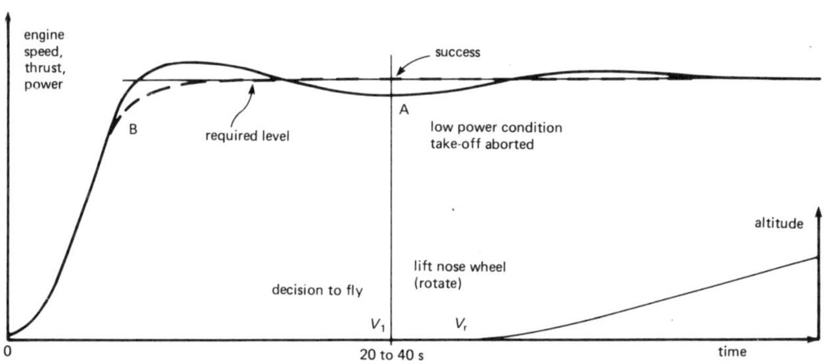

Figure 12.7 *An example of the effect of a control system transient response interacting adversely with the timing of an aircraft take-off*

redesigning the temperature signal amplifier, so that it anticipated the approach towards the temperature limit, and resulted in curve B.

Operational problems also occur with *electronic* control systems, even to the extent of comprising the majority of all the engine's problems. Dirt, moisture and corrosion of all connectors cause trouble. The infant mortality rate, of solid-state components which fail very early in life, can be overcome by a few hundred hours of operation in a laboratory. Analogue devices are prone to drift, and require carefully controlled atmospheric temperature and humidity. Signals from nearby radio and TV systems, or interference from other electrical equipment may confuse the system, and need careful screening. Welding without prior isolation can severely damage the controls.

To these physical problems should be added inappropriate system design, in which the sequence of starting activities or safety interlocks fails to cater for all possibilities. Such a case may occur when one engine of a group trips out for any reason: this sort of sudden load change should not cause the others also to trip out.

Finally, the fuel itself may cause damage to the units. Components which are silver plated, to reduce wear or avoid seizure, lose the plating prematurely if excessive sulphur is present in the fuel. The lighter fuels may lack adequate lubricating properties to protect sliding parts, particularly in the fuel pump. The fuel to be used may therefore dictate the choice of materials used in the fuel control system.

13

Gas Turbine Engine Noise Reduction

Noise is unwanted sound, generated by a vibrating surface or a turbulent flow; transmitted by synchronised pressure fluctuations through a medium (section G.1); and received by the listener's ear or, indirectly, via a microphone and a recording and playback system.

Open-cycle gas turbine engines satisfy all the conditions for noise generation. The actions of compressor and turbine blading in accelerating and decelerating the air flow inefficiently generate sound waves which are propagated along the inlet and outlet ducts into the surrounding air. To this is added noise from the combustion process. Aircraft engines have the additional source of the exhaust jet, which is dissipated turbulently in the surrounding atmosphere as heat and noise. The listeners include the residents near an airport, the public at large and the employees working near an industrial engine plant.

Noise is harmful. It causes deafness by excessive level and duration. Even moderate levels of background noise reduce the ability to hear speech intelligibly and increase fatigue, necessitating the use of preventative and protective measures.[52, 80, 81] High levels of noise also damage machinery, by inducing fatigue cracking in aircraft panels and structures, and compressor and turbine blades (appendix F). The continual vibration also induces wear and frettage between mating surfaces, especially those in the combustion system which are free to slide to accommodate differential expansion.

The excessive noise of the jet aircraft of the 1950s inspired the activity which led to the much quieter turbofan engines of the 1970s. Some of the technology used is applicable to industrial engines, and particularly to fans, compressors and turbines. This chapter discusses engine noise and its main sources, the means to avoid blade noise by design, and the ways of reducing the effects of the noise of existing engines. For readers unfamiliar with noise terminology, an introduction is included in appendix G. The later parts of that appendix consider more deeply the generation of blade noise.

13.1 The Noise Field Round an Engine

An engine running in the open generates a narrow band noise spectrum, as shown in figure 13.1. This example would apply to a turbofan engine running at a medium speed, with contributions as shown from the compressor, combustion, turbine and exhaust jet. The turbine noise is shown to be of higher frequency than that of the compressor, because it usually has more blades. The internal noise propagates almost entirely through the inlet and outlet, with very little emerging through the engine casings.

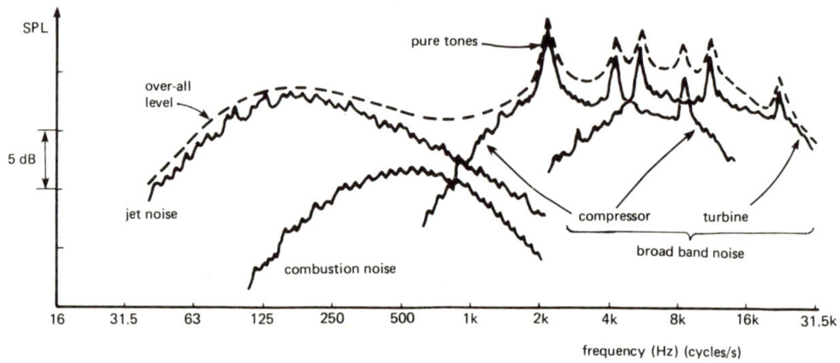

Figure 13.1 *Typical noise spectrum for the main components and their combined level, at medium engine speed*

The level of the separate contributions would change markedly with changes of engine configuration, running speed or microphone position. Each signal radiates directionally, as shown in figure 13.2*a*, which shows only a few noise contours. The letter L indicates the position of contours of equal level from the different sources, and contours of different levels could be drawn for all sources, as shown for the compressor and fan noise. While compressor and combustion noise is seen to radiate forward and rearwards, turbine noise can only propagate rearwards. This is because the nozzle guide vanes are choked in most circumstances; the sonic velocity of the gas prevents the upstream passage of sound waves (section B.2).

If the engine moves past the observer, the direction in which each contribution sounds loudest will be where the contours reach lines parallel to the engine axis. Figure 13.2*b* shows the variation of level along a line parallel to the axis, which could represent the signal measured as an aircraft flies past. This is the region of maximum noise: the level obviously falls as the horizontal scale is extended in either direction.

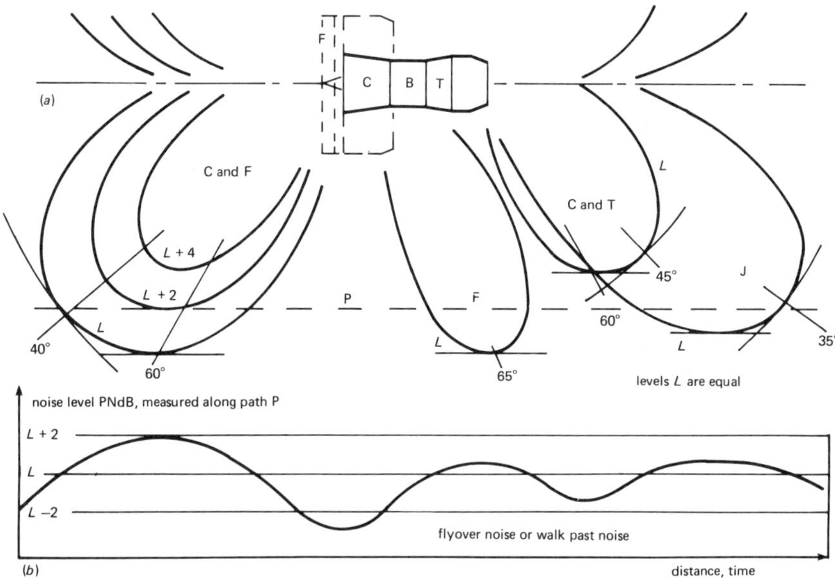

Figure 13.2 *(a) Contours of constant level showing the directivity of noise radiation from the various sources, and (b) the consequent variation of flyover noise with time*

13.2 Compressor and Turbine Blade Noise

The noise of fans, compressors, turbines and sirens is generated by air flowing through one or more stages, each comprising a multi-bladed rotor and stator. In this discussion, the terminology used will be that applying to a single-stage fan, but the mechanisms of noise generation will also apply to compressors and turbines. The siren case is different, because it is designed to make as much noise as possible.

Fan noise results from four main factors. The first is *broad-band noise*, at low levels throughout the frequency spectrum, which results from turbulent flow in the blade boundary layers. It correlates well with blade incidence angle (section E.3) and tip Mach number (section B.4.3), and is generated near the blade trailing edges.[82A] It therefore appears in the noise spectrum for a rotor running without stators, as in figure 13.3. The increased noise at lower frequency results partly in this case from the jet noise of the swirling outlet flow (section 13.4).

The engine order frequency is the multiple of the rotational speed, N. The peaks at blade passing order and its second harmonic result from the second causative factor, *atmospheric turbulence* in the inlet air flow. The main flow distortions arise from the intake wall boundary layer, blow in doors (section 11.4.4), ingestion of ground vortex (section 6.4.2), cross winds and general

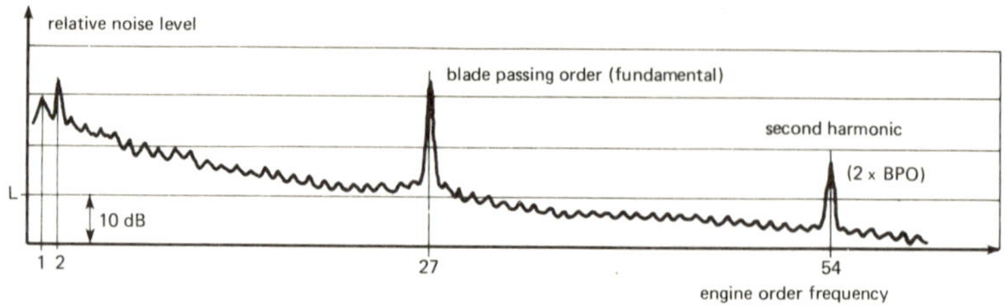

Figure 13.3 *The noise spectrum for a 27-bladed fan rotor with subsonic tip speed, measured at 45° to the outlet axis*

gustiness. Flight experience shows that these tones disappear once there is significant forward velocity[83]: the noise level increase in static operation may be as high as 20 dB as a result of this turbulence. Causes which may additionally affect industrial engines include eddies and uneven flow distribution from bends and corner vanes in the inlet ducting, and eddies generated by objects in the inlet plenum or region.[84]

The third factor is *rotor-stator interaction noise*, which is generated in smooth flow conditions as a result of adverse combinations of blade and vane numbers and blade tip speed. This effect is demonstrated simply by figure 13.4. The addition of a set of stator vanes close behind the rotor reduces the broad-band outlet swirl noise, but drastically increases the blade passing order tones and harmonics and introduces most engine order tones from the fundamental upwards.[85] The underlying cause of this noise is the interruption of the low-velocity blade wakes (figure E.6) by the following stators. The stators feel intermittent changes of lift, which affect their pressure field and make them radiate noise. The signals emitted by the adjacent vanes in a set may be out of phase, in which case they cancel and decay. If they are in phase, they combine to propagate as sound waves at blade passing order frequency, with additional harmonics (section G.3). This phasing of the signals may be avoided by the appropriate choice of the relative numbers of blades (B) and vanes (V) (section G.7). In simple terms, high-speed subsonic fans do not generate these tones if

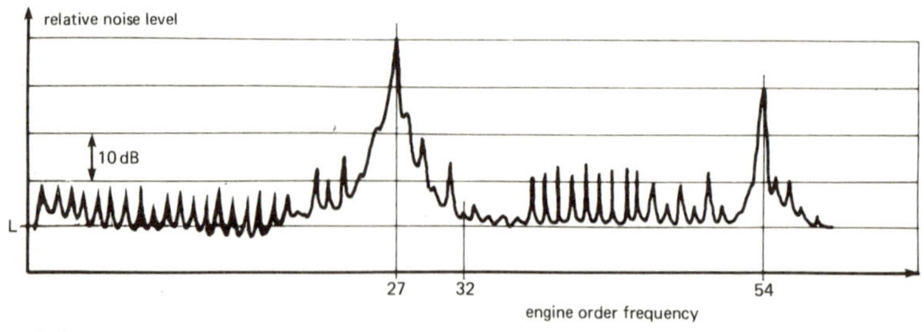

Figure 13.4 *The considerable increase in noise level relative to figure 13.3, caused by the placement of 32 stator vanes spaced at one-third of the rotor blade axial chord downstream of the fan. (The levels marked L in both figures represent the same noise level)*

V/B exceeds 2, while lower speed fans are quiet at a lower value of V/B. The large number of vanes required by this ratio of 2 may well introduce problems of structure, cost, annulus blockage or performance deficit when maintaining an appropriate space/chord ratio (section E.6).

The last factor is operation at *supersonic* blade tip speeds. Each blade produces its own shock wave, which propagates along the inlet duct, but variability between the blades creates shock waves at different angles and strengths. Adjacent waves at converging angles merge until only a few shock waves emerge, generating a relatively low frequency sound known as *buzz-saw noise.*[82A] All the engine order tones are also increased by supersonic operation.

Blade noise in a new engine cannot be predicted from first principles, but there is sufficient understanding and experience to permit a good estimate. A simple rule of thumb is that it varies with the sixth power of flow velocity or blade tip velocity,[80A] because of its dipole nature (section G.4). In practice, several factors (section G.9) combine to make the noise vary from between the square and the ninth power of the velocity, for different fans. The combination of several stages causes complication by making the rotor blades interrupt the wakes from the preceding stators, causing secondary disturbances to the following stators.

13.3 Combustion Noise

The passage of air through the numerous sharp-edged orifices, slots, holes and chutes induces considerable turbulence. This is augmented by the turbulent combustion process, and transmits noise via the engine exhaust to radiate as broad-band 'white' noise of a moderate level.

There may also be noticeable discrete tones generated at specific frequencies, all of which are undesirable. The lower frequency noises result from various resonant couplings and are known as chugging and rumble. *Chugging* results from the time delay between fuel entry and its subsequent combustion. If a momentary air-pressure drop causes a brief increase in the fuel flow, the increased heat may increase the pressure in phase with an organ pipe resonant, cyclic pressure increase. This sort of self-sustaining cycle generates noise at a few hundred hertz: changing the fuel nozzle pressure drop relative to flow may effect a cure.[32C] *Rumble* also occurs at low frequency: it is believed that a temperature fluctuation reaching the choked turbine NGVs causes a pressure wave to be reflected back to the fuel nozzles. If this changes the fuel flow condition with adverse phasing relative to the temperature fluctuations, then resonance may ensue. These two phenomena may differ only in name.

Higher frequency noises such as howl and screech are usually associated with afterburners. The *howl* mechanism involves standing waves in which a half wavelength of the oscillation corresponds to the diameter or radius of the chamber. Another cause may be the frequency of vortex shedding from the flame holder.[32C] *Screech* occurs sometimes when the exhaust from an afterburner nozzle is heavily overchoked. A series of expansions and contractions forms a periodic shock wave pattern along the jet. Should this pattern be set into an oscillatory motion along the axis of the jet, the result is a high-frequency screech of very high intensity.[80A]

13.4 Jet Noise

Jet noise arises from extremely turbulent mixing of the jet with the surrounding atmosphere. Two zones can be defined, as shown in figure 13.5: the primary core in which the flow continues as it left the nozzle, and the turbulent mixing zone. The boundaries of these zones do not form the neat cones indicated, because the turbulent region makes large excursions into the core and the atmosphere, but the cones indicate the mean boundaries of the regions.

The noise is generated mainly at the shear zone with the atmosphere, where the randomness of the turbulence generates noise over a wide frequency spectrum. The spectrum itself shifts to lower values with distance along the jet [80A] : the frequency of the peak noise within the spectrum falls from an initial high value (at the intense shear zone near the nozzle) to a lower value downstream where the mixing is more gentle (figure 13.5). The temperature also falls as atmospheric air is entrained. Reference 80A quotes the non-dimensional Strouhal number fD_e/V_e as falling from about 0.7 to 0.3 over the jet length shown, but there are many side effects and non-linearities which prevent the frequency spectrum for all jets from being represented by a single non-dimensional line. [82B]

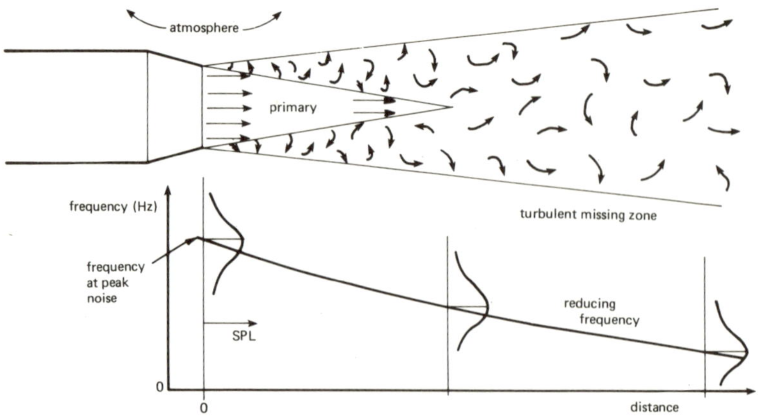

Figure 13.5 *The reduction in frequency as the scale of the turbulence increases with distance along the jet*

The noise from the jet is also difficult to non-dimensionalise, because the maximum level is measured at an angle from the jet axis which increases slightly with velocity, the noise contours (figure 13.2*a*) become longer and narrower with increased velocity, and gas density and atmospheric pressure also affect it. A correlation for jet noise at any angle is given in reference 82B (from SAE Aerospace Recommended Practice (ARP)876), but a moderately accurate expression for the peak noise level or sound-pressure level, SPL, which occurs at between 20° and 40° to the exhaust axis, is

$$SPL_{30} = 10 \log_{10} \left[\left(\frac{V_e}{c}\right)^9 \times \left(\frac{A_e}{R^2}\right) \times \left(\frac{p_a}{1.013}\right)^2 \right] + 150 \text{ dB} \qquad (13.1)$$

where R is the radius to the measuring station. The equivalent expression for noise measured at 90° to the axis is

$$SPL_{90} = 10 \log_{10} \left[\left(\frac{V_e}{c}\right)^{7.5} \times \left(\frac{A_e}{R^2}\right) \times \left(\frac{p_a}{1.013}\right)^2 \right] + 140 \text{ dB} \quad (13.2)$$

The indices 7.5 and 9 are close to the value 8 which is predicted by the standard theory for quadrupoles (section G.4.[80A] An approximate correction for exhaust temperature or density could be added to the calculated SPL, in the form

$$+ [47 \log_{10}(V_e/c) + 7.7] \log_{10}(T_a/T_e)$$

Bypass and fan engines have *coaxial exhaust jets*. Here the most significant variables affecting noise are the area ratio and velocity ratio of bypass to core exhausts.[82B] Optimally, the introduction of a bypass flow around an existing core flow, with $A_b/A_c = 4$ and $V_b/V_c = 0.4$ to 0.6, reduces the noise radiated at 90° by 5 dB, and reduces the high-frequency content of the noise. If the bypass flow velocity is raised to that of the core jet, the effect is to enlarge the nozzle by a factor of 5 and increase the noise level: equation 13.2 indicates an increase of 7 dB.

Another design of outlet involves the use of *corrugated nozzles*. These might reduce the noise by between 3 and 10 dB by dividing the exhaust flow into narrow strips: the primary zone becomes much shorter, so the size of the turbulent mixing zone is reduced (figure 13.6). These were common on aircraft

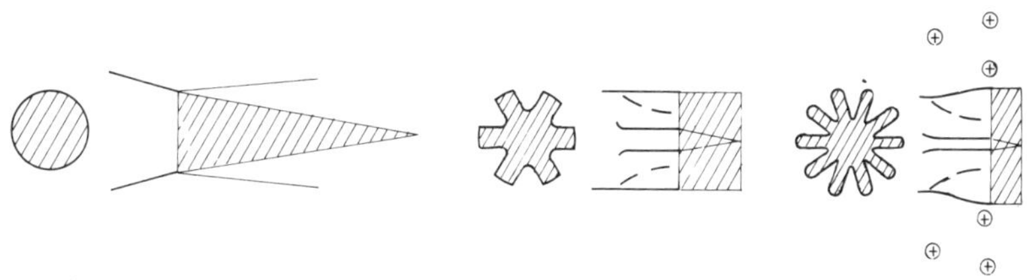

Figure 13.6 *The reduction in noise at constant flow area when the size of the primary zone is reduced by changing from circular to six-corrugated or ten-lobed nozzles*

of about 1960, before high bypass ratio fans were introduced. The reduced dimension across the nozzle opening raised the frequency spectrum at approximately constant Strouhal number.

13.5 Engine Noise Reduction

Engine noise levels are now limited by law. For *aircraft*, the noise permitted varies with the number of engines, the maximum weight and the stage of the flight cycle. These are tabulated in the latest edition of reference 86 and will be progressively reduced. As an example, aircraft with four engines and weighing over 272 t, and certificated between 1969 and 1977, were not to generate an effective perceived noise level exceeding 108 EPNdB outside a defined rectangular area on the ground. This area started 2 km before landing touch-down, extended to 6.5 km beyond the start of ground roll for take-off, and was 0.7 km wide. Plotting this area on an airfield map, with the larger areas for lower noise levels and including all runways and aircraft types, helps planners determine quiet zones for new suburbs and industry.

Industrial engines are covered by factory regulations and noise codes. A typical requirement for a packaged engine generator set is that it will not generate more than 80 dBA beyond a metre from its case.

Quietening by *design*, of blade noise and jet noise, has been introduced in sections 13.2 and 13.4. The blade noise techniques also apply to *centrifugal compressors*, which are hard to quieten with pipe or channel diffusers because the ratio of vane to blade numbers is always wrong. Multi-vane cascades or vaneless diffusers hold more promise, but small rotors may run fast enough to leave only rotor speed orders within the audible range. In *axial compressors*, it may be reduced approximately as follows: remove inlet guide vanes (if present) – 10 dB; double the axial spacing between the rotor blades and stator vanes – 4 dB,[82A] although with length, weight and whirl speed penalties; space the stator vanes unevenly – 8 dB (section G.9); or reduce speed by 10 per cent – 2.5 dB. In *turbines*, the higher gas temperature reduces the Mach number relative to the compressor, so permits a reduced vane-to-blade ratio.

The use of *acoustic linings* in ducts, with Helmholtz resonators formed by honeycomb material sandwiched between the outer wall and an inner, perforated skin, may provide noise reduction of 10 dB if properly applied. They should preferably be several duct diameters long, and the radial thickness may vary to cover a range of frequencies.[87] These linings, appropriate blade numbers and high bypass ratio all contribute to the trend shown in figure 13.7. It should be noted that the noise has been scaled to identical thrust levels in this figure; a direct comparison may be obtained approximately by adding $10 \log_{10}$(bypass ratio). An alternative duct lining is the bulk absorber type, using fibrous material to dissipate the acoustic energy. All linings need drainage to avoid fire risk (section 14.5) from trapped fuel, or lining damage from ice formation.

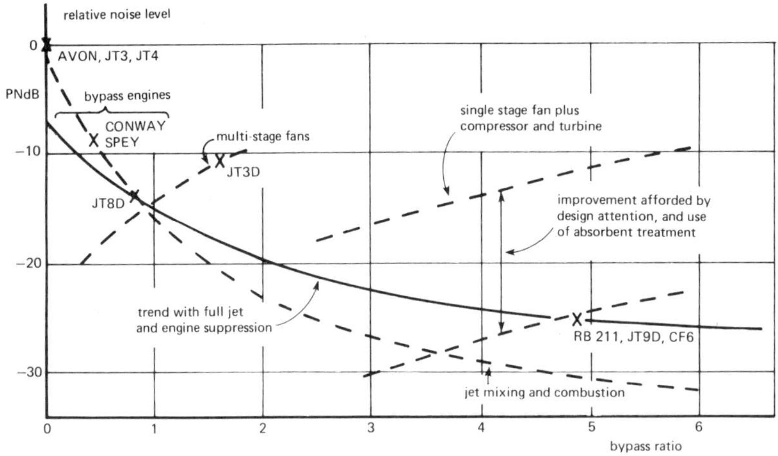

Figure 13.7 *Chronological trend of noise reduction with later aircraft engine designs (by courtesy of Rolls-Royce Ltd)*

Other *duct treatment* may involve long ducts to take the noise to a remote place, a very smooth inlet flow to minimise turbulence (section 13.2), a sonic intake with a throat which chokes (section B.2) to prevent upstream sound transmission at a fixed flow limit, or reactive mufflers.

Often an industrial engine will be housed in a building having some of the characteristics of a test cell (figure 4.3). The intake and exhaust are silenced and remote from people, and an engine intake drawing air from inside the room needs plenty of clear space. The walls and doors need to be massive, and without leaks as noise escapes very easily through gaps. The mass law indicates that doubling the mass of a wall causes the sound to attenuate by 4.4 dB at constant frequency, while the sound is attenuated by an extra 6 dB for each doubling of frequency.[80B] As sound-power levels in the room may reach 170 dB,[80C] the mass of the test cell is very significant. Double or triple glazing may be needed for the control room window.

For an engine operating in the open, distance from people helps attenuate the higher frequencies (section G.2). Deflectors or walls may help reduce the noise by reflecting it away or upwards. The engine may be oriented with its inlet or exhaust pointing towards the nearest group of people, to give them the advantage of the directivity of noise radiation (figure 13.2*a*).

14

Mechanical Design Considerations

14.1 Internal Loads and Jet Engine Thrust

The thrust of a jet engine is the reaction to the rate of change of momentum imparted to the working fluid.

The reaction can only be applied by pressure differences acting across the area of the various surfaces within the engine. To take the simplest analogy, a released balloon accelerates away because the internal pressure on the skin opposite the nozzle is greater than the atmospheric pressure on the front of the skin. If the nozzle were to be capped, the pressure difference across the skin would be balanced by that across the cap, and there would be no thrust.

The turbo-jet engine has no simple surface corresponding to the unbalanced skin area of the balloon. Instead, it has numerous blades and vanes, tapered casings and the ends of the l-p rotor, of which the axially projected areas contribute to the axial thrust. The highest internal pressure occurs at the combustion system: in general, the components ahead of the combustor are pushed forwards and the rearward components are pushed rearwards by the internal pressure. The *gross thrust* is therefore the difference between all the forward thrusts and all the rearward thrusts.

The *forward thrusts* are applied against the back of the compressor rotor blades and stator vanes, at the walls of the compressor outlet diffuser and against the hub of the l-p turbine rotor. The *rearward thrusts* are applied to the combustion discharge nozzles, the turbine nozzles and blades and the inside of the convergent nozzle. The axial thrusts on the compressor and turbine rotors are transmitted to the casing via a thrust bearing (section 14.2.1).

The loads are applied to these components by the changing pressure and velocity of the gas, which may be assessed by the axial *gas load*. The forward thrust on the compressor is equal to the increased gas load imparted, and so on. Figure 14.1 shows the changes through the engine. At any station n along the

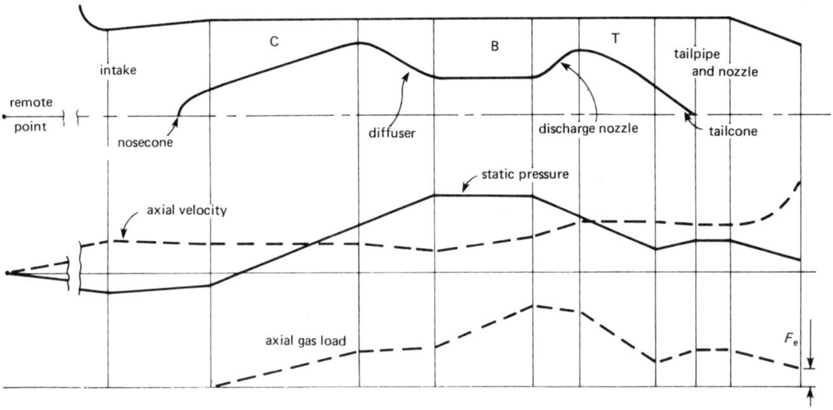

Figure 14.1 *Variation of static pressure, axial velocity and axial gas load through the engine*

engine, the gas load is measured as $\dot{m}V_n + (p_n - p_a)A_n$. This also applies at the exit plane e, where the gas load is the engine thrust

$$F = \dot{m}V_e + (p_e - p_a)A \tag{14.1}$$

The term $(p_e - p_a)$ applies if the nozzle is overchoked (section B.2). Then, the stagnation pressure p_{0e} is so high that p_e is above the ambient level (equation B.17) across the nozzle exit. Effectively, it implies an increased pressure on the inside of the balloon skin. If the nozzle is not choked, then the residual pressure term disappears and the thrust is

$$F_e = \dot{m}V_e \tag{14.2}$$

Dimensionless forms of these equations are given in section B.5.

The final, 'propelling' *nozzle* clearly exerts a rearward thrust, but its small area maintains the high internal pressure which accelerates the exhaust gas to its highest possible velocity. The thrust increase from an *afterburner* or *reheat* system (sections 3.5.3, 7.4, 11.4.1) arises because it can maintain the pressures in the engine while using a larger nozzle area to accommodate the greater gas temperature; the rearward thrust on the nozzle is therefore reduced or, in momentum terms, the outlet velocity is increased. If *thrust reverser* doors are closed across the nozzle, the high, rearward pressure load is matched by the forward flow of the outlet gas as it exerts a rearward thrust.

If the engine is moving forward, its thrust is still determined by the rate of increase of rearward momentum of the flow. Ram effect (section B.5) increases all the internal pressures and the gross thrust, but the approaching airflow must be subtracted, the intake momentum drag. This yields the net thrust, which falls as flight velocity rises, in the form

$$F_{net} = F_e - \dot{m}V_{aircraft} \tag{14.3}$$

To give scale to the enormous mass flows involved in thrust generation, typical values range from 10 to 700 kg/s for the smallest jet engine to the largest turbo-fan (section 2.1). These figures correspond to about 8 to 570 m^3/s of free air, or about 17 000 to 1.2×10^6 ft^3/min in older units. The exhaust velocity ranges from about 250 m/s for the fan outlet to about 800 m/s for the exhaust of a reheated engine. Various combinations of $\dot{m}$, V_e and p_{0e} result in a range of aircraft engine thrusts from about 2000 N (450 lbf) to 250 000 N (56 000 lbf) at the time of writing.

14.2 Structural Design

The design of the engine structure, both the static casings and the rotating parts, must satisfy a number of operational requirements, many of which may occur simultaneously.

(1) The internal passages are shaped to guide and contain the working fluid.

(2) The casing aligns the rotating assemblies and reacts the radial and axial (section 14.1) bearing loads.

(3) The casing contains the internal pressure as it varies throughout the engine (section 14.1), with minimal oval distortion at the split lines.

(4) The casing reacts to the axial gas loadings (section 14.1) and, in aircraft applications, transmits the resultant engine thrust to one or more mounting points.

(5) The casing withstands stresses due to movement, for instance earthquakes, vehicle motion or aircraft landings (typically five to ten times gravity).

(6) In transport engines, the casing reacts gyroscopic loads applied by the rotors through the bearings, as the direction of travel is changed.

(7) The casing may be required to contain internal components which, if they break and are then released, could cause severe damage to the surroundings; for instance, to an aircraft and its passengers.

(8) The internal load paths ensure maximum carcass rigidity and strength, with a minimum weight.

(9) The casing dismantles in a number of sections to facilitate engine build and strip, or access for routine maintenance.

(10) The rotating assemblies withstand all centrifugal and torsional stresses at maximum speed every time that speed is reached, within a permitted cyclic life (4.4 and 14.6).

(11) The rotors also withstand the stresses at an allowable overspeed condition for a prescribed short period (section 12.1.1).

(12) The rotors transmit the axial gas load applied at the blade radius (section 14.1) to the thrust bearing at their centre without excessive axial deflection.

(13) The components of the rotor locate during assembly to facilitate rotor balancing, and do not shift during operation so as to cause loss of balance.

(14) The shafts joining the rotors withstand the applied torque, provide axial location for the rotor and are easily uncoupled for removal.

(15) The bearings locate the rotors but also provide some flexibility, so that the rotors may pass through any critical vibration speeds and assume their natural centre of rotation (section 14.3).

(16) Seals, usually of finned labyrinth type minimise the loss of high pressure fluid from all junctions between the static and rotating components.

(17) The rotor structures incorporate provision for cooling flows through internal passages.

In practice, many of these requirements are ideals that cannot be realised in every engine design. The shortcomings may then result in operating problems or add to the maintenance or repair procedures, which will vary according to the detail design methods and construction philosophy of the manufacturer.

14.2.1 Labyrinth Seals

Labyrinth seals (figure 8.1) are formed by a cylinder with external fins, spaced axially at one pitch on a disc, running closely inside fins at a different pitch inside a cylindrical static member. The pitch difference ensures that at least two opposed fins are in close proximity in any relative axial position. This type of seal adequately separates zones of different pressure by permitting a small leakage flow through the small radial gap round the seal circumference: the flow itself generates the required pressure drop (section 3.2.4). This works well at a steady speed, but changes of speed may cause relative axial and radial movements between the fixed and moving seals (section 14.2.2).

The size of the labyrinth seals at the ends of the compressor and turbine rotors determines the end load on the rotor. A large diameter seal leaves a small area of the rotor exposed to the static pressure at that station through the engine, and provides small opposing forces on both the rotor and the casing. The seal diameters are chosen so that the axial tensile loads of a joined compressor and turbine rotor balance out to a small residual end load, to be reacted by the thrust bearing. This load is usually arranged to act always in one direction, at all engine operating conditions, thus avoiding bearing problems (section 14.4). The seals used to adjust rotor end loads are sometimes called balance pistons.

14.2.2 Transient Dimensional Changes

The clearance of labyrinth seals at the ends of rotor and stator blading varies between the different conditions of engine operation. During a start, the turbine blades and nozzles heat up and grow radially before the casings and discs. The radial growth of the rotor blades may necessitate that the sealing gap at the blade shrouds be oriented axially (figure 8.1). Then, depending on the axial growth of the casings and the turbine shaft as they warm up, and relative to the reference position of the thrust bearing which couples them, the turbine blade sealing fins may move forward or backwards relative to the seal segment. This is only one critical clearance. The others, which mostly involve radial gaps, involve all the labyrinth seals and the compressor blade and stator tip clearances.

The effect of these variable gaps is mainly to reduce the *performance* of the engine when excessive leakage takes place. The radial gaps are affected by material stress and strain which change with engine speed, as against thermal growths and contractions which take longer. The effects are therefore different during acceleration relative to deceleration. The absolute value of the radial gap is set by avoiding a rub at minimum clearance, or by the use of wearaway coatings or feather-edged rotor blade tips which determine their own clearance. These are particularly important with certain combinations of blade and casing materials, where light rubs may generate sparks and debris, or even cause a fire in the case of titanium rubs. The wearaway coatings also allow for casing distortion resulting from design (section 14.2.3) or a transient carcass deflection.

The determination of the optimum labyrinth seal clearance may involve the use of soft wearaway probes on the static member: subsequent strip shows the range of movement of the seal, but not the phase of operation in which its extremes were reached. For measurements at specific conditions, tangential X ray or gamma ray photography may be used, as shown in figure 14.2 [88]

Variation of the compressor blade tip clearance has a more serious effect on operation, particularly in engines with nickel alloy compressor discs. These take longer to heat than steel, so may match the slow growth of cold steel casings quite well during acceleration. In prolonged operation, the differential thermal expansion lets the casings expand away from the discs and blades, with adverse effects on *surge margin* (section 11.1). Provision is made to control the tip clearance by air-cooling the compressor and turbine casings when required, for instance at steady speed and during deceleration but not during acceleration.

The last stage of operation to cause concern is the *shut-down*. Again, the components lose heat at different rates. An emergency shut-down from a high speed condition may well cause the turbine NGV inner shroud to shrink too fast for the hot rotor discs, closing up the clearance and causing temporary or permanent seizure. It is normal to idle an engine for a specified period before shutting down, to avoid this condition, and to measure the time taken to stop as a regular check to monitor any change in internal friction and drag.

14.3 Engine Vibration Problems

At first glance, it would appear that the gas turbine's smooth, continuous operation and lack of reciprocating components would render it quite free of

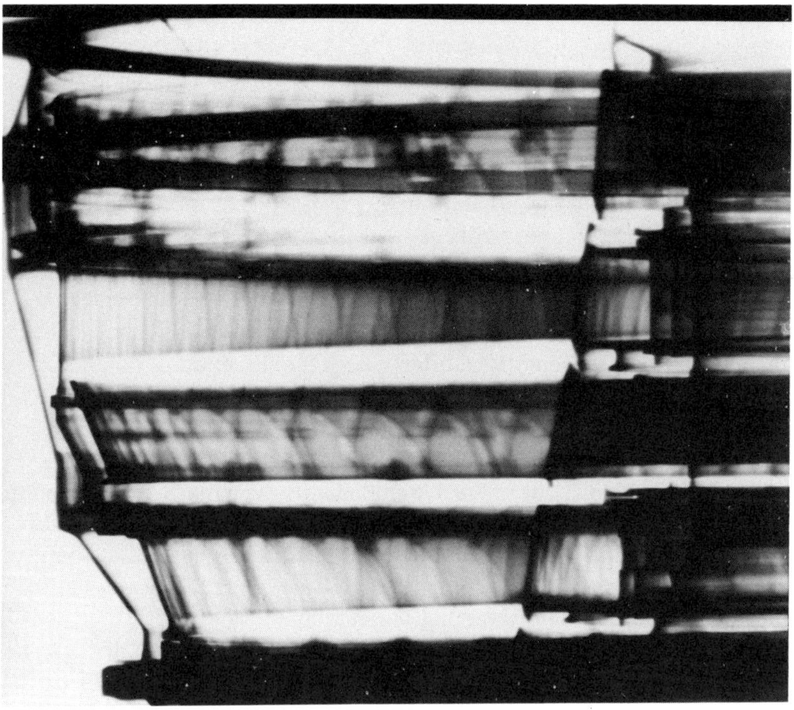

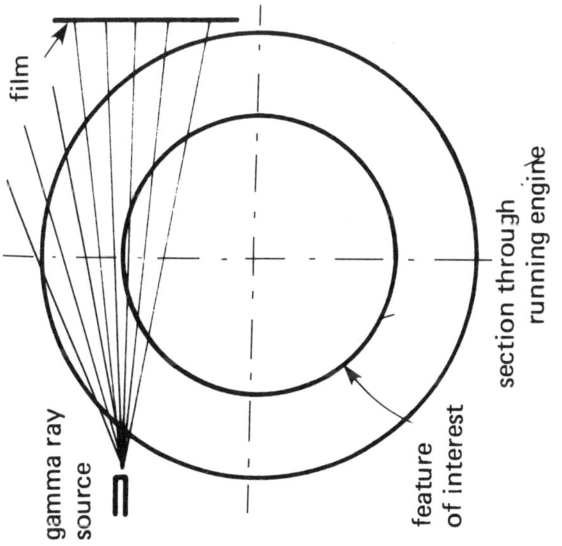

Figure 14.2 *Tangential gamma-ray photography to inspect the running clearances of seals (by courtesy of Rolls-Royce Ltd)*

vibration problems. They are indeed smooth in comparison with many other engine types but components can fail as the result of excessive vibration stresses. This section considers the movements of the main rotating assemblies which cause the engine carcass to shake, transmitting a vibration to the supporting structure. The main problem is the vibration of blades and other internal parts which cannot be felt externally: these are caused by blade passing orders and internal flow conditions, and are discussed in appendix F.

Each compressor and turbine rotor has its own imbalance, mass and bearing support stiffness. The natural frequencies of the rotors, at which the vibration amplitudes are large, are therefore usually different, so the identity of the rotor causing a particular vibration problem is indicated by the speed at which the vibration peak occurs. The bearing support structure is usually flexible enough to keep the natural frequency low, often below the idle speed. It may involve a spring-mounted housing, a housing mounted in a lightly clamped plate which provides damping as it orbits, or a squeeze film which provides oil damping.

In its turn, the casing vibrates in response to the force applied by the orbiting bearings, its own mass, the flexibility of its attachment to the mounting structure and the mass of the mounting structure. The effect on the mounting structure is a vibration which can be felt, possible with audible noise. The character of this transmitted signal may become harsher if the bearings orbit sufficiently to contact their travel limiting snubber.

The allowable magnitude of engine casing vibration is often below 0.1 mm (4 thou or mils), to avoid the development of long-term breakage problems. It may be measured by vibration pickups bolted to the casing, which feed the combined signals from all rotors into an appropriate meter. The limits of rotor imbalance during assembly are of the order of up to 0.5 kg mm (0.7 oz in.), depending on the size of the rotor and the type of engine. The achievement of such fine balance requires the individual balancing of each disc with its blades, assembly of the rotor with the heavy side of consecutive discs on opposite sides, and careful balancing of the final assembly. Long compressor rotors of this construction are particularly hard to blance, and are prone to balance shifts as components move or settle (section 14.2, item 13).

The assembly of the engine is thus a very skilled task. Apart from individual rotor imbalance, vibration problems can be encountered by connecting a compressor and turbine with their heavy sides opposite, creating a vibration couple. Alignment of the casings and bearing assemblies is also critical, as a misaligned splined shaft connection may generate hard-to-trace, non-integral vibration frequencies. The engine is therefore assembled standing with its shaft vertical in some cases, to ensure alignment of the casings.

14.4 Bearings and Lubrication System

The bearings used include both ball or roller anti-friction bearings and plain journal bearings. Most aircraft engines use combinations of ball and roller bearings, heavy industrial units normally use plain bearings and other engines may use either type.

Ball and roller bearings may encounter severe problems in gas turbine engines as the speed and load vary. The load is transmitted radially through the roller, but at an inclined radius through balls which carry axial as well as radial load (figure 14.3*a* and *b*). At steady speeds, both types ideally run like epicyclic gear sets, with no slip between the rolling element and race at the loaded contact area. During an engine speed change, the rotor may reach its next steady speed before the caged rollers reach theirs. The inner race therefore skids against the rollers, leading to a wear problem. A solution in this case is to provide viscous oil drag in a narrow gap between the cage and the inner race (figure 14.3*c*), to help the cage follow the changes of rotor speed. Skidding is also a problem with ball bearings, particularly if the direction of thrust changes. From spinning about the axis shown in figure 14.3*b*, the ball is required very quickly to take up the different rotation shown in figure 14.3*d*. This requires the design of air seals to keep the rotor load always in one direction (section 14.2.1).

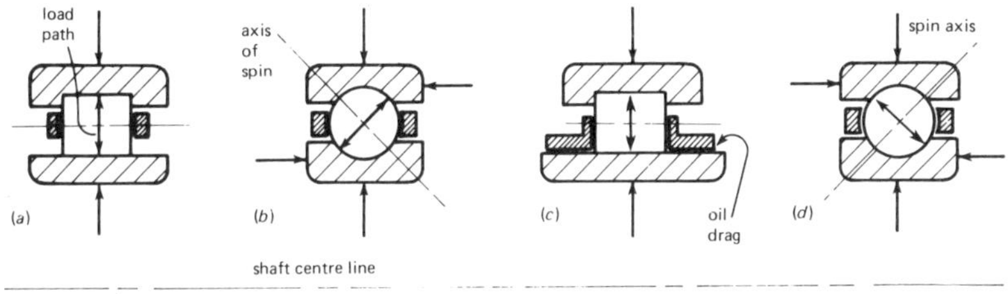

Figure 14.3 *Some aspects of ball and roller bearing operation which affect the design of smaller gas turbine engines*

Bearings of both types generate heat, the plain bearings by viscous-friction and the rolling bearings by the continual slight deformation and release of all the rolling surfaces. A major function of the lubricant is to *cool* the bearings, below the melting or softening point of the journal material or the tempering temperature of the anti-friction bearings (often 393 K, 120 °C). The oil is therefore supplied to and scavenged from the bearing chambers at a considerable rate. The chambers are sealed to contain the oil, by rubbing mechanical seals or metal ring or labyrinth seals with small clearances. With the last two types, oil is kept in by an inward flow of engine-cooling air, which also surrounds and insulates the bearing chambers from their hot surroundings.

Oils for gas turbine engines are developed for operation at high temperatures. A major cause of oil failure is oxidation at high temperature, causing increased acidity and viscosity, and chemical changes. This depends on the temperatures reached and the time spent at that condition, including the time in a hot engine after shut-down. In really severe conditions the lighter fractions are vaporised out, leaving oil varnish deposits and eventually, coke deposits. The high-temperature coking propensity is measured by the percentage of carbon residue left after the oil is permitted to evaporate off a hot surface. Such deposits may be serious if they form outside the bearing chamber as the result of leakage, affecting cooling air flows and possibly raising the metal temperature to the oil's ignition point. The deposits may also form in pockets in the hotter chambers, hindering the heat removal, blocking passages, and clogging the chamber seals with resultant wear and leakage problems.

The oils are of several types, dependent on their viscosity and chemical additives. The normal viscosities for aircraft engines are 7, 5 and 3 cSt, measured at 373 K (100 °C). The lower viscosity oils tend to cause less trouble with bearing skidding and with coking. The load carrying capacity and effect on rubber and synthetic seals varies between brands. Seals swell by different percentages in different oils and may be an important consideration when changing brands; changing from a high swell to a low swell type may let the seal contract sufficiently to leak, particularly if it is old and hardened. Each brand is therefore approved for an engine only after extensive tests.

The lubrication system conditions the oil as it circulates. The scavenge pump capacity may need to be of three to five times the feed pump capacity to cater for any air entering via bearing chamber seals. Air bubbles are removed from oil returned to the tank, and are vented from the oil system through an oil mist trap of some type. The oil is cooled by heat exchange with external air or water or the engine fuel supply (section 12.2) and it is filtered before returning to the bearings. A small amount of oil consumption is normal, but *high oil consumption* (HOC) requires investigation for the causes of leakage. Any oil which is consumed is likely to appear as smoke or vapour in the cooling air outlet (section 3.3) or exhaust. The regular topping up of the oil system is beneficial in diluting the older oil to negligible quantities, before its oxidation life limit is reached. The system should contain at least enough oil for a month's usage in industrial applications.

The oil pressure is limited for safety by including a pressure relief valve, and sensors may also be included to warn of low pressure or of high or low temperature (section 12.1.2). For starting at very low temperatures, it is sometimes advisable to pre-heat the oil to reduce its viscosity and drag. Below a minimum level, perhaps –40 °C, it is essential to heat the oil, as the viscous drag may shear the oil pump driving shaft or other highly stressed components.

14.5 Safety and Reliability

Gas turbine engines are designed for long life and reliability with high standards of control (chapter 12) and systems to warn of serious impending faults. Extensive precautions are taken to prevent serious failures, by techniques such as the determination of safe cyclic lives (section 14.6). Where the safety of people is involved, there are secondary systems such as shields to contain parts which may break (section 14.2, item 7).

Some parts could not be contained by any structure light enough to fly. Turbine discs are an obvious example, either complete or in pieces following an overspeed (figure 14.3). Such an event could occur if a shaft broke, but only in certain circumstances. In a split-shaft engine, if the h-p shaft broke in front of the thrust bearing, the compressor would rapidly slow down but there might be enough remaining energy in the combustion to overspeed the turbine. It could be that, if the fuel control system is driven from the front of the compressor, its slowing might reduce the fuel supply in time, or it might attempt to supply more to correct for the loss of speed (section 12.3). The safer way is to make the h-p shaft deliberately weaker behind the thrust bearing, so that the released turbine is blown back against the static structure and braked to a halt.

Consider the alternative failure of the l-p shaft, possibly as the result of an ingestion (section 6.4.2). The h-p system may now run happily as a self-contained engine, and even the frictional drag of the stators may not save the released l-p turbine in time. In this case, an *emergency fuel shut-off* device may be required. It may work by detecting abnormal axial movement of the l-p compressor or turbine rotor, or it may be required to detect a relative angular rotation between them. Alternatively, the shaft speed could be measured for governing from the turbine end. A more likely failure is overspeeding of the power turbine after the generator trips off line (12.5.3); the speed limit is given in section 9.4.

Such thought processes come under the heading *failure analysis*. The engine maker may write a document to demonstrate the safety of his engine, showing predictions of all possible consequences of any primary failure of the engine or its accessories. An example of the application of this forethought, and as well as experience, is in the precautions against fire.

Fire is most likely to occur outside the engine, in the region of its fuel and oil systems. In aircraft, the design precautions taken include: a firewall to separate the hot end of the engine from the accessories; adequate drainage, and ventilating airflows to purge oil or fuel vapour; earthing of all fuel and oil pipes to the engine to avoid static electricity generation; wire locking or the use of vibration proof fasteners in the fuel and oil systems; the use of protective materials, for covering the fuel and oil system plumbing and fittings, which will not burn or melt easily even in a 15 minute nacelle fire of 32 kW (110 000 Btu/h); fire detection systems which sense an excessive temperature level or a sudden change; and fire-extinguishing plumbing and nozzles.

Operational safety is also maintained by the inclusion of *anti-icing* systems at the engine intake. If ice were to form on the nacelle lip, or on the intake casing struts or IGVs, it might break off intermittently in lumps, and cause severe damage or so change the blading profile as to affect the performance. Internal supplies of hot air to warm these areas, external electric resistance mats and pulsating pneumatic covers are among the devices to control and limit ice build-up.

Reliability of the engine, with high confidence levels that there will be no failure or malfunction, affects safety in some cases, cost in many cases, or the likelihood of successful completion of a military mission or other activity. Failure within a peak lopping operation could have severe effects on cost control (section 12.6). The efforts made to ensure reliability include extensive development testing (section 14.6) and, in subsequent operations, monitoring and analysis. It is demonstrated in reference 89 that thorough analysis of the failure patterns is of assistance in identifying an engineering problem.

14.6 Development Testing

Engines which will be produced in some numbers usually undergo a development programme as a back-up to the design process. Many facets of the engine operation are explored, of which some are listed in section 4.3. It is desirable that each facet should be explored throughout its operating envelope and even beyond. For instance, in testing for blade vibration problems, it is desirable that the test engine should be matched to run well beyond 100 per cent speed in case a flutter problem occurs at 102 per cent: some production variations or rare operating conditions may bring this condition within the running range and cause a failure. Similarly, aircraft engines may be tested in high-altitude test cells, at greatly reduced inlet pressure and temperature, to explore their behaviour before risking flight. Another aspect of bench testing is the determination of material temperatures throughout the engine at extreme conditions, and verification that the loss of strength with temperature remains within safe limits. One of the techniques used is *thermal painting*, with paints which change colour permanently at known different temperatures, to indicate the peak temperatures reached. The experience of all development engines is accumulated, and fed back to determine design improvements.

Many components are also separately rig tested, to explore their characteristics and cover certain destructive activities without risking a whole engine. One case is the determination of the safe *cyclic life* for the discs and shafts of aircraft engines.[90] The life is only a few thousand cycles (section 4.4) because the components are stressed beyond their fatigue limit. New compressor and turbine discs are accelerated and slowed cyclically in a pit to failure (figure 14.4) to establish a statistically safe life. Used discs are subsequently withdrawn from service at increasing life intervals to verify their remaining life when spun to failure.

Less critical parts are also given accelerated tests to failure, with vibration or stress or thermal cycles which duplicate engine operation. These cyclic tests include bearing loads, casing pressures and loads, and the thermal fatigue of turbine blading. Systems are also tested for function and life, including fuel, oil,

Figure 14.4 *The collected pieces of a turbine disc spun to failure in the sand-bagged pit of a cyclic test rig.*

ignition, starting, airbleed, electrical and others. The management and content of such a development programme is well described in reference 91.

Some engines may be modified externally for a new application, without altering the major internal parts. It is then not necessary always to type test (section 4.3) from scratch, but only to verify that the new pieces and the revised use for old parts, such as a different loading of the casings, are acceptable. In the case of engines for experimental rather than production or passenger-carrying use, a reduced length qualification test may suffice. An example of the schedule for such a test of 48 hours' duration is shown in table 14.1: it is like a shortened type test schedule and was applied to the engines discussed in reference 12.

Table 14.1 Qualification Test Schedule

The 48-hour test comprises eight stages, each being as follows.

Cyclic Testing 1.5 h

Six cycles of: accelerate from idle to take-off

run 2.5 min at take-off

decelerate to approach ($89\% \, N_2/\sqrt{T_{01}}$)

run 5 min at approach

accelerate to take-off

run 2½ min at take-off

decelerate to idle

run 5 min at idle ($58\% \, N_2$)

Five of these cycles to go to normal take-off, $97\% \, N_2/\sqrt{T_{01}}$, while the sixth goes to emergency take-off, $100\% \, N_2/\sqrt{T_{01}}$, $104\% \, N_1/\sqrt{T_{01}}$

Endurance Testing 2 h

Run	30 min at normal take-off	$97\% \, N_2/\sqrt{T_{01}}$	$99.4\% \, N_1/\sqrt{T_{01}}$	$F_e = 27.1$ kN
	60 min at max. continuous	94%	95%	$85\% \, F_e$
	30 min at approach	89%	84%	$50\% \, F_e$

Incremental Testing 2.5 h

Run for 9 min at each of the following percentage values of N_1:

98.2, 95.3, 91.7, 88.1, 84.5, 81.0, 79.8, 76.2, 75.0, 70.3, 64.5, 59.5, 53.6, 47.7, 41.7, 35.7 and idle.

15

The Selection of a Gas Turbine Engine

Aircraft makers, oil companies, power utilities and a few other industries already have extensive experience in selecting engines to suit their requirements. Some other industries with power and heat requirements have not yet adopted gas turbine engines but could possibly benefit from their use. This chapter is written therefore for engineers without direct experience of gas turbines, and considers industrial applications only. Further information on engine selection can be found in reference 4D.

15.1 Diesel Engine, Steam Turbine Plant or Gas Turbine?

This question has already been answered for many industries, each of these power plants having the clear balance of advantage in certain applications. The gas turbine has found its niche in aircraft and naval ship propulsion, many oil industry applications, intermittent electricity generation and total energy systems (chapter 2). To some extent, the nature of the power plant selects it for a given duty; for instance, diesel engines to drive reciprocating pumps and compressors; gas turbines to drive high-speed machinery such as centrifugal or axial flow pumps or compressors, and steam plant to provide process steam from the turbine exhaust. Not all are so distinct: in some applications the diesel engine or steam turbine may compete strongly with the gas turbine: the choice then depends on several factors, some of which are listed in very general terms in table 15.1. Several items listed may be contentious, and merit further discussion.

The *capital cost* data are very approximate, and apply to engine/generator sets in 1978/79 – see section 15.2.3. For gas turbines, the price depends very much on the maker or supplier, the country of origin, the size range, the duty or rating and the point of delivery: the capital cost data are FOB while the installed prices are for a turnkey contract, ready to switch on and start. Combined cycle plant costs about $ US 150–300/kW FOB.

Table 15.1 Comparison between Different Types of Power Plant

Parameter	Diesel engine	Gas turbine	Steam turbine plant
Usual power range	10 kW to 10 MW	100 kW to 100 MW	1 MW to 1000 MW
Capital cost (approx)	$U.S. 110–180/kW	$U.S. 80–150/kW (aero) –280/kW (heavy)	$U.S. 140–240/kW (turbine only)
Installed cost (approx)		$U.S. 150–300/kW	$U.S. 800–1200/kW
Size and weight/kW	Medium	Small	Large
Specific installation requirements	Heavy foundations Vibration isolation Quietening enclosure	Intake air filters Intake and exhaust mufflers	Ash handling facility Copious supplies of water Exhaust de-pollution
Plant life (typical)	10 to 20 years	15 to 30 years	25 to 35 years
Inspection/overhaul periods	Annual	Several years	Annual
Efficiency (typical)	0.35 to 0.40	0.20 to 0.40	0.30 to 0.40
η factor at half load	1.00	0.8 to 0.85	0.91 to 0.96
Fuel grade and type	High Liquid (diesel grade)	High to medium Liquid or gas (see section 2.7)	Medium to low Gas, liquid or solid
Staffing requirements	A few	None (extra to plant)	Many
Lubrication needs	Frequent oil and filter changes	Oil topping up	Negligible
Maintenance periods	Weekly and monthly	Monthly and annual	Daily

The *installation* requirements depend on the power range. Smaller diesel and gas turbine generator sets are often pre-assembled as a transportable package. The gas turbine sets in particular may then just rest conveniently on the ground, or attached by three-point mounting within the plant or on the roof. Very large gas turbine units tend to be installed as separate units, individually mounted on the floor and connected by ducting as a permanent installation.

The *efficiency* figures are not strictly comparable between the three power plants. The diesel engine figure applies to most engines and operating conditions. The gas turbine values cover a wide range of makes and models: the highest values apply to very high pressure ratio engines or those with regeneration. A normal operating efficiency is about 0.25 to 0.3, and this will fall at part load. The steam plant works at only slightly better efficiency: 0.35 to 0.38 is the limit set by the typical pressure of 170 bar (2500 lb/in.^{2}) and temperature of 810 K (1000 °F).[92] In practice this only applies to very large plants, which operate long term as low as 0.3. Smaller plant has lower peak efficiency, even below 0.3. Combined cycle installation is not tabulated here, but its efficiency reaches towards 0.5 peak (section 3.6.7) while total energy systems lie between 0.6 and 0.85.

No mention is made of *variable costs*. These may be inferred from the efficiency, fuel cost, extra staff, maintenance and life data. The *staffing* requirement for the gas turbine is usually so low on a day-to-day basis that, if it is associated with other plant or a factory, no extra personnel are required. The main feature of steam plant which offsets its high capital cost is the low cost of its relatively low-grade fuel, but it is suggested [8] that, when environmentalists demand the removal of smoke and sulphur dioxide from its exhaust gas to make it as clean as the gas turbine, there is little operating cost difference (per kW h) between the two types. The maintenance costs for a gas turbine depend on many factors, which will be discussed separately, in section 15.2: a typical low cost may be $ 15 per kW year.

The *maintenance tasks* for diesel and steam plant are well known. Some of the tasks to be expected with gas turbines are cleaning of the inlet filter (section 15.4) and compressor (section 6.4.1) to maintain as new performance at the governed turbine entry temperature limit, and monthly topping-up of the oil system. The oil in many engines is never changed, and the filter life is long. The control system requires little servicing unless operational problems are encountered (section 12.6). The turbine blades may be inspected and a few replaced after several years, or they may all be recoated for another life (section 8.4). After such an overhaul, the performance should be restored to its original level and no running-in period is required.

15.2 Gas Turbine Type and Duty

This choice is influenced by numerous factors. The competitive engines which could perform the required function (section 15.3), and are of comparable capital cost, may offer substantially different operating costs in terms of fuel usage and maintenance. Table 15.2 summarises the parameters which affect the long-term maintenance and repair costs: some of these are discussed below.

Table 15.2 Typical Maintenance Cost Factors – currency units/kWh

Parameter	*Operating Condition*	*Factor*[1]
Rating	Base load	1.0
	Mid-range	1.5
	Peak load	2.5
	Reserve peak	5.0
Fired hours per start[2]	1000	1.0
	100	1.25
	10	2.0
	1	3.0
Fuel	Dry clean gas	1.0
	Class A distillate oil	1.5
	Treated heavy oil[3]	3.0
Installation space	Large building	1.0
	Close fitting enclosure	1.2
Experience at rating offered	Proven	1.0
	Limited development	2.0
	Prototype	4+
Design philosophy[4]	Heavy industrial	1.0
	Medium industrial	2.0
	Aero-derivative	2.5
Standard of operation and maintenance	Good	1.0
	Average	2.0
	Poor	4+

1 Multiply all factors.
2 Double for fast start.
3 Low sodium and vanadium.
4 Factors reduce with experience.
(Adapted from reference 4D by courtesy of the Council of the Institution of Mechanical Engineers).

15.2.1 Engine Rating and Life

The main influence on maintenance costs and engine life is the condition of the turbine blades, which is strongly dependent on the engine rating. The rating basically refers to turbine entry temperature, but may also involve rotational speed. These affect the blade creep life: typically the blade creep life is halved for every 20 °C increase in temperature near its rated operating point. An engine rated for severe short-term operation may have a turbine blade life of only 2000 or 3000 hours (section 2.3).

The turbine life is also strongly dependent on *starting time*. Although the gas turbine is nominally a quick starting engine, the full use of this capability reduces its life and, in peak lopping duty, will introduce thermal fatigue as the limiting factor (section 8.3). *Fuel quality* also has a marked effect, even a relatively clean, heavy oil being more damaging to the turbine than gas. Poorer oil and gaseous fuels are several times more destructive (section 8.4), and basically require the use of relatively low turbine entry temperatures with reduced efficiency and increased capital cost per kW (section 1.5). The use of external combustion for a closed-cycle engine permits the use of cheaper fuels (section 2.8). Operating experience in connection with different fuels is given in references 53 and 93.

15.2.2 Aero-derivative or Industrial Origin?

Table 15.2 may be initially misleading, in that the aeroderivative engine has an extensive development background from which it is de-rated for gas generation duty in industrial use, while the industrial engine and power turbine often have little development. Internal technicalities, such as the number of shafts or turbine blade cooling, have little effect: for instance, the gas temperature in an engine with blade cooling is raised to bring the metal temperature up to that of the uncooled blade. The aero-derivative engine is likely to cost a little less to buy, but more to maintain. The aero-derivative engine is likely to be mechanically safe because of its passenger-carrying background, while the heavy industrial engine is safe by conservative design. The former may have an overhaul life of 10 000 to 40 000 h, the latter 25 000 to 100 000 h. The aero-derivative engine's higher efficiency may be matched by regeneration in the industrial engine.

The main functional difference is in the *repair procedure*. In the event of an internal failure, or scheduled internal inspection, the aero-derivative engine is usually removed and shipped to the maker or an approved overhaul facility. This is often necessitated by the special tooling, the delicate components and critical settings during reassembly. A replacement engine may be installed within a few hours, either from the user's spares or as a rental unit. Plant downtime is minimised but the cost may be high.

The industrial engine is usually subject to *on-site repair*, but is likely to need such attention less often than the aero-derivative. The features and construction permit handling by normal maintenance personnel. A half casing may be lifted for inspection, or a component such as the turbine may be removed for re-blading. The time involved may be less than a day, or several weeks, which may involve plant downtime. Smaller industrial gas generators may also be exchanged if necessary, and returned to the factory for repair. The remoteness of the site, personnel accommodation, cost of downtime and standby arrangements (section 15.3) may well determine the acceptable repair procedure and, therefore, the engine origin. This is discussed further in reference 4E.

Also of great importance to this choice is the *reputation* of the makers and engines on the short list. Reputations are variable among engines from each origin, and the outstanding reputation of one of the engines offered may determine the choice above all other considerations.

Finally, these two origins may be further subdivided. The aero-derivative engine may stem from a *civil* or *military* background. The former may be more reliable, with fewer random failures, because long life and passenger safety are more important to the civil role. The industrial engine may be of *heavy* or *medium* design, the former probably being longer lived and more reliable, costing more to but and install but less to maintain, and having a less efficient operating cycle.

15.2.3 Sources of Information

It may be a problem for the newcomer to determine the makers' names, let alone their reputation. A few are named in this book, but a fuller list would be impractical as it would inevitably miss some and might soon go out of date. There are several useful journals, of which some are (alphabetically)

(1) *Diesel and Gas Turbine Progress*
(2) *Gas Turbine World*
(3) *International Power Generation*
(4) *Offshore*
(5) *Oil and Gas Journal*
(6) *Turbomachinery International*

They have up-to-date information and advertisements, and some issue an annual catalogue which lists the makers and their products. Journal 2 regularly lists competitive bids and orders, which indicate the range of current prices, as in table 15.1. Some of these cost data were also obtained from reference 92.

The makers' and engines' reputations may be gauged from other users, whose names the makers should be proud to supply, and at meetings and conferences such as the Annual International Gas Turbine Conference. There is also the Diesel Engineers' and Users' Association (London), which publishes information on

Table 15.3 Standard Ratings specified in ISO 3977–1978(E)

Class	Annual running hours	Name
A	up to 500	Reserve peak
B	up to 2000	Peak-load
C	up to 6000	Mid-range
D	up to 8760	Base-load

Range	Annual average number of starts
I	over 500
II	up to 500
III	up to 100
IV	up to 25
V	Continuous operation without planned shut-down for inspection and/or maintenance within a specified period

Examples
ISO standard peak-load rating (2000 h, 500 starts) Class B: Range II.
ISO standard base-load rating (8760 h, 25 starts) Class D: Range IV.

operation and costs. For a serious study, and any subsequent engineering or contract work, there are many engineering consultants who specialise in gas turbine work, and who are listed in some of the catalogues.

It has been assumed above that engine comparisons relate to the same *power level*. Until recently, powers have been quoted on different bases, but ISO 3977 (BS 3863) [52] rationalises ratings into classes of the operating hours per year, and into ranges of the number of starts required. These are listed in table 15.3, and two common examples are included. The performance ratings are determined by acceptance test procedures laid down in ISO 2314.[35]

ISO 3977, Gas Turbines – *Procurement*, is also a very useful source of general information on definitions, fuels, control and environmental pollution. It is of assistance with the contract side, specifying the technical information to be supplied by both the purchaser when enquiring and the manufacturer when tendering. Many countries issue their own version of it.

15.3 Planning a Gas Turbine Installation

The first requirement is to satisfy the installation's *functional* and *performance* needs. The relevant parameters to be chosen or specified may include power (or thrust), shaft speed, fuel consumption and fuel type, air mass flow and exhaust gas temperature: the relative importance of these may differ widely between applications. The parameters may be determined by preliminary cycle calculations (chapter 3). It is advisable to err on the conservative side as regards losses and to allow a generous performance margin for plant tuning or uprating. Excessive generosity may reduce the plant efficiency, but considerably increases the engine life and reduces maintenance costs. A margin which is too small and is lost, so that the engine finally proves too small, is a far worse penalty.

The nominal installation power may range from considerable excess of *standby capacity* to zero. Adequate standby could avoid the loss of production time involved in engine repair (section 15.2.2). Where personal safety is concerned, such as on off-shore drilling platforms, the standby could be 100 per cent, involving the use of two engines at half power where one could suffice. Lower standby levels, which may keep a plant at full production, may be obtained by using several engines as gas generators, all running to supply the power turbine. If one should fail, or be shut down for scheduled maintenance, the others may be raised from part load to base-load level, or perhaps from continuous base-load to a higher, peak rating.

Where the standby capacity is inadequate to maintain full operation with one engine out, the concept of *availability* is used. It also allows for the loss of more than one engine; it is defined in several ways, a typical one being

$$\text{availability} = \frac{\text{period time} - \text{planned and unplanned downtime}}{\text{period time (time when operation is required)}}$$

Typical availability for gas turbine plant ranges from 0.9 to 0.99. Analysis in reference 94 showed that, with the likely failure rate of the engines (see also [89]) planning for 0.97 availability would result in 1 per cent loss of gas transmission delivery, while 0.91 availability would lose 7 per cent. It would be worth while to run an engine at a very high rating to avoid such a loss. Such analysis also shows the importance of being able to plan the downtime, and avoid unexpected problems.

Plants designed for *cogeneration* typically extract about 10 per cent less of the exhaust gas energy than could feasibly be extracted. This permits a reduction of cost in the steam plant section of about 40 per cent, and forms a reasonable compromise.

There are many further considerations, including contracts and finance, which are beyond the scope of this book. The engineering analysis should, however, be balanced by liaison with the local *power authority*. Some plants generate surplus electricity, for which the authority may pay a reasonable rate and which may help it at peak times. A plant which generates most of its power needs in-house, but which may call upon the local authority for a large emergency supply in the event of an engine failure, may find it pays a heavily penal rate. This again emphasises the desirability of spare capacity.

15.4 Ancillary Requirements

A range of ancillary equipment is available, and some items are essential to ensure a satisfactory plant operation and life. Engine manufacturers can supply or arrange for these units, and some supply them as plug-in modules which are shipped separately as complete systems. Examples may include the air filters, exhaust silencers, fuel treatment plant, starting system, control room and oil-cooling plant.

Air filters are used for silencing, for atmospheric dirt and sand, and for airborne salt. Salt is important in marine environments but may also be present in air hundreds of kilometres from the sea [4E], and is more serious in hotter engine types (section 8.4). Inertial separators remove the heavier particles, but an oil-wetted and washed nylon mesh coalescer is often used to remove sand and salt after an inertial stage. Paper and cloth bag filters are also used, but need considerable area for the high mass flow. Filter systems may use several stages of different types to ensure safety from the corrosive elements, which are only permitted in minute quantities (section 8.4). Some filters are self-cleaning but may cost more, offsetting the saving in maintenance.

Fuel treatment includes filtering and temperature control for liquid fuels (section 12.2) and gas. Wet gas, with drops or larger slugs of liquid fuel, can result in turbine burn-out from a sudden, excessive rate of heat release. Traps, demisters and filters are essential to avoid slugs and to reduce droplets to a very low, finely dispersed level. Heating the gas to 20 °C above its dew point may prevent condensation. Other aspects of fuel control, and the control system design itself, also need careful planning (section 12.6).

Oil cooling is likely to be the only need for cooling water, unless compressor intercooling is used. Only small quantities are required [92], and air cooling may be used if the environment is cool.

The location of working areas for personnel, and the *control room*, should be out of the area where the very rare case of a major shaft, disc or casing failure could result in injury.

Appendix A

The Use of SI Metric Units

SI (Système Internationale) units may be applied very conveniently to gas turbine calculations, and avoid the confusion, pitfalls and numerous constants of earlier systems of units. SI units use prefixes to establish size ranges in multiples of power of one thousand. Taking the watt (W) as an example, one thousandth (10^{-3}) of a watt is a milliwatt, mW; a millionth (10^{-6}) of a watt is a microwatt, μW; a thousand (10^3) watts is a kilowatt, kW; and a million (10^6) watts is a megawatt, MW. Further prefixes are 10^9 = giga (G), 10^{12} = tera (T) and 10^{15}= peta (P): these larger multiples are most commonly used with power and energy.

An exception to this prefix rule is the kilogram (kg), but the tonne (t) follows the rules for larger masses. Some other units, such as the metre, do not make use of all the prefixes or have other, special names which are not included here.

Parameter, Symbol	*Unit*	*Conversion factors*
Length, L	metre	1 m = 39.370 in. = 3.281 ft
Area, $A = L^2$		1 m^2 = 1550 $in.^2$ = 10.764 ft^2
Volume, $v = L^3$		1 m^3 = 61024 $in.^3$ = 35.315 ft^2
Angle, α	radian	1 rad = $(180/\pi)$ degrees
Radius, R		1 m
Time, t	second	1 s = 1 min/60 = 1 h/3600
Angular velocity, $\omega = \alpha/t$		2π rad/s = 1 rev/s = 60 rev/min
Frequency, $f = 1/t$	hertz	1 Hz = 1/s = s^{-1}
Velocity, $V = L/t$		1 m/s = 3.281 ft/s = 3.6 km/h = 2.237 mile/h
Acceleration, $a = L/t^2$		1 m/s^2 = 3.281 ft/s^2
Mass, M	kilogram	1 kg = 2.205 lb = 0.0685 slugs = 10^{-3} tonne
Mass flow, $\dot{m} = M/s$		1 kg/s = 2.205 lb/s
Momentum, $\dot{m}V$		1 (kg/s) (m/s) = 1 kg m/s^2
Force, $F = Ma = \Delta\dot{m}V$	newton	1 N = 1 kg m/s^2 = 0.2248 lbf = 10^5 dyne
Torque, $\tau = FR$		1 N m = 0.7376 lbf ft
Energy (work), E	joule	1 J = 1 N m = 0.7376 ft lb

Parameter, Symbol	*Unit*	*Conversion factors*
Heat, Q	joule	1 J = 1 Btu/1055 = 1 Chu/1899
Power, P	watt	10^3W = 1 kW = 1 kJ/s = 1.341 hp (British)
Density, $\rho = M/v$		1 kg/m^3 = 0.0624 lb/ft^3
Dynamic viscosity, μ		1 Ns/m^2 = 10^3 cP (centipoise)
Kinematic viscosity, $\nu = \mu/\rho$		1 m^2/s = 10^6 cSt (centistoke)
Pressure, p	bar	1 bar = 10^5 N/m^2 = 14.504 lb/in.^{2} = 2088.5 lb/ft^2
Stress, σ	pascal	10^6 Pa = 1 MPa = 1 N/mm^2 = 145.04 lb/in^2
Temperature, T	kelvin	$K = {}^\circ C + 273 = ({}^\circ F + 460) \times \frac{5}{9} = {}^\circ R \times \frac{5}{9}$
Specific heat, C_p or C_v		10^3 J/kg K = 0.239 Btu/lb °F (Chu/lb °C)
Efficiency, η	1.0 =	100% = 0.293 W/(Btu/h) = 0.000393 hp/(Btu/h)
Specific fuel consumption, s.f.c.		1 kg/N h = 9.807 lb/lbf h (thrust) 1 kg/kW h = 2.956 lb/hp h (power)

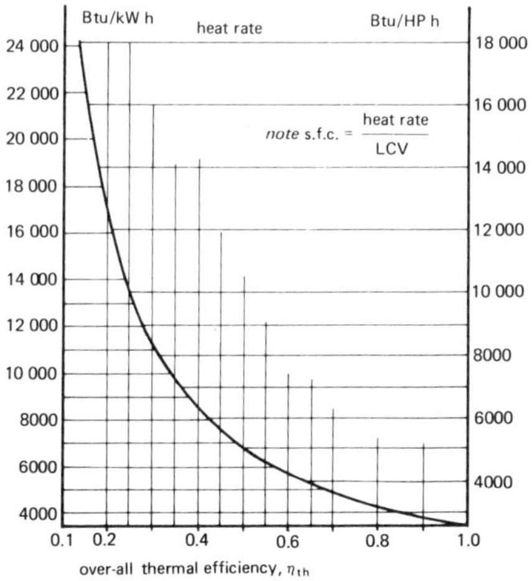

Figure A.1 *Conversion between engine thermal efficiency and heat rate*

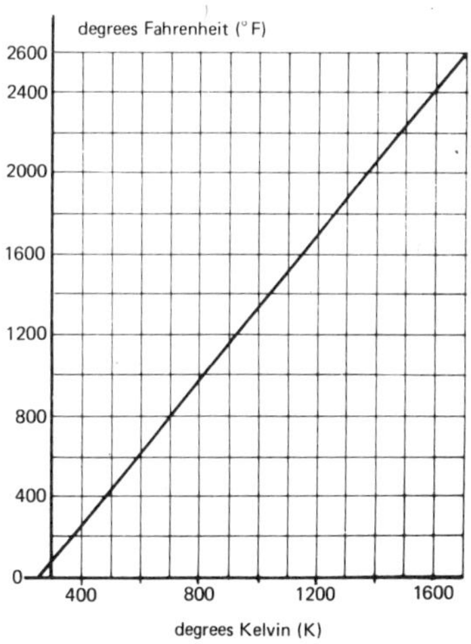

Figure A.2 *Conversion between Fahrenheit and Kelvin units of temperature*

Appendix B

Compressible Fluids and Dimensionless Parameters

The properties of gases are fundamental to the operation of the gas turbine engine, because their compressible nature enables them to store and release energy as required. This appendix presents some basic concepts and physical explanations as background to the subject, with a summary of the mathematics.

B.1 The Gaseous State – Static

A gas is essentially a fluid superheated to a temperature well above its boiling point. Whereas a liquid lies in the bottom of a container, because of the predominance of gravitational and inter-molecular forces, the molecules of a gas fill the container uniformly. Heating of the fluid increases the internal energy of the molecules, and therefore their velocity, so that the predominant forces in the gaseous state occur as the molecules collide and rebound off other molecules and the container walls.

The *internal energy* of the gas is stored in translational, rotational and vibrational modes. Translation represents the kinetic energy of the molecules and involves three degrees of freedom, being motion of each molecule along the mutually perpendicular x, y and z axes. Rotational energy is stored by rotation of each molecule as a flywheel about the three axes, representing three more degrees of freedom: Maxwell showed that the energy stored in a molecule is equally divided between its translational and rotational degrees of freedom. Vibrational modes involve the oscillation of the atoms within the molecule, relative to each other, but these store little energy except at very high temperatures (section B.4.2).

The *molecular structure* determines its number of degrees of freedom. A polyatomic molecule generally has all six degrees of freedom (for example, CH_4, NH_3). Diatomic molecules (for example, H_2, O_2, N_2) and linear polyatomic molecules (for example, CO_2, O=C=O) may be considered as a dumbell aligned with one axis: while they may store significant levels of energy by rotating about both the other axes, spinning the dumbell about its own axis stores negligible energy. A monatomic molecule can store no energy by spinning (and has no possible vibratory mode). Diatomic and monatomic molecules therefore store energy with five and three degrees of freedom respectively, which has an effect on their values of specific heat and γ (section B.4.2) and on the turbomachinery to handle them (section E.9).

Of the energy stored in the molecule, the fraction which is stored in translational modes is related to the gas *pressure*. As molecules strike and rebound from the container walls, the forces resulting from their momentum change generate the pressure applied over the area of the wall. Taking V_m to be the mean (r.m.s.) molecular velocity, the kinetic theory of gases shows that

$$p/\rho = V_m^{\,2}/3 \tag{B.1}$$

where p is pressure and ρ is density. As $V_m^{\,2}$ is proportional to the internal energy and the temperature, raising the temperature of a given mass (number of molecules) of gas in a fixed volume increases its pressure. Inserting more molecules into the volume increases density and pressure as, alternatively, does reducing the volume of a fixed mass of gas. These relationships are linked by the characteristic *equation of state*

$$p/\rho = pv = RT \tag{B.2}$$

where v is the specific volume $(1/\rho)$ and R is the characteristic gas constant (appendix D and section B.4.2). This applies to an ideal gas, where only the collision forces are significant and the volume of the molecules themselves is negligible relative to that of the container. For a real gas, with high density or low temperature, the compressibility factor Z is introduced in the form $pv = ZRT$.

B.2 The Gas in Motion – Stagnation Conditions and Flow Limit

So far, the gas has been considered in a static condition: that is, with no net motion or flow in any direction even though all the molecules are in motion.

Consider now the flow of gas at velocity V along a duct. The static conditions still apply normal to the flow: that is, the duct walls are exposed to some reference static pressure, p_r. An observer moving with the flow also senses the pressure to be at this static level, but an observer facing the flow senses a higher pressure, as in a strong wind. The Pitot probes shown in figure B.1*a* sense the random molecular velocity plus the flow velocity (dynamic head), so show a higher pressure on the gauge. This is the *stagnation pressure*, denoted by p_{0r}, and named because it locally stops the flow where it repels the approaching molecules: it is also known as total pressure. The terms total and static apply also to *temperature*.

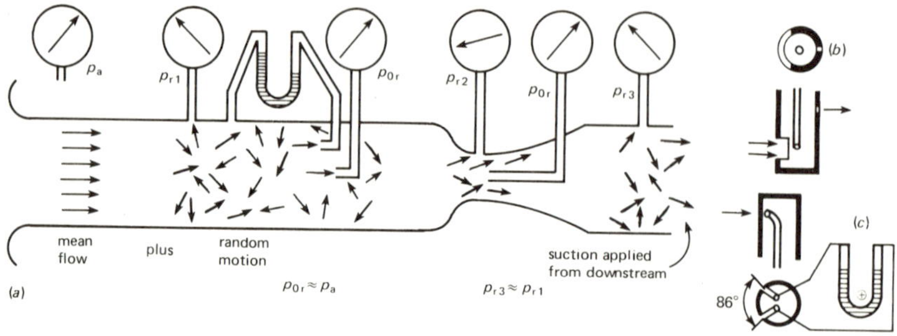

Figure B.1 *(a) Pressure measurements of a flow induced to enter a variable area duct by suction from a downstream pump; (b) a stagnation temperature measuring thermocouple probe and; (c) a direction-sensing double static pressure measuring probe*

The practical effect of this upon flow in a duct is as follows. If the entry is from the atmosphere (or a very large chamber), then the total pressure in the duct is the chamber static pressure. There is a slight loss of total pressure resulting from frictional loss at entry and along the duct, but T_{0r} remains constant (at the ambient static value) because no energy is lost (section C.4). Variations in duct cross-sectional area have a major effect on p_r and T_r, however. Reducing the area reduces p_r and T_r because the flow velocity increases (section B.4.1 and 3).

If the pressure downstream of an area constriction or throat is steadily reduced, the inlet flow will increase until it reaches a fixed limit where p_r is a fixed fraction of p_{0r}. This is shown mathematically in section B.4.3 and is called the *choking* condition. It arises because the flow reaches the velocity of sound at the throat. The message that the downstream pressure has been reduced still further could only be transmitted upstream, to induce the flow to accelerate even further, by a pressure signal travelling against the flow. As such a signal can only travel at sonic velocity, the message fails to get through.

The *measurement* of T_{0r} is nearly as easy as for p_{0r}. A probe with an opening facing the flow contains a thermocouple to sense the temperature of the arrested flow. A small bleed hole (figure B.1*b*) permits a slight flow to speed the response to temperature changes. Static pressure is measured from a flush drilling in the wall, sharp edged but without a projecting burr. Two holes in a cylindrical probe may also be used to sense static pressure if properly spaced (figure B.1*c*). If the two holes are connected to separate gauges or across a manometer, they will show equal levels when the probe is rotated to face the flow, and permit the assessment of flow direction. The hardest parameter to measure is static temperature for, if the thermocouple is immersed in the flow, it must inevitably pick up a component of the flow effect. Alternative methods of measuring temperature, particularly at high levels, are discussed in section 12.1.4.

B.3 Working Processes

Energy, heat and work may be interchanged by working processes but the total energy of the system is conserved. The *first law of thermodynamics* states that

$$dQ = dE + dW + d(KE) + d(PE) \tag{B.3}$$

where the prefix 'd' indicates a change in value, dQ is heat input, dE is the increase in internal energy of the gas (section B.1) and dW is the work done by the gas on its surroundings. The kinetic energy (KE) and potential energy (PE) terms allow for change in flow velocity or altitude: the former applies to gas turbines which use a steady flow of working fluid and the latter applies to the atmosphere (section B.4.5), but is negligible in a normal-sized engine. These quantities are measured per unit mass of gas.

Work involves a force times a displacement, or a pressure times a volume change. For instance, a blade displacing gas against a resisting pressure does work. The compressible nature of gas implies the inherent work content which originally compressed it, and which may be recovered by controlled expansion. This work content enables any element of gas to exclude the surrounding gas from the space it occupies.

Heat addition is measured by *specific heat*, the quantity which raises unit mass of material through one degree of temperature. Two values are used for gases, C_v for heat supplied at constant volume and C_p for heat supplied at constant pressure. Heat input to a constant volume involves no work even though the pressure rises, so it only increases the internal energy ($C_v\ dT = dE$). Thus, the *internal energy* at any temperature is given by C_vT. Heat input at constant pressure does involve work input, as the volume must increase, and it also involves energy input as the temperature must rise to permit the same number of molecules to occupy the larger space. That is, $C_p\ dT = C_v\ dT + p\ dv$ and, at any given temperature, $C_pT = C_vT + pv$. Combination of this with equation B.2 shows that

$$C_p - C_v = R \tag{B.4}$$

Now, by definition

$$C_p/C_v = \gamma \tag{B.5}$$

Therefore

$$C_p = R\gamma/(\gamma - 1) \tag{B.6}$$

and

$$C_v = R/(\gamma - 1) \tag{B.7}$$

These are standard relationships which will frequently be used.

The term *enthalpy* is used to describe the total heat or energy content of the working fluid, and is also used in specific form (enthalpy per unit mass). The static enthalpy is (B.3)

$$H = E + W = C_v T + pv = C_p T \tag{B.8}$$

and the stagnation enthalpy is

$$H_0 = E + W + KE = C_v T + pv + \tfrac{1}{2} V^2 = C_p T_0 \tag{B.9}$$

A change of enthalpy fully describes the change of condition of the gas during heat addition or removal, or during work input (compression) or extraction (expansion). In a flow process between states 1 and 2, the work involved is therefore

$$W = H_{02} - H_{01} = C_p(T_{02} - T_{01}) \tag{B.10}$$

Thermodynamics texts also derive this from the work shown as the shaded area in figure B.2.

The efficiency (appendix C) of a process is often assessed by an increase in *entropy*(s). A change of entropy is defined by $\mathrm{d}s = \mathrm{d}Q/T$, so it has the same units as R, C_p and C_v. It represents energy which is dissipated by the generation of eddies and turbulence in the flow during compression or expansion or within a duct. Without such losses, the flow could be reversed to return to the starting point. If, in addition, no heat was transferred between the gas and its container during the process (an adiabatic process), then there would be no change of entropy. Such a reversible adiabatic process is called an *isentropic* process (same entropy) and is the ideal striven for by the machine designer. Putting $\mathrm{d}Q = 0$ in the equation B.3, substitution from several of the other relationships above, and mathematical manipulation as shown in thermodynamics texts yields the basic equations for isentropic compression or expansion processes

$$pv^{\gamma} = \text{constant} \tag{B.11}$$

and

$$\frac{p_{02}}{p_{01}} = \left(\frac{T_{02}}{T_{01}}\right)^{\gamma/(\gamma-1)} \tag{B.12}$$

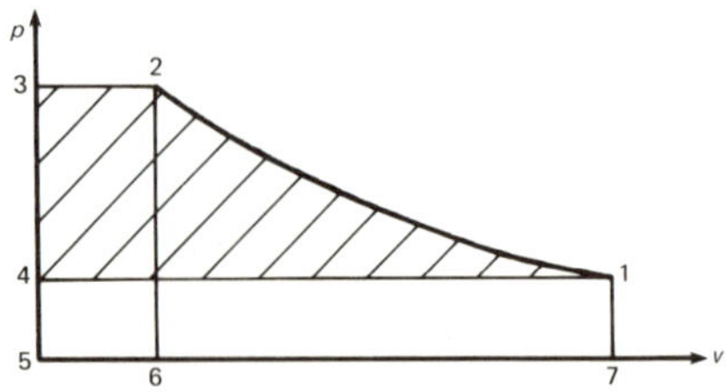

Figure B.2 *Pressure–volume diagram showing the work done on the gas during a flow process. Areas: 1457 = work done by working fluid when inserted into machine; 1267 = work done on working fluid during a compression process; 2356 = work done on working fluid to force it out of machine; 1234 = net work done on fluid*

For a real process, allowance must be made for inefficiency, as covered in appendix C.

B.4 Miscellaneous Derivations

B.4.1 The Velocity Equivalent of Temperature

Subtracting equation B.8 from B.9 permits the flow velocity to be linked to the static and total temperatures. In its standard form, the velocity equivalent of temperature is shown by

$$T_{0\mathrm{r}} - T_\mathrm{r} = V^2/2C_p \tag{B.13}$$

Thus, as the flow accelerates in a nozzle (section B.2), for instance, the stagnation temperature remains constant while the random motion V_m is progressively exchanged for bulk motion V in the direction of flow.

B.4.2 Values of R, γ, C_p, C_v and c

As one kg mol of any gas occupies 23.6444 m^3 at ISA conditions, it may be calculated from equation B.2 that R = 8314.3/M.W., where M.W. is the molecular weight. The constant values of R for several gases are tabulated in appendix D.

The value of γ may be derived simply by consideration of the internal energy of the gas. From section B.3 and equations B.1 and B.2, the internal energy of unit mass of gas is given by

$$E = C_v T = C_v V_\mathrm{m}^{\ 2}/3R$$

Now, the kinetic energy stored in a unit mass of gas is $V_\mathrm{m}^{\ 2}/2$, which is also the total internal energy of a monatomic gas. As this is divided (section B.1) between its three degrees of freedom (DF), the internal energy stored in gas of any molecular form may be expressed as

$$E = (\mathrm{DF}/3)\,(V_\mathrm{m}^{\ 2}/2)$$

As E is the same quantity in both these equations, it follows that

$$C_v = R(\mathrm{DF}/2)$$

Substitution of this into equation B.7 yields values for γ of 1.67, 1.40 and 1.33 respectively for gases with monatomic, linear and non-linear molecules.

In practice, the value of γ and of the specific heats is not constant because the vibratory modes (section B.1) are excited increasingly as temperature rises. Bond bending modes start from moderate temperatures while bond stretching modes start at higher temperatures and may not be fully excited until the the temperature reaches several thousand kelvins. Chemical thermodynamics shows that, with n atoms per molecule, there are $(3n - 5)$ vibratory modes for a linear molecule, and $(3n - 6)$ modes for a non-linear molecule: each fully developed mode adds another R to the value of C_v and C_p, and reduces the value of γ accordingly. These variations may be determined by empirical formulae fitted to measure data: sample formulae for C_p alone are given in appendix D to permit γ and C_v to be calculated, and the values of γ at ISA are also shown as calculated from these data and equation B.6.

The values of γ and R determine the velocity of sound waves, which cause minute isentropic compressions and expansions as they pass. Physics texts and some fluid mechanics texts demonstrate that the *sonic velocity* is given by

$$c = \sqrt{(\gamma RT)} = \sqrt{(\gamma p/\rho)} \tag{B.14}$$

Insertion of the standard values of γ in this expression shows that, for any gas, the value of c is clearly less than the mean molecular velocity, $V_m = \sqrt{(3p/\rho)}$ (equation B.1). The flow velocity of c (section B.2) is therefore not an absolute limit, and the flow may be induced to accelerate further by a divergence downstream of the throat.

B.4.3 Mach Number and Critical Ratios

The Mach number of a flow is the ratio of its velocity to the velocity of sound

$$M = V/c = V/\sqrt{(\gamma RT_r)} = V/\sqrt{(\gamma p_r/\rho)} \tag{B.15}$$

Manipulating this with equations B.6, 12, 13 and 14 yields

$$\frac{T_{0r}}{T_r} = 1 + \frac{(\gamma - 1)}{2} M^2 = \left[\frac{p_{0r}}{p_r}\right]^{(\gamma-1)/\gamma} \tag{B.16}$$

When the flow is choked or flight is at sonic velocity, $V = c, M = 1$, and

$$\frac{T_{0r}}{T_r{}^*} = \frac{\gamma + 1}{2} = \left[\frac{p_{0r}}{p_r{}^*}\right]^{(\gamma-1)/\gamma} \tag{B.17}$$

where $T_r{}^*$ and $p_r{}^*$ are the critical, minimum values at the throat.

B.4.4 Comparison of Compressible with Incompressible Flow

In incompressible flow, the pitot tube measures the first two terms of the Bernoulli equation ($p + \frac{1}{2}\rho V^2 + \rho gz$ = constant) where z allows for altitude. Therefore

$$p_{0r} = p_r + \tfrac{1}{2}\rho V^2 \text{ (incompressible)} \tag{B.18}$$

where $\frac{1}{2}\rho V^2$ is the dynamic head. With compressible flow, the expansion of

$$p_{0r}/p_r = (1 + (\gamma - 1) M^2/2)^{\gamma/(\gamma-1)}$$

and the use of

$$p_r \gamma M^2/2 = \rho V^2/2$$

yields the equivalent expression for compressible flow

$$p_{0r} = p_r + \tfrac{1}{2}\rho V^2(1 + M^2/4 + (2 - \gamma)M^4/24 + (2 - \gamma)(3 - 2\gamma)M^6/192 + \ldots \tag{B.19}$$

The error in assuming that air behaves as an incompressible fluid may be found by putting various values of M in equation B.19. Up to Mach 0.2, the error in calculating total pressure is less than 3 per cent. This means that low-speed calculations, for instance involving ventilating fans and windmills, may be done simply by using incompressible flow theory. It has often been used also in cascade work (section E.4).

B.4.5 The International Standard Atmosphere (ISA)

The atmosphere may be considered as a static compressible fluid in which the variation of altitude and potential energy is significant (equation B.3). It is taken as the annual average condition at 40° latitude and is defined by a temperature of 288.15 K at sea level, which falls by 0.0065 K per metre of increasing altitude up to 11 km and is then constant at 216.65 K to 20 km. The sea-level pressure is taken as 101325 N/m^2, and the pressure at any altitude up to 11 km may be calculated from

$$\frac{p_{(\text{required})}}{101325} = \left(\frac{T_{(\text{at same altitude})}}{288.15}\right)^{5.25} \tag{B.20}$$

This is similar to equation B.12, except that the value of $\gamma/(\gamma - 1)$ for air is 3.5. The higher index results from heat input by radiation from the Earth. If a mass of air were to rise, expanding isentropically (index 3.5) as the pressure reduced, then it would cool faster than the actual temperature lapse rate, would be heavier than the surrounding atmosphere and would sink to its stable level. When the atmosphere becomes unstable, as in a thunderstorm, so that its actual index falls below 3.5, significant vertical flow occurs.

B.5 Dimensionless Ratios

Performance parameters are often used in non-dimensional form, as introduced in chapter 3. A well-known one is *efficiency*, which relates output to input.

Mach number and the ratios of *pressure* and *temperature* expressed in equations B.12, B.16 and B.17 are also in common use.

Flow *velocity* is made dimensionless by Mach number and also by its relationship to inlet stagnation temperature. From equation B.13

$$\frac{V}{\sqrt{T_{0r}}} = \sqrt{\left[2C_p\left(1 - \frac{T_r}{T_{0r}}\right)\right]} \tag{B.21}$$

and, at choking, the critical velocity is given by

$$\frac{V^*}{\sqrt{T_{0r}}} = \sqrt{\left[2C_p\left(\frac{\gamma - 1}{\gamma + 1}\right)\right]} \tag{B.22}$$

Rotational speed is expressed as $N/\sqrt{T_{0r}}$, which is $V/\sqrt{T_{0r}}$ divided by the circumference at which V is measured. Both $N/\sqrt{T_{0r}}$ and $V/\sqrt{T_{0r}}$ are related to Mach number, and are therefore parameters describing the flow condition relative to the adjacent hardware. They are used as dimensionless numbers although they actually include the dimensions of $\sqrt{C_p}$.

Mass flow is given by $\dot{m} = \rho A V$, where A is the duct cross-sectional area. Its dimension less form is given by

$$\frac{\dot{m}\sqrt{T_{0r}}}{Ap_{0r}} = \frac{\rho V\sqrt{T_{0r}}}{p_{0r}} = \frac{p_r}{RT_r} \times \frac{T_{0r}}{\sqrt{T_{0r}}} \times \frac{V}{p_{0r}} = \frac{1}{R} \times \frac{p_r}{p_{0r}} \times \frac{T_{0r}}{T_r} \times \frac{V}{\sqrt{T_{0r}}}$$

from which

$$\frac{\dot{m}\sqrt{T_{0r}}}{Ap_{0r}} = \frac{\gamma}{\gamma - 1}\sqrt{\left\{\frac{2}{C_p} - \left(\frac{p_r}{p_{0r}}\right)^{2/\gamma}\left[1 - \left(\frac{p_r}{p_{0r}}\right)^{\gamma - 1/\gamma}\right]\right\}} \tag{B.23}$$

or

$$\frac{\dot{m}\sqrt{T_{0r}}}{Ap_{0r}} = M\sqrt{\left[\frac{\gamma}{R\left(1 + \frac{\gamma - 1}{2}M^2\right)^{(\gamma+1)/(\gamma-1)}}\right]} \tag{B.24}$$

and in the choking case

$$\frac{\dot{m}\sqrt{T_{0r}}^*}{Ap_{0r}} = \gamma\sqrt{\left[\frac{\left(\frac{2}{\gamma + 1}\right)^{(\gamma+1)/(\gamma-1)}}{C_p(\gamma - 1)}\right]} \quad (\approx 0.04 \text{ for air and combustion gas}) \tag{B.25}$$

These also are not truly dimensionless, containing C_p or R, but these are substantially constant or determinable while the value of γ is being obtained (equations 3.4 and B.6). The value of A is also constant so is often ignored, leaving the non-dimensional form $\dot{m}\sqrt{T_{0r}}/p_{0r}$. Note that pressure is in units of N/m^2.

Using the same type of manipulations as for mass flow, the dimensionless equations for *thrust* may be found. For a choked nozzle, equation 14.1 yields

$$\frac{F_e}{A_e p_a} = 2\left[\left(\frac{2}{\gamma + 1}\right)^{1/(\gamma-1)}\right]\frac{p_{0e}}{p_a} - 1 \tag{B.26}$$

while, for an unchoked nozzle, equation 14.2 yields

$$\frac{F_e}{A_e p_a} = \frac{2\gamma}{\gamma - 1}\left[\left(\frac{p_{0e}}{p_a}\right)^{(\gamma-1)/\gamma} - 1\right] \tag{B.27}$$

If A_e is unknown, it is calculated from equation B.25 for use in B.26 in the choked case, but equation 14.1 is used in the unchoked case.

The object of these dimensionless forms is to permit the plotting, by one unique line for each parameter, of the engine performance which will apply regardless of variations of inlet temperature and pressure. These equations of engine thrust are not dimensionless relative to variations in flight velocity, as p_{0e} rises along with all other internal pressures in proportion to *ram pressure ratio* R_{01}, while p_a remains constant at a given altitude. The value of R_{01} is found from the stagnation pressure rise resulting from aircraft forward velocity

$$R_{01} = \frac{p_{01}}{p_a} = \left[1 + \left(\frac{\gamma - 1}{2}\right)M^2\right]^{\gamma/(\gamma-1)} \tag{B.28}$$

The fully dimensionless thrust is then

$$\frac{1 + (F_e/A_e p_a)}{R_{01}} = 2\left[\left(\frac{2}{\gamma + 1}\right)^{1/(\gamma-1)}\right]\frac{p_{0e}}{p_{01}} \tag{B.29}$$

which applies whether the nozzle is choked or unchoked.

Fuel flow is treated as a rate of energy flow, $\dot{m}_f \times LCV/Ap_{0r}\sqrt{T_{0r}}$. This ratio also includes the units of C_p, which is ignored, as is A, which is constant in a given engine. The parameter may therefore conveniently be used in the form $\dot{m}_f/p_{0r}\sqrt{T_{0r}}$.

Appendix C

Assessment of Efficiency and Performance

C.1 Efficiency of Compression or Expansion

The process of compression or expansion of a gas involves an energy input to compress, or an energy recovery from an expanding gas. The level of energy transfer associated with the process depends very much on the type and quality of the hardware which controls the motion of the gas. The ideal is to compress with minimum energy input, or expand with maximum energy recovery. When the process falls short of the ideal, then its efficiency is less than unity. Sections C.2 and C.3 present two means of assessing the efficiency of a process, relative to the ideal, but a physical explanation is needed initially.

The energy input during compression is measured in terms of the rise of stagnation temperature (see section 1.1, equation B.10). If the process is ideal, the temperature rise is minimal, being related to the pressure rise (equation B.12). If less than ideal, the temperature rise is increased relative to the pressure rise, meaning that the energy input is increased relative to the pressure rise obtained. The energy recovered during expansion is also measured in terms of the change in stagnation temperature, in this case a drop in temperature. For a given pressure drop, the ideal case yields a maximum temperature drop (equation B.12) and, therefore, energy output. If less than ideal, the temperature drop and energy recovery are reduced relative to the pressure drop.

The effect of inefficiency in both the compressor and turbine, therefore, is to leave a higher final temperature at the end of the process relative to the pressure change involved. Alternatively, the effect could be described as leaving a lower pressure at the end of the process than the energy change would suggest. Thus the inefficient compressor generates only a low pressure, and the inefficient turbine drops the pressure excessively from its starting point. In practice, efficiency is assessed in terms of the effect on the final temperature of the gas.

C.2 Isentropic Efficiency

This method of performance assessment considers the change in gas condition between the beginning and end of a process, and is based on the irreversibility of the process in terms of entropy (section B.3). The word isentropic means constant entropy, and is therefore applicable to the ideal compression and expansion processes shown in figure C.1. These processes are reversible (no losses) and adiabatic (no heat transfer), as discussed in section B.3. The heat addition and heat rejection processes are discussed also in section 3.1.

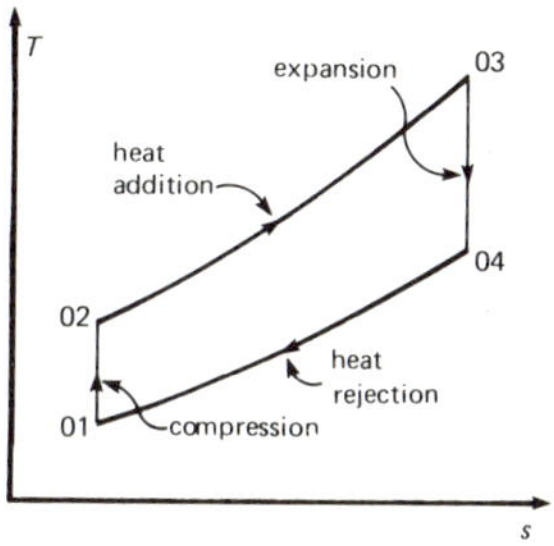

Figure C.1 *The ideal engine cycle*

If the process is not reversible, the entropy increases. This is demonstrated in figure C.2, using the station numbers 1 and 2 for the beginning and end of each process whether it involves compression or expansion. The process can be divided into two sections; the isentropic part from condition 1 to 2′ (two prime), followed by an internal heating process 2′ to 2 in which the turbulence and eddies generated are dissipated by viscous friction in the form of heat. Obviously, the more turbulence the greater the heat input. The isentropic efficiency assesses this by using stagnation temperature ratios in the form

$$\text{compressor:}\quad \eta_c = \frac{T_{02}' - T_{01}}{T_{02} - T_{01}} = \frac{\Delta T_0'}{\Delta T_0} \tag{C.1}$$

$$\text{turbine:}\quad \eta_t = \frac{T_{01} - T_{02}}{T_{01} - T_{02}'} = \frac{\Delta T_0}{\Delta T_0'} \tag{C.2}$$

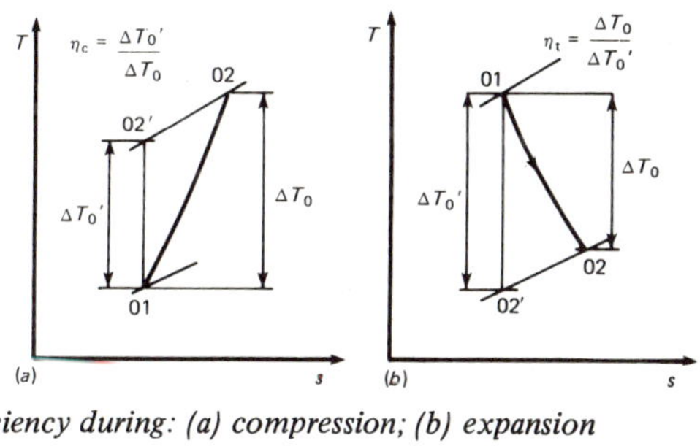

Figure C.2 *Inefficiency during: (a) compression; (b) expansion*

C.3 Polytropic Efficiency

While the causes of inefficiency remain the same, the use of polytropic efficiency is a common alternative means of assessment (section 1.5). If the compression or expansion process is considered as an infinite number of very small, consecutive compressions or expansions, then the polytropic efficiency is effectively the isentropic efficiency of each small stage of the process.

From equation B.12, an isentropic process can be defined by

$$p = \text{constant} \times T^{\gamma/(\gamma-1)}$$

Taking logs

$$\ln p = \frac{\gamma}{\gamma - 1} \ln T + \text{constant}$$

Differentiating

$$\frac{\mathrm{d}p}{p} = \frac{\gamma}{\gamma - 1} \frac{\mathrm{d}T'}{T} \tag{C.3}$$

where $\mathrm{d}T'$ represents the isentropic temperature change of a small part of the process. If the process is not fully isentropic, then the polytropic efficiency η_∞ may be applied with the real temperature change $\mathrm{d}T$, thus

compression: $$\frac{\mathrm{d}p}{p} = \frac{\gamma}{\gamma - 1} \frac{\eta_{\infty\mathrm{c}}\mathrm{d}T}{T}$$ expansion: $$\frac{\mathrm{d}p}{p} = \frac{\gamma}{\gamma - 1} \frac{1}{\eta_{\infty\mathrm{t}}} \frac{\mathrm{d}T}{T}$$

integrating

$$\ln p = \frac{\eta_{\infty\mathrm{c}}\gamma}{\gamma - 1} \ln T + k_1 \qquad \ln p = \frac{\gamma}{\eta_{\infty\mathrm{t}}(\gamma - 1)} \ln T + k_2$$

taking anti-logs

$$p = k_1 T^{\eta_{\infty\mathrm{c}}\gamma/(\gamma-1)} \qquad p = k_2 T^{\gamma/(\gamma-1)\eta_{\infty\mathrm{t}}}$$

or

$$p \propto T^{\eta_{\infty\mathrm{c}}\gamma/(\gamma-1)} \qquad p \propto T^{\gamma/(\gamma-1)\eta_{\infty\mathrm{t}}} \tag{C.4}$$

Whereas the isentropic index of compression is γ, as used in pv^γ = constant (equation B.11), the polytropic index n may now be determined from

compression: $$\frac{n}{n-1} = \frac{\eta_{\infty\mathrm{c}}\gamma}{\gamma - 1}$$ expansion: $$\frac{n}{n-1} = \frac{\gamma}{\eta_{\infty\mathrm{t}}(\gamma - 1)} \tag{C.5}$$

The major advantage of using polytropic efficiency is that it may be considered constant throughout the process and is a measure of the quality of the machine.

If a constant value of η_∞ is chosen, for instance 0.90, the equivalent value of isentropic efficiency is found to vary with compression or expansion ratio, as shown in figure C.3. In particular, it shows how increased pressure ratio reduces the isentropic efficiency of a compressor, regardless of its aerodynamic quality. The reason is shown by the divergence of the lines of constant pressure on the temperature–entropy diagram in figure C.4. The over-all change in stagnation temperature is the sum of many small changes, and is also the sum of the isentropic temperature changes of each small stage divided by its polytropic efficiency. Thus

$$\Delta T_0 = \Sigma \Delta T_{0\infty} = \Sigma \Delta T_{0\infty}'/\eta_\infty$$

But, by definition, $\Delta T_0 = \Delta T_0'/\eta_c$. Therefore

$$\frac{\eta_\infty}{\eta_c} = \frac{\Sigma \Delta T_{0\infty}'}{\Delta T_0'}$$

which can be seen to be greater than unity. Hence, η_C is less than η_∞. At each small stage of the process, the *preheat* or increase in temperature due to friction in the preceding stages calls for a relatively greater work input at the next stage.

The reason why turbine isentropic efficiency increases with pressure ratio may be similarly explained. In this case, the *reheat* is the higher temperature of the gas at any stage than isentropic expansion would yield. This raised temperature relative to pressure enables the subsequent stages of expansion to yield greater work output. It is thus natural for a turbine to have a greater isentropic efficiency than a compressor of comparable design quality.

C.4 Ducting Efficiencies

The gradual loss of stagnation pressure in the flow along a duct, or through an inlet or a nozzle, is not matched by a loss of stagnation temperature, as no work

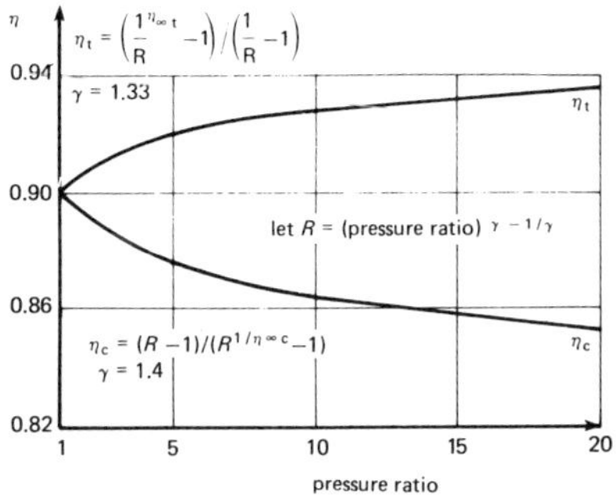

Figure C.3 *Variation of isentropic efficiency at constant polytropic efficiency*

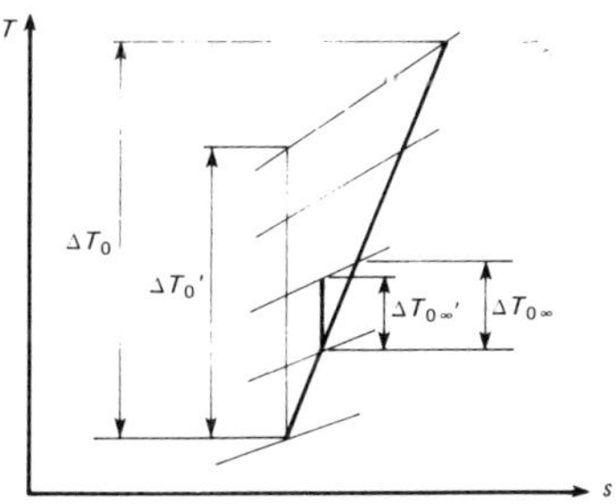

Figure C.4 *Relationship between isentropic and polytropic efficiencies*

is done (section B.2). The state point on a temperature–entropy diagram simply moves to the right (increasing entropy at constant temperature) for the stagnation conditions, while the static state point also moves vertically as the velocity changes.

Intake and duct losses may be considered simply as percentage losses of total pressure, while a nozzle is often considered to be loss free if it is short because it involves an accelerating flow. These losses are considered in greater detail, especially for aircraft engines in flight, in reference 31.

C.5 Assessment of Engine Ideal Performance

The over-all efficiency is assessed in terms of net power output divided by energy input

$$\eta_{th} = \frac{\text{turbine power} - \text{compressor power}}{\text{heat energy input}}$$

$$= \frac{\dot{m}_{34} C_{p34}(T_{03} - T_{04}) - \dot{m}_{12} C_{p12}(T_{02} - T_{01})}{\dot{m}_{23} C_p (T_{03} - T_{02})}$$

If $\dot{m}$ and C_p are taken as constant and the compression and expansion processes are performed at 100 per cent efficiency, then the efficiency for this simple, ideal case is

$$\eta_{th} = \frac{T_{03} - T_{04} + T_{01} - T_{02}}{T_{03} - T_{02}}$$

$$= 1 - \frac{T_{04} - T_{01}}{T_{03} - T_{02}} = 1 - \frac{T_{04}/T_{01} - 1}{T_{03}/T_{02} - 1} \times \frac{T_{01}}{T_{02}}$$

Now, between two lines of constant pressure

$$\frac{T_{02}}{T_{01}} = \frac{T_{03}}{T_{04}}$$

so

$$\frac{T_{04}}{T_{01}} = \frac{T_{03}}{T_{02}}$$

Therefore

$$\eta_{th} = 1 - \frac{T_{01}}{T_{02}} \tag{C.6}$$

or

$$\eta_{th} = 1 - \left(\frac{p_{01}}{p_{02}}\right)^{\gamma - 1/\gamma} \tag{C.7}$$

This represents the ideal maximum efficiency obtainable at a given pressure ratio, and is plotted in figure C.5. In practice, it could be higher because C_{p34} is higher than C_{p12}, but this is more than cancelled out by the various inefficiencies, pressure losses and other effects discussed in section 3.2.

It is also worth while to consider the optimum relationship between compression ratio, p_{02}/p_{01}, and over-all temperature ratio, T_{03}/T_{01}, in order to obtain the maximum power output. No power can be obtained in two extreme cases of the value of pressure ratio. One case is that where the pressure ratio is unity. The other is where the pressure ratio is so great that the temperature after compression reaches the allowable turbine entry temperature: that is, $T_{02} = T_{03}$, so no heat can be added.

The power output is given by

$$P = \dot{m}_{34}C_{p34}(T_{03} - T_{04}) - \dot{m}_{12}C_{p12}(T_{02} - T_{01})$$

If $\dot{m}$ and C_p are taken as constant and the processes are again 100 per cent efficient, then the following can be derived

$$\frac{P}{\dot{m}C_pT_{01}} = \left(1 - \frac{T_{01}}{T_{02}}\right)\frac{T_{03}}{T_{01}} - \left(\frac{T_{02}}{T_{01}} - 1\right)$$

Differentiating with respect to (T_{02}/T_{01}) and equating to zero, it can be shown that the power output reaches a maximum when

$$\left(\frac{T_{02}}{T_{01}}\right)^2 = \frac{T_{03}}{T_{01}} \qquad \text{(C.8)}$$

or when

$$\frac{p_{02}}{p_{01}} = \left(\frac{T_{03}}{T_{01}}\right)^{\gamma/2(\gamma-1)} \qquad \text{(C.9)}$$

Plotting power output against the log of the pressure ratio yields the curves of figure C.6.

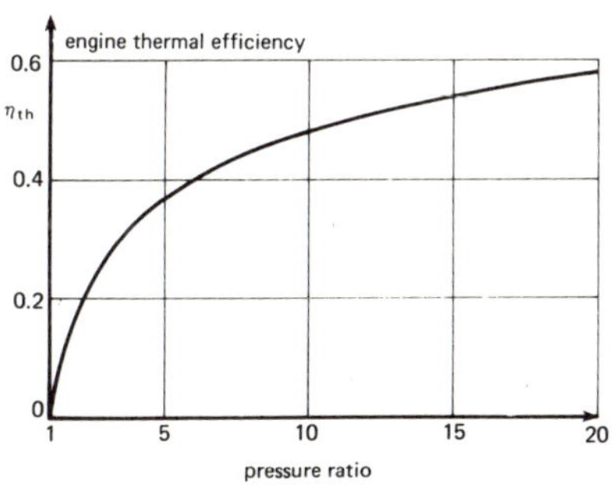

Figure C.5 *Increase in maximum possible thermal efficiency with pressure ratio*

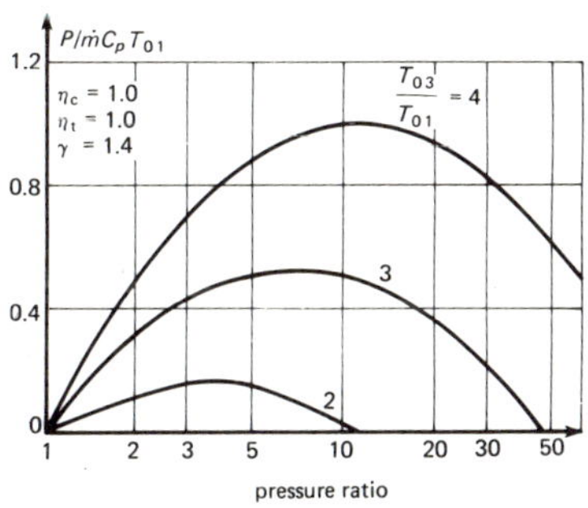

Figure C.6 *Effect of over-all temperature ratio on work output (simplified)*

C.6 Propulsive Efficiency and Work

The net thrust of a jet engine in flight is $\dot{m}(V_e - V_a)$, from equation 14.3. The power is therefore equal to $\dot{m}(V_e - V_a)V_a$, whereas the rate of energy input is given by the change in kinetic energy of the flow through the engine, $\frac{1}{2}\dot{m}(V_e{}^2 - V_a{}^2)$. The propulsive efficiency is given by

$$\eta_p = \frac{\text{power output}}{\text{energy input}} = \frac{m(V_e - V_a)V_a}{\frac{1}{2}m(V_e - V_a)(V_e + V_a)}$$

so

$$\eta_p = \frac{2}{1 + V_e/V_a} \qquad \text{(C.10)}$$

Assuming constant mass flow and exhaust velocity, the aircraft velocity at which the most work is done can be found, by differentiating the power output relative to aircraft velocity, to be where $V_a = V_e/2$. The efficiency and work can be plotted against velocity ratio, as shown in figure C.7. The lack of work at $V_a/V_e = 1$ results from the fact that there is no thrust, and the efficiency of 1.0 is meaningless because there is no addition of kinetic energy either. The optimum use of the engine occurs somewhere above the point of maximum work, where the efficiency is 0.67, but the practical selection of an engine for a particular aircraft must also consider the engine thermal efficiency, as shown in figure C.8. If the nozzle is unchoked (section B.2), the parameter specific net thrust/flight speed reduces to $(V_e/V_a) - 1$.

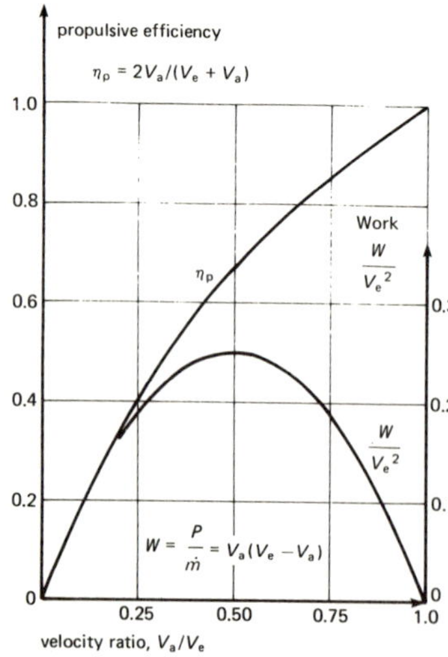

Figure C.7 *Variation of aero-engine propulsive efficiency and work with flight/jet velocity ratio*

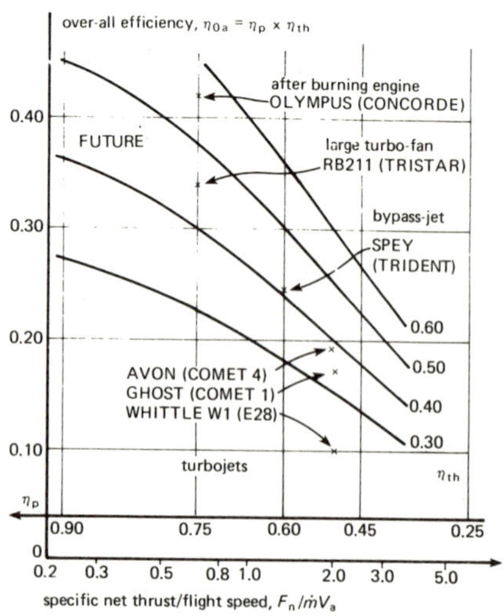

Figure C.8 *Improvement in over-all aero-engine efficiency with time and changes of configuration*

Appendix D

Chemical and Thermodynamic Properties of Working Fluids and Fuels

Table D.1 Properties of Typical Working Fluids and Process Gases

	Chemical formula	Molecular[a] weight M.W.	Boiling[a] point @ 1 atm (K)	Density[a,b] @ ISA (kg/m^3)	Gas const $R(=pv/T)$ 8314.3/M.W.	(C_p/C_v) @ ISA γ	Specific heat capacity[c,d] $C_p = a + b\Upsilon + c\Upsilon^n$ (J/kg K) where Υ = temperature/100 a	b	c	n	Dynamic[a] viscosity @ ISA, μ ($cp \times 10^3$)	Thermal[a] conductivity @ ISA (J/s m K)
Acetylene	C_2H_2	26.04	189	1.1019	319.3	1.112	2188	376.7	–11.6	+2	9.90	0.0200
Air	–	28.964	83	1.2256	287.1	1.399	947	21.3	–0.31	+2	18.00	0.0252
Ammonia	NH_3	17.03	240	0.7308	488.2	1.310	1746	147.4	–909	–2	9.65	0.0233
Argon	A	39.93	88	1.6908	208.2	1.666	521	0	0	–	21.87	0.0173
n-Butane	C_4H_{10}	58.17	273	2.4580	143.0	1.094	277	527.9	–16.3	+2	–	0.0149
Carbon dioxide	CO_2	44.01	195	1.8761	189.0	1.296	1005	20.0	–1959	–2	14.57	0.0157
Carbon disulphide	CS_2	76.13	319	–	109.2	1.227	684	8.8	–989	–2	9.67	0.0074
Carbon monoxide	CO	28.01	82	1.1853	296.9	1.400	1015	14.6	–164	–2	17.20	0.0242
Chlorine	Cl_2	70.91	236	3.0466	117.2	1.326	522	0.9	–401.2	–2	13.07	0.0083
Ethane	C_2H_6	30.09	184	1.2720	276.4	1.192	313	531.5	–15.4	+2	8.94	0.0201
Ethylene	C_2H_4	28.07	169	1.1860	296.3	1.238	422	426.6	–13.0	+2	9.60	0.0194
Helium	He	4.00	4	0.1692	2078.6	1.666	5199	0	0	–	19.20	0.1473
Hydrogen	H_2	2.016	20	0.0853	4124.1	1.405	13532	161.9	+2491	–2	8.65	0.1816
Hydrogen bromide	HBr	80.92	206	3.4241	102.7	1.401	331	6.4	+77.6	–2	17.95	0.0083
Hydrogen chloride	HCl	36.46	188	1.5428	228.0	1.397	720	14.2	+344.3	–2	13.98	0.0135
Hydrogen cyanide	HCN	27.03	299	–	307.6	1.307	1457	41.8	–2229	–2	–	0.0117
Hydrogen sulphide	H_2S	34.08	212	1.4410	244.0	1.325	959	36.3	–565	–2	12.33	0.0140
Methane	CH_4	16.05	109	0.6785	518.2	1.318	882	470.7	–11.2	+2	10.71	0.0327
Neon	Ne	20.18	27	0.8535	411.8	1.666	1030	0	0	–	30.76	0.0472
Nitric oxide	NO	30.01	120	1.2704	277.1	1.387	980	12.8	–195	–2	18.50	0.0250
Nitrogen	N_2	28.02	77	1.1854	296.7	1.401	1020	13.4	–179	–2	17.25	0.0252
Nitrogen dioxide	NO_2	46.01	294	1.9469	180.7	1.295	916	20.7	–1519	–2	–	–
Nitrous oxide	N_2O	44.01	185	1.8748	188.9	1.281	1038	19.6	–1939	–2	14.25	0.0164
Oxygen	O_2	32.00	90	1.3530	259.8	1.399	936	13.1	–523	–2	19.90	0.0258
Propane	C_3H_8	44.14	231	1.8650	188.4	1.129	229	542.5	–16.6	+2	7.90	0.0166
Steam	H_2O	18.03	373	0.7620	461.1	–	1695	57.1	0	–	–	–
Sulphur dioxide	SO_2	64.06	263	2.7107	129.8	1.269	721	12.3	–1202	–2	12.10	–

Table D.2 The Properties of Some Gaseous Fuels

	Chemical formula	Molecular weight	Density @ ISA (kg/m^3)	Lower calorific value @ 288 K (MJ/kg)	(MJ/m^3)[b]	(Btu/scf)
n-Butane	C_4H_{10}	58.17	2.458	45.71	112.41	2917
Carbon monoxide	CO	28.01	1.185	10.10	11.97	311
Ethane	C_2H_6	30.09	1.272	47.48	60.43	1584
Hydrogen	H_2	2.016	0.085	119.95	10.22	265
Methane	CH_4	16.04	0.678	50.01	33.95	881
Propane	C_3H_8	44.14	1.865	46.35	86.42	2290

Table D.3 The Properties of Some Liquid Fuels

Pure liquids	Chemical formula	Molecular weight M.W.	Boiling[b] point @ 1 atm (K)	Density[b] @ 228 K (kg/litre)	Lower (net) calorific value @ 298 K (MJ/kg)[b,e]	(Btu/lb)
Benzene	C_6H_6	78.11	353	0.879	40.14	17255
n-Butane (L.P.G.)	C_4H_{10}	58.17	273	0.575	45.34	19490
Ethanol	CH_5OH	34.06	352	0.794	27.23	17705
n-Heptane	C_7H_{16}	100.21	371	0.684	44.56	19155
Methanol	CH_3OH	32.04	338	0.796	19.94	8570
n-Octane	C_8H_{18}	114.23	399	0.703	44.43	19100
Propane (L.P.G.)	C_3H_8	44.14	231	0.505	45.98	19765

Miscellaneous fuels	Density ρ @ 288 K (kg/litre)	Boiling range (K)	Kinematic viscosity @ temp. (cSt)	(K)	Lower (net) calorific value @ 288 K (MJ/kg)	(Btu/lb)
Diesel[b]	0.84	453 to 613	5.0	288	42.9	18440
Fuel oil[b,f]	0.80 to 0.97	453 to ?	1.4 to 70	355	$53.1 - 12.3\rho$	–
Gasoline[b]	0.72 to 0.76	301 to 473	0.5 to 0.65	288	41.9 –44.0	{18010 18915
Kerosine [36]	0.73 to 0.83	–	2.1 to 3.0	288	$51.7 - 10.5\rho$	–
Sulphur [23]	1.789 (410 K)	718	4.58	410	9.20	3955
Tar (coke oven)[b]	1.138 to 1.18	372 to 645+	–	–	36.0 to 37.7	{15475 16205

Table D.4 The Properties of Some Solid Fuels

	Density[b] (no pores) (kg/m^3)	Bulk density[b] (with pores) (kg/m^3)	Higher (gross) calorific value* (MJ/kg)	Higher (gross) calorific value* (Btu/lb)
Carbon (to CO)	1800 to 2300	–	9.20[e]	3955
(to CO_2)	–	–	32.76[e]	14080
Charcoal	1400 to 1700	–	34.30[e]	14745
Coal	1250 to 1700	580 to 830	32 to 36[b]	13755 15475
Coke	1750 to 2000	300 to 510	33.60[b]	14445
Peat	1150 to 1250	–	19.70[b]	8470
Pine bark	–	220 to 320	19.77[g]	8500
Wood	500 to 1100	–	17.30[b]	7435

* These calorific values apply to the combustible content only. They may be degraded by content of ash and water approximately in proportion to their fraction of the weight.

The data in this appendix are adapted and calculated from the following sources.

[a]R. C. Weast (Ed.), *Handbook of Chemistry and Physics*, 56th ed (CRC Press, 1975–6)

[b]J. W. Rose and J. R. Cooper, *Technical Data on Fuel,* British National Committee, World Energy Conference, 7th edn. (1977)

[c]K. K. Kelly, U.S. Bureau of Mines, Bulletin 584 (1960)

[d]H. M. Spencer, *Ind. Engng Chem.*, **40** (1948) 2152–4

[e]G. J. Van Wylen and R. E. Sonntag, *Fundamentals of Classical Thermodynamics*, 2nd edn. (Wiley, New York, 1976)

[f]Mobil Technical Bulletin, *Heavy Fuel Oil*

[g]I. Gilmour, private communication

Appendix E

Axial Compressor and Turbine Blading

This appendix presents aspects of the aerodynamics, design and performance of blading as additional background to chapters 6 and 8. Compressor and turbine blading are treated together where possible, with their differences highlighted. The discussion of compressor blades generally refers to subsonic blading, as appropriate to industrial and most aircraft use, but the supersonic and transonic (subsonic root, supersonic tip) blading of modern aircraft engines is specifically mentioned in a few places.

E.1 Air Flow and Blade Angles

The common nomenclature used for compressor and turbine blades can be seen in figure E.1, which represents a section taken through the blading at a chosen diameter. This nomenclature, although numbered for the rotor blades, can be applied to both the stator and rotor blade rows, as depicted in figures 6.2 and 8.3. In chapters 6 and 8, the power input to or output from the rotor is expressed in the form

$$P = \dot{m}UV_a(\tan\alpha_1 - \tan\alpha_2) \tag{E.1}$$

where α_2 for the turbine has a negative value.

The angles are thus important in determining the performance of the machine, but it should be noted that the airflow angles (α) are seldom equal to the blade angles (β) at the leading and trailing edge. The measurement and effects of these angular discrepancies are discussed under cascade testing (section E.4), from which one set of results is shown in figure E.2, indicating typically how the flow angles determine whether the machine will act as a compressor or turbine.[63A]

The difference between compressor and turbine blading can be seen by comparing figures 6.2, 8.3 and E.1. The direction of blade curvature and the

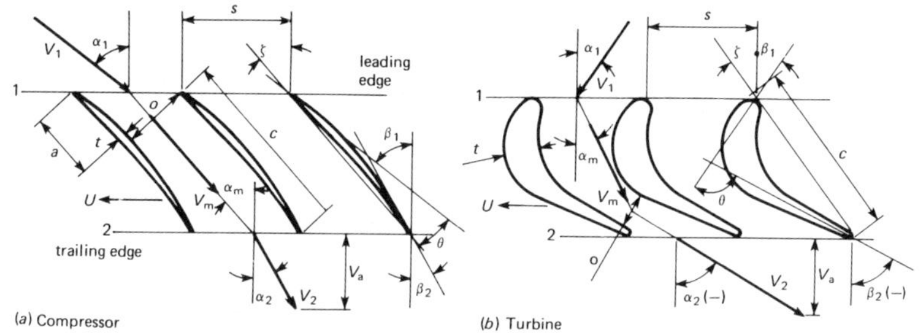

Figure E.1 *Blading dimensions, angles and nomenclature: s space; c chord; a chord to maximum camber; H blade height (not shown); K tip clearance (not shown); θ camber angle ($\beta_1 - \beta_2$); i incidence angle ($\alpha_1 - \beta$); t maximum thickness; α air or fluid flow angles; β blade camber line angles; ζ stagger angle; ε deflection angle ($\alpha_1 - \alpha_2$); δ deviation angle ($\alpha_2 - \beta_2$), o throat opening*

position of the throat (the narrowest part of the passage) clearly differ: in the compressor the flow diffuses (slows and gains static pressure) in both rotor and stator and work has to be put in by each successive rotor to maintain the flow, whereas, in the turbine, the passages in stator and rotor expand (accelerate) the flow, which is slowed by the removal of energy at each rotor stage. The relationship of the inlet and outlet flow angles can be assessed quantitatively by the parameter 'reaction', which is discussed next.

E.2 Impulse and Reaction

The 'degree of reaction' (D) is a parameter which defines the proportion of the stage's static temperature change (rise in the compressor, drop in the turbine)

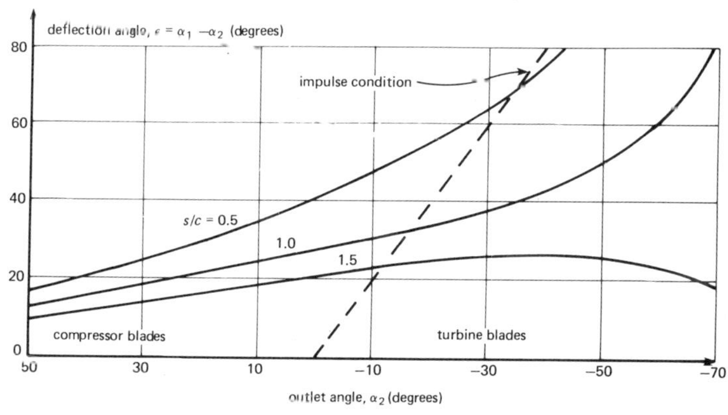

Figure E.2 *Deflection angle versus outlet flow angle for compressor and turbine blades (redrawn from reference 63A)*

occurring in the rotor. Assuming that the absolute flow velocity is the same at entry to and exit from the stage, so that $\Delta T_s = \Delta T_{0s}$, then

$$D = |(T_2 - T_1)/\Delta T_{0s}| \quad \text{(E.2)}$$

Using the fact that the work of the stage ($\dot{m}C_p\Delta T_{0s}$) results solely from the motion of the rotor $[\dot{m}C_p(T_3 - T_0) + \frac{1}{2}\dot{m}(V_3{}^2 - V_0{}^2)]$, it can be derived that the reaction for compressor and turbine blading is defined by

$$D = 1 - \tfrac{1}{2}(V_a/U)(\tan\alpha_0 + \tan\alpha_3) \quad \text{(E.3)}$$

where V_a is assumed constant. From this equation can be derived the normal expressions for compressor and turbine reaction

$$D(\text{comp}) = \tfrac{1}{2}(V_a/U)(\tan\alpha_1 + \tan\alpha_2) = \tfrac{1}{2}\,\frac{\tan\alpha_1 + \tan\alpha_2}{\tan\alpha_0 + \tan\alpha_1} \quad \text{(E.4)}$$

and

$$D(\text{turb}) = -\tfrac{1}{2}(V_a/U)(\tan\alpha_1 + \tan\alpha_2) = 1 - \tfrac{1}{2}\,\frac{\tan\alpha_0 + \tan\alpha_3}{\tan\alpha_0 - \tan\alpha_1} \quad \text{(E.5)}$$

Substitution of the appropriate values 0, $\frac{1}{2}$ or 1 for D in these equations yields, with further manipulation, the flow-angle conditions shown in figure E.3. Zero reaction implies impulse rotor blading, with passages of constant area and no change in flow velocity. 50 per cent reaction implies the same temperature change in rotor and stator, and the blading is called 'symmetrical' because of the mirror-image similarity of the blade shapes. 100 per cent reaction implies an

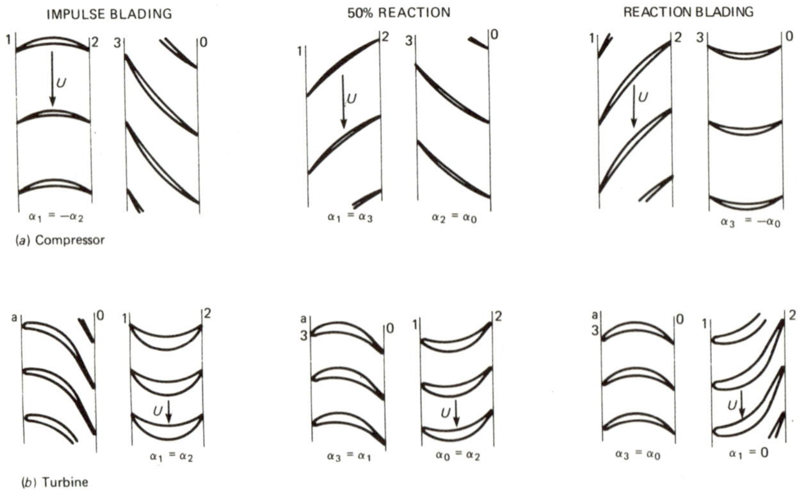

Figure E.3 *Blading angles for different degrees of reaction*

impulse stator. However, in turbine practice, blades are twisted, using impulse and reaction angles, while the nozzles have $\alpha_a \approx 0$, the impulse configuration only. Typical values of reaction are quoted in sections E.6 and E.7.

E.3 Forces on Blading

The fluid flowing through a moving blade row experiences a change in whirl velocity, which involves useful work (sections 6.2 and 8.2), and a change in axial gas load (section 14.1) which affects static pressure and axial velocity. These changes result from forces exerted by the blades upon the flow.

Figure E.4 shows how the flow momentum vector is changed by the blade row from $\dot{m}V_1$ at angle α_1 to $\dot{m}V_2$ at angle α_2: constant axial velocity (V_a) is assumed, although this is not always the case (section E.6). An impulsive force F_i, approximately normal to the blade chord, represents the change in flow direction while F_r(reaction) is exerted by the pressure change to accelerate or decelerate the flow. The combination of F_i and F_r makes the resultant F_w, the tangential force which is applied by the torque at the rotor shaft.

F_r is aligned with the mean momentum vector $\dot{m}V_m$ at angle α_m, as used in cascade testing (section E.4), where

$$V_m = V_a \sec \alpha_m \tag{E.6}$$

and

$$\tan \alpha_m = \tfrac{1}{2}(\tan \alpha_1 + \tan \alpha_2) \tag{E.7}$$

F_i is normal to F_r. The force exerted by the flow on the blade is F_b, which is equal in length but opposite in direction to F_i and acts against the concave side of the blade. It is the resultant effect of the pressure difference acting across the area of the blade, but may be considered as a simple force with the approximate

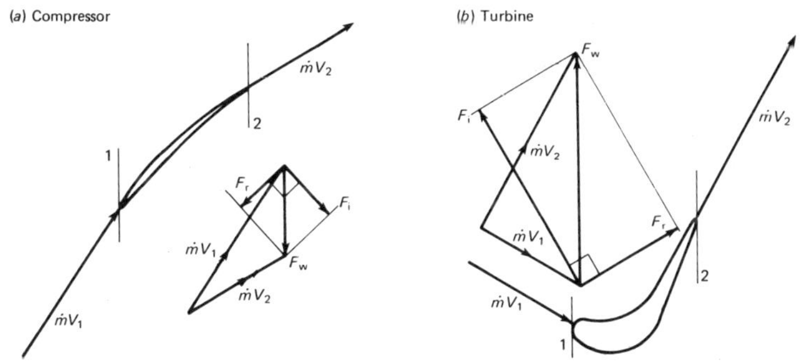

Figure E.4 *Tangential force exerted on the flow by the combination of blade impulsive force and pressure change reaction force*

value

$$F_b = \frac{\dot{m}}{n}\left(\frac{V_1 + V_2}{2}\right)(\alpha_1 - \alpha_2) \tag{E.8}$$

where n is the number of blades in the stage and the angles are measured in radians.

The effect of F_b is to induce a bending stress in the cantilevered blade and its attachment to the disc. This is minimised by making the blade *lean* away from the gas load. The centrifugal load then acts along a line offset to the convex side of the blade root, and provides a bending moment opposing that of the gas load. The two moments do not always balance, but the lean angle may be chosen to equalise the peak bending stresses which occur in both directions. For instance, the positive moment with high gas load at take-off in an aircraft engine may match the negative moment when the centrifugal moment exceeds the gas load in the reduced pressure atmosphere at high-altitude cruise. The blades must nevertheless be strong enough to withstand the complete loss of gas load which may occur briefly during surge.

These aspects of loading and stress are covered more deeply in many other books, such as references 31, 62, 72 and 95. A simple assessment of centrifugal stress will, however, be presented in section E.8 to give some indication of the important parameters.

E.4 Cascade Testing – Two-dimensional Flow

The relationship between the blade profile and the air-flow angles at any diameter is determined by cascade testing. Several airfoils of the chosen profile are set in parallel, as in figure E.1: the whole row of blades can be rotated on a turntable to different *incidence* angles relative to an approaching calibrated flow. The direction of the flow behind the blades is measured to determine the *deflection* of the flow from its approach path, and this can be related to the known blade angles, chord and spacing [63A]. The forces exerted on the turntable are also measured, yielding the lift and drag forces on the airfoils under test. The back pressure behind the cascade may be controlled, particularly in the case of supersonic compressor blades which considerably increase the static pressure [63B].

The blade's characteristics are determined by its profile, which is defined by a simple code. For instance, a British NGTE 12C2/25P40 airfoil is defined as

> maximum thickness/chord ratio: $t/c = 0.12$
>
> base profile C2, defines variation of thickness, which is disposed equally along each side of the central camber line, and the radii at leading and trailing edges.

camber angle $\theta = 25^\circ$

P indicates a parabolic shaped camber line;
C would indicate the use of a circular arc

the maximum camber is at 40 per cent of the chord from the leading edge: $a/c = 0.4$.

American NACA profiles are also in extensive use, [95] and other shapes are listed by European research establishments.

The *incidence* angle $i(\alpha_1 - \beta_1)$ is an input parameter, and is varied over a wide range of values to perform the cascade testing. This permits the blade performance to be assessed at design and off-design conditions in the engine, as the flow coefficient V_a/U and, therefore, incidence vary through the running speed range.

The *deviation* angle δ arises from the reluctance of the flow to turn through the full camber angle of the blades and is defined as $(\alpha_2 - \beta_2)$ for compressor blades, and as $(\cos^{-1}(o/s) - \alpha_2)$ [63C] for turbine blades. $\mathrm{Cos}^{-1}(o/s)$ is the geometric angle normal to the throat, relative to the axial direction, and is a more realistic datum to use than β_2 as the trailing edge can be well downstream of the throat. Deviation increases with θ, s/c, a/c and ξ [63A] but, in the case of the turbine, is reduced by increasing the throat angle $(\cos^{-1}(o/s))$ and by increasing the Mach number. [96] It is still expressed empirically, rather than by a derived equation, but it is very important to relate the deviation accurately to the blade geometry as variation of the outlet flow angle has a major effect on performance.

The measured axial and tangential forces are used to determine the airfoil's lift and drag forces, normal and parallel to the mean flow vector (equations E.6 and E.7). [95] These forces in turn yield the lift and drag coefficients, the former being

$$C_L = \text{lift force}/\tfrac{1}{2}\rho V_m{}^2 c = 2(s/c)(\tan\alpha_1 - \tan\alpha_2)\cos\alpha_m \qquad \text{(E.9)}$$

This is clearly related to the work equation E.1 and to the *deflection* or turning angle $\epsilon(\alpha_1 - \alpha_2)$. Often C_L is based on V_1 for a compressor or V_2 for a turbine, instead of V_m. Increasing Mach number has little effect on C_L for the compressor but, in the turbine, the deviation reduces, so the deflection increases with M. Typical variation of C_L with blade geometry is shown for compressor and turbine blades in figure E.5, α_2 and s/c being the most significant influences. It has been found [63A] that, for a compressor, the design value of $(\tan\alpha_1 - \tan\alpha_2)$ can be related empirically to s/c by the equation

$$\tan\alpha_1 - \tan\alpha_2 = 1.55/(1 + 1.5\,s/c) \qquad \text{(E.10)}$$

The deflection angle (ϵ) is not constant along the cascade because the deviation varies cyclically across each blade space. Figure E.6*a* illustrates for the compressor how the flow follows the blade surface on the concave side but

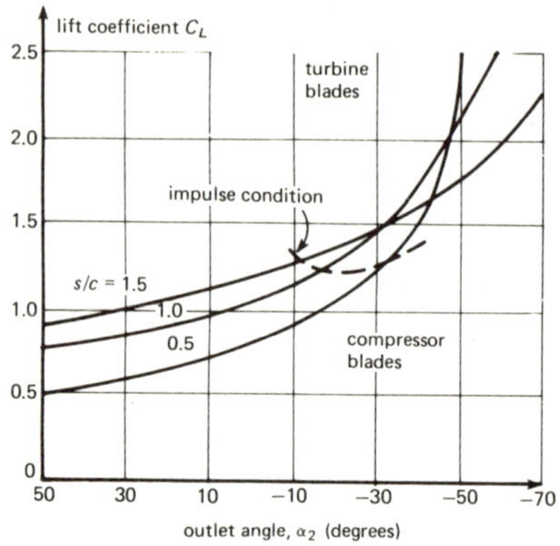

Figure E.5 *Lift coefficient versus outlet angle for compressor and turbine blades (redrawn from reference 63A)*

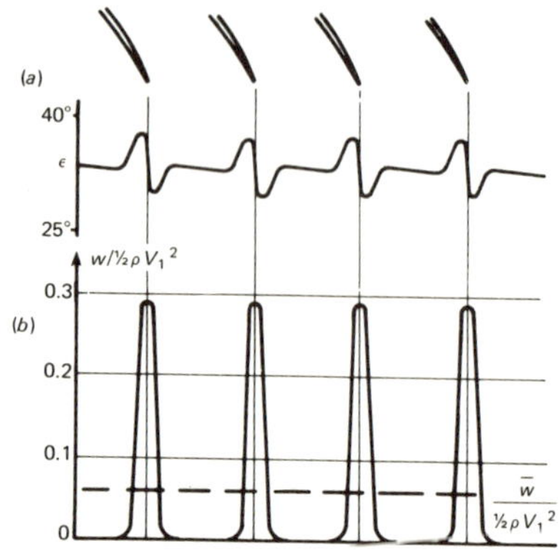

Figure E.6 *Typical variation of deflection angle and pressure loss across blade space, for compressor blading*

deviates from the convex side. Figure E.6*b* shows the lower level of total pressure in the disturbed wake behind each blade, the loss (w) being the result of frictional drag over the blade surfaces. This loss is averaged to $\bar{w}$ across the blade spacing, representing the mean loss of total pressure through the cascade. This is the profile drag, for which the coefficient is

$$C_{DP}(\text{comp}) = \text{drag force}/\tfrac{1}{2}\rho V_m{}^2 c = (s/c)\,(\bar{w}/\tfrac{1}{2}\rho V_1{}^2)\cos^3\alpha_m/\cos^2\alpha_1 \tag{E.11}$$

The profile drag coefficient for turbine blading is expressed in terms of the pressure loss coefficient, $Y = (p_{01} - p_{02})/(p_{02} - p_2)$, as discussed in section 8.2, where Y is equivalent to $\bar{w}/\tfrac{1}{2}\rho V_2{}^2$. Thus

$$C_{DP}(\text{turb}) = (s/c)\,Y\cos^3\alpha_m/\cos^2\alpha_2 \tag{E.12}$$

Profile losses are affected by flow velocity, rising considerably at a Reynolds number below about 10^5, and are also greatly increased by a thicker trailing edge.[63C] The typical variation of these losses with incidence is shown in figure E.7.[63A] The apparently high losses of the turbine blading have to be considered with the turbine's high work capacity in assessing the efficiency of the machine.

The deflection and C_L can usefully be plotted with C_{DP} against incidence, as in figure E.8 (for compressor blading). C_L and ϵ are shown as one curve with different scales, and the C_{DP} curve is akin to the $\bar{w}/\tfrac{1}{2}\rho V^2$ curve of figure E.7. Such is the similarity of the shapes of the curves that the non-dimensional scales ϵ/ϵ^* and $(i - i^*)/\epsilon^*$ may also be attached to this figure, to demonstrate the characteristic at off-design conditions. The asterisk is used to denote the nominal design values of incidence and deflection. ϵ_s is the value of deflection at stall,

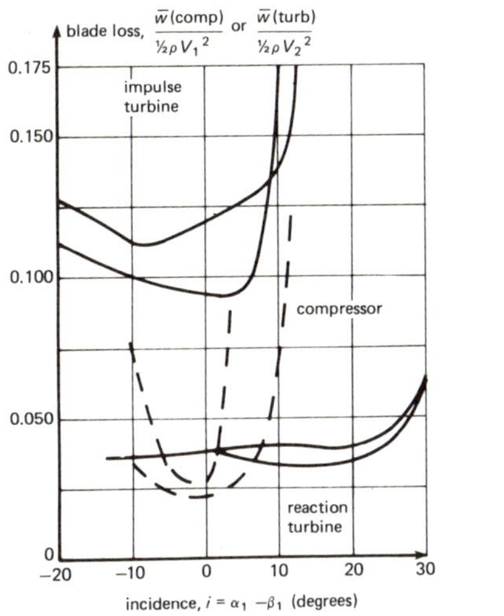

Figure E.7 *Typical compressor and turbine blade loss versus incidence (redrawn from reference 63A)*

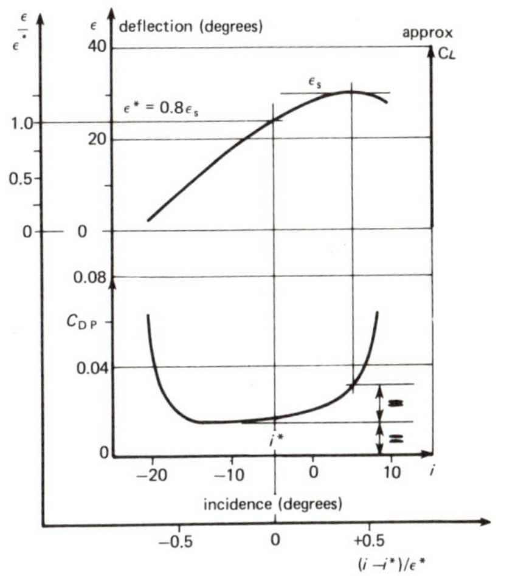

Figure E.8 *Typical variation of lift and drag versus incidence for compressor blades (redrawn from reference 63A)*

which is defined as being where the drag rises to double its minimum value. ϵ^* is chosen to be $0.8\epsilon_s$, and corresponds with the design condition given by equation E.10. At any off-design value of i, the appropriate values of C_{DP} and ϵ can be determined from curves of this type, using the non-dimensional scales.

E.5 Three-dimensional Effects – Secondary Flows

The main differences between real machines and the cascade rig are

(1) the blade row is bent to form a complete circle, so the s/c ratio varies from hub to tip;

(2) each blade row is closely preceded and/or followed by another, each disturbing the ideal flow pattern of the others (section F.3);

(3) every second blade row moves with velocity U varying from hub to tip, affecting the approach angle α_1 (to rotor and stator);

(4) there are real boundaries to the flow passage at hub and tip, with associated annulus wall boundary layers.

These differences require the use of several adjustments to make cascade results applicable to the real compressor or turbine. The first is the use of a blade twist from hub to tip, as discussed in sections E.6 and 7 to cater for differences (1) and (3). The effect of twist is to vary the flow geometry and blade profile with radius: a cascade test therefore applies to only one diameter and at least three cascade tests are required to determine the blade shapes at, for instance, root, mean and tip diameters.

The machine geometry imposes three departures from the mean, nominal flow condition. The increase in whirl velocity in the compressor rotor causes a slight radially outward flow which is reversed in the stator: conversely the turbine blade causes a slight inward flow. Secondly, the annulus wall drag causes a considerable distortion in the axial velocity profile, as illustrated in figure E.9*a* for the condition at a late stage in a compressor, this being where the worst distortion occurs. The low velocity at root and tip amounts to an effective *blockage* to the flow, [6M] which has to be allowed for by increasing the blade length at the later stages to maintain constant the effective mean flow velocity. The third departure is the tendency to *free vortex flow.*

The peaky velocity profile induces a very important further disturbance to the flow pattern, called secondary flow vortices. They result from the turning of the flow in the curved blade passages. As with any turning flow vector, a Coriolis force is experienced and, in section 5.2, it was quoted as being $2\omega V_r$. In axial flow machinery with blade passage curvature of a given radius, say R_b, the term $2\omega V_r$ is equivalent to $2V_m^2/R_b$. The force therefore varies with the square of the flow velocity, and is clearly much greater near the blade mid-height. Two main vortices are set up in which the flow is outwards (relative to the centre of blade passage curvature) near the mid-height and inwards against the weaker Coriolis force at the hub and tip. These vortices are shown in figures E.9*b* and *c* for the compressor and turbine.

The third vortex in the compressor blading results from the scrubbing effect of the adjacent, moving, outer annulus wall and flow leakage through the blade tip clearance *k*. This effect is negligible in turbine blading because of the opposite

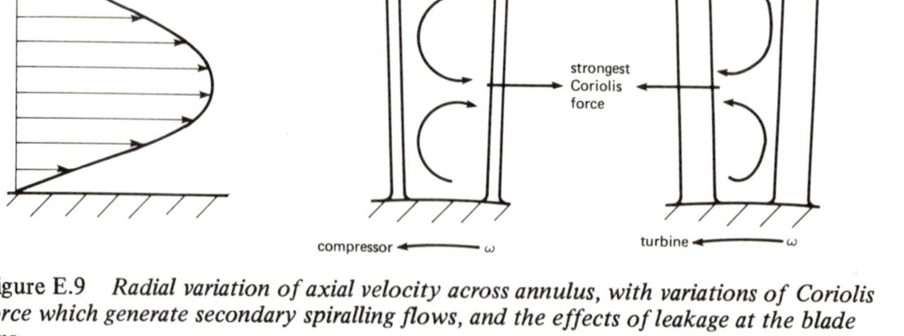

Figure E.9 *Radial variation of axial velocity across annulus, with variations of Coriolis force which generate secondary spiralling flows, and the effects of leakage at the blade tips*

sense of wall motion relative to blade curvature and/or because turbine blading may be shrouded, eliminating the moving wall. Nevertheless, axial leakage at the blade tip occurs because of the turbine's high level of pressure differences, and this is also considered as a significant loss.

The $C_{D\mathrm{P}}$ value from cascade testing requires to be increased by drag coefficients allowing for the parasitic effects of the annulus drag, secondary flows and tip clearance, to permit estimation of the performance of real machines. These drag coefficients are determined empirically and are added together in the form

$$C_D = C_{D\mathrm{P}} + C_{D\mathrm{a}} + C_{D\mathrm{s}} + C_{D\mathrm{t}} \tag{E.13}$$

Of these, $C_{D\mathrm{a}}(=0.02\ s/h)$ and $C_{D\mathrm{s}}(=0.018\ C_L{}^2)$ are used for compressor blading and, together, have a similar value to $C_{D\mathrm{P}}$.[31] The C_D curve is similar in shape to the $C_{D\mathrm{P}}$ curve of figure E.8. For turbines, only $C_{D\mathrm{s}}$ and $C_{D\mathrm{t}}$ are used, as discussed in section E.7.

This method is superseded by the deeper study of the three-dimensional flow patterns, as described in reference 97 for compressors. Such study integrates all these effects, quite reasonably as the secondary flows are intrinsically caused by annulus drag.

E.6 Typical Compressor Blade Data

Airfoil profiles differ significantly between subsonic and supersonic compressors. Subsonic blading has profiles similar to the wings of gliders or light aircraft, with the maximum thickness t of $0.1c$ to $0.15c$ placed typically at $0.3c$ from the large leading edge radius and joined by a slow taper to the small trailing edge radius. Supersonic blading has small leading and trailing edge radii and the shape may consist of two circular arcs (lenticular), a straight wedge followed by circular arcs, or simply two wedges set back to back.[63B]

The position of maximum *camber* (a/c) in subsonic compressor blades affects the profile drag and operating behaviour. With a low value, about 0.4, the blades are tolerant of a wide range of incidence angles and will pass a large flow without choking; this suits operation at the rear stages.[6L] Conversely, with $a/c = 0.6$, the advantages of high work capacity and a high Mach number before the drag rises significantly, match it to the requirements of the front stages.

The value of *reaction* in subsonic compressors is commonly 50 to 60 per cent at mean diameter, letting the slightly more efficient rotor do a little more than half of the diffusion.[95] Supersonic or transonic (supersonic flow at tip, subsonic at root) aircraft compressors, which commonly have no inlet guide vanes at the first stage, are designed to work with zero approach and outlet whirl, $\alpha_0 = 0$(6.1). This requires a high value of reaction, with shock formation in the rotor passages, and the stators approach the impulse condition.

Compressors handling light gases may use low reaction to reduce blade velocity and stress (section E.9).

The blade *twist* is chosen by the designer to match the real flow conditions as well as possible. Several assumptions may be made to simplify the calculations, one of which is free vortex flow ($V_w \times R$ = constant). In compressors, it would call for excessive deflection at the rotor blade root, where V_w and ($V_{w2} - V_{w1}$) would be higher than at the tip, with a corresponding increase in C_D. An alternative would be to assume and design for constant reaction at all radii, but then V_a and V_w vary with radius and radial equilibrium is not satisfied (the radial variation of static pressure has to balance the centrifugal force on the air flowing along its spiral path through the compressor). Other possibilities are available [95] but the common choice is 'half-vortex', using the arithmetic mean of the whirl velocities calculated by the free vortex and constant reaction methods. This yields compressor rotor and stator blades with moderate twist and variation of reaction from root to tip.

A high value of blade *aspect ratio* ($h/c > 3$) yields a high lift-to-drag ratio and an efficient compressor. As h is fixed by the mass flow, the mean radius and the value of V_a or Mach number, a high value of h/c implies a low value of c. In turn, this permits a short, rigid, lightweight compressor. In practice, however, such blades are more prone to vibration failures (appendix F) from reasons of mechanical excitation or aerodynamic flutter. Current aircraft compressors achieve robustness and high performance by using lower aspect ratios (from 1.0 to 2.5), higher axial velocity, and drum construction rather than separate discs on a shaft. The larger chord permits larger blade spacing at constant s/c with fewer blades and consequent lower costs. An exception is in the fan design for aero engines, which use higher aspect ratios (up to about 4.0) and use one or more snubber or shroud rings (clappers) within the blade height to control vibrations and impact damage (section 6.4.2).

The *hub/tip* diameter or radius *ratio* (R_h/R_t) affects performance in several ways. A low value permits a slightly smaller over-all diameter for a given mass flow, but increases the blade height. This increases the problems of designing for radial equilibrium and, if the chord is also increased to keep a constant aspect ratio, then the losses will increase. For a multi-stage compressor, the lowest value of R_h/R_t at the first stage is reasonably about 0.5, but it may fall as low as 0.3 for the fan of high bypass ratio turbofan engines. A practical limit here is the strength of the fan disc, which is small, has many blade root fixing slots and must withstand the loads imposed by long, heavy blades (section E.3). There must also be space within the hub radius to incorporate the shaft and a bearing chamber (section 14.4). At the outlet end of the compressor, the boundary layer and tip clearance losses may become excessive relative to the region of effective, unblocked flow (section E.5) if R_h/R_t rises above about 0.92; with a resultant loss of efficiency and performance. As R_h/R_t rises above about 0.85, increasing the *tip clearance* above 0.05 per cent of diameter also severely causes the efficiency to drop [33].

The compressor performance is dependent also on variations of the *mean radius* of the annulus. Three simple cases are shown in figure E.10, with *a* constant tip radius, *b* constant mean radius and *c* constant hub radius. For a given area contraction ratio between compressor inlet and outlet, geometry *a* gives a shorter last stage blade than geometry *c* and must necessarily have fewer stages, if the hub/tip ratio of the last stage is not to exceed 0.92. Thus, geometry (*c*) can generate a higher pressure ratio, although its work input per stage may be penalised by the reduced mean blade speed. Denning and Jordan [4B] show that, with the Mach number reducing from 0.5 to 0.25 through the compressor and R_h/R_t equal to 0.92 at outlet the compressor rotor of constant tip radius is restricted to a pressure ratio of 14; with constant mean radius it could achieve a ratio of 24 and it could exceed 50 to 1 with constant hub radius. The choice of geometry *a* in many early designs therefore imposed a substantial handicap, while geometry *b* offers a good compromise.

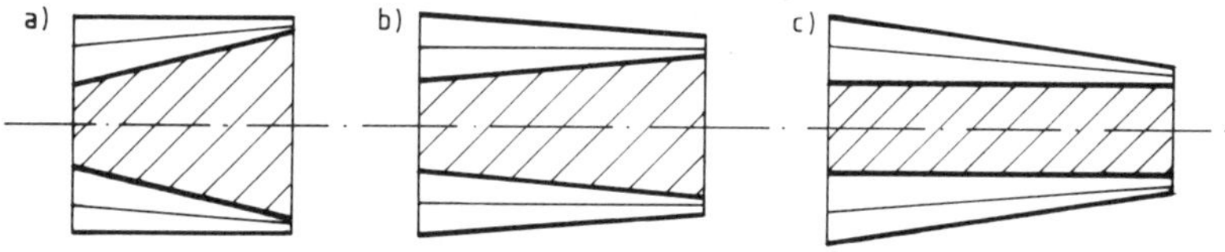

Figure E. 10 *Three compressor annulus configurations showing the variation of hub/tip ratio with changing mean radius*

The Mach number of the flow is also important to the geometry. In the above case, the considerable drop in Mach number towards the compressor outlet benefits the combustion system inlet. If the axial velocity is kept constant, the Mach number falls slowly as the temperature rises. M may be restricted to 0.4 to minimise the losses, particularly for industrial use, but a higher value is appropriate to aircraft use to reduce the engine diameter and weight and the nacelle size and drag. For a large turbofan of transonic design, the axial Mach number may approach 0.6, with a relative value of 1.8 at the blade tips.

E.7 Typical Turbine Blade Data

Turbine blading is normally designed for the free vortex condition, $V_w \times R$ = constant. Thus V_0 decreases while U increases at larger radius, and the velocity triangles vary from root to tip. This calls for the use of twisted airfoils; in practice the nozzles are often not twisted, for manufacturing reasons, while the blades have a high degree of twist.

The degree of *reaction* is normally about 50 per cent at mean diameter, the blade twisting towards impulse form at the root and reaction form at the tip (figure E.3). This limits the minimum value of hub/tip ratio, R_h/R_t: if the hub radius is too small, the blade angles may go beyond the impulse condition, causing the flow accelerated by the nozzle to be recompressed by the blade root with considerable losses. This limits the minimum value of R_h/R_t to about 0.65. A reasonable maximum value is about 0.85; if the blade is any shorter, the end effect losses from the boundaries and tip clearance may overshadow the useful work done.

The losses are usually determined by the methods of references 61, 98 and 99. From equation E.13, the tip clearance and secondary flow (including annulus drag) losses are assessed in the form

$$C_{Dt} = nC_L{}^2 \frac{k/h}{s/c} \tag{E.14}$$

and

$$C_{Ds} = C_L{}^2 \frac{\lambda}{s/c} \tag{E.15}$$

The value given to n is 0.5 for unshrouded blades or 0.25 for shrouded blades, while λ is the secondary loss parameter shown empirically in figure E.11.

The *hub/tip ratio* affects the secondary flow losses, because λ is expressed in terms of $(A_2/A_1)^2/(1 + R_h/R_t)$. The losses are clearly reduced by increasing R_h/R_t, that is, by the use of a blade that is short relative to its mean radius. However, the hub/tip ratio also interacts with the flare: this provides the necessary rate of area increase to maintain constant V_a while the pressure decreases, and generally results in a reducing value of R_h/R_t towards the later stages. It is sometimes necessary to increase the mean flow radius, and both R_h and R_t, to maintain a satisfactory ratio. This is most appropriate if the later stages are attached to a shaft rotating at lower speed; the increase in radius is matched by a raised blade speed U.

The turbine operates with accelerating flows, but diffusion processes occur in two regions. The flare itself is commonly limited to an included angle of 25 to 30°, to reduce the rate of boundary layer growth. The main diffusion problem is at the rear part of the blade's convex surface. The flow accelerates round the tighter initial radius while, being irrotational, it is quite slow along the concave surface (figure E.12). Towards the passage outlet, the flow velocities tend to balance, requiring diffusion and risking separation from the convex surface, with increased losses. This can be avoided by keeping the NACA *diffusion parameter* $(V_{max} - V_2)/V_{max}$ down to about 0.2.[6F] The use of a parabolic camber line with a/c at about 0.35 to 0.4 assists this: earlier blade shapes tended to involve circular arcs and straight lines for ease of manufacture, but operated with lower efficiency.

There is an optimum blade spacing, or *pitch/chord ratio* (s/c) which will result in minimum profile loss, the other losses being independent of s/c. [98] The values of s/c which appear to suit most researchers' criteria are shown [63C] to be approximately 0.4 to 0.6 for impulse blades and 0.9 to 1.0 for reaction blades. Thus, the normal twist of rotor blades suits the spacing requirements for low loss

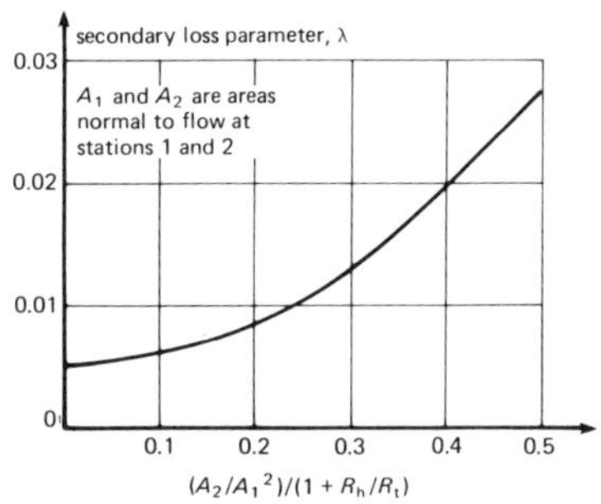

Figure E.11 *Secondary loss parameter for turbine blading, varying with annulus geometry (Redrawn from reference 6F by permission of the Council of the Institution of Mechanical Engineers)*

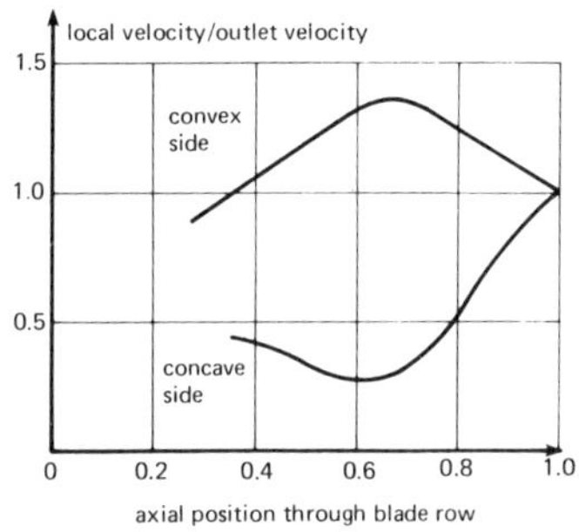

Figure E.12 *Typical two-dimensional velocity distribution round turbine rotor blades (Redrawn from reference 6F by permission of the Council of the Institution of Mechanical Engineers)*

at all radii. However, the blade thickness/chord ratio (t/c) is relevant to s/c as it affects the width of the passage between the blades. The maximum value of t/c which may be used without causing the passage to diverge is equal to $(s/c) \times (1 - o/s)$, and t is commonly between $0.15c$ and $0.25c$. The ratio s/c, together with the operating radius and the hub/tip ratio, yields the blade aspect ratio h/c. Although commonly 3 to 4 in value, this may range from 1 to 6. Larger values imply a shorter chord and reduced secondary losses for a given annulus.

Typical blade angles are $-50°$ to $-70°$ for α_2, and α_3 is normally below $20°$. At the final stage, α_3 is lower still, to minimise flow losses in the downstream ducting.

E.8 Blade Radial, Tensile Stress

Centrifugal loading is the main constant force applied to the blade roots. The material stress at the root may be calculated very simply if the blade is assumed to be parallel along its length and has no tip shroud. Using m for the airfoil mass acting at mean radius, ρ_b for the material density and A_b for the area of the blade section (as seen in figure E.4), the stress at the root is

$$\sigma = m\omega^2 R_m/A_b$$

$$= \rho_b A_b(R_t - R_h)\omega^2 \tfrac{1}{2}(R_t + R_h)/A_b$$

$$= \tfrac{1}{2}\rho_b(R_t{}^2 - R_h{}^2)\omega^2$$

$$= \tfrac{1}{2}\rho_b \left[1 - \left(\frac{R_h}{R_t}\right)^2\right] U^2 \qquad \text{(E.16)}$$

$$= \pi\rho_b A N^2 \qquad \text{(E.17)}$$

where A is the annulus area in the plane of the blades.

E.9 The Use of Alternative Working Fluids

In closed-cycle engines it is sometimes advantageous to use working fluids other than air (section 2.8). Inert gases (helium, argon, carbon dioxide) offer freedom from corrosion and, in the event of leakage, they are neither toxic nor explosive, although they may be expensive. Their properties differ in general according to the number of atoms per molecule (section B.1), and specifically in relation to the molecular weight, M.W. Their different properties call for different blading design and may affect the size and cost of the turbomachinery. The optimum working fluid is that which offers the required power output at minimum cost.

In an ideal cycle (section C.5), the *power output* is the difference between the compressor and turbine power and may be expressed in the form

$$P_{out} = P_{34} - P_{12} = \dot{m}C_p(T_{03} - T_{02})(1 - T_{01}/T_{02}) \qquad \text{(E.18)}$$

The mass flow is $\dot{m} = \rho A V_a$, which contains density and volumetric flow terms. The density is linked with the specific heat (ρC_p) when comparing different gases: whereas a light gas has a high specific heat relative to mass (section B.4.2), its low density gives it no advantage relative to volume. Gases with the same molecule type therefore have ideally the same value of ρC_p, as shown in table E.1.1.

The comparison could be made on several different bases: the same turbine entry temperature, the same pressure ratio, the same maximum pressure, etc. Keeping the pressure ratio and levels constant and, therefore, the volumetric flow rates $A_1 V_{a1}$ and $A_2 V_{a2}$ constant for all gases, yields the comparison shown in table E.1.2. The monatomic gas is clearly at a disadvantage, but the different levels of T_{02} mean that these cycles are working at different efficiency levels.

Consider instead a cycle optimised for *power output* by the relationship $T_{02}{}^2 = T_{01} \times T_{03}$ (equation C.8). Working between fixed temperature limits will require different pressure ratios for the different types of gas (table E.1.3). This provides the further choice of whether to make the low pressure p_{01}, the high pressure p_{02}, or an intermediate pressure equal for all gases. If p_{01} is constant, then the ratios of $(P_{out}/A_1 V_{a1})$ for the three gas types will equal the ρC_p ratios in table E.1.1, still disadvantageous to the monatomic gas.

The more obvious choice is to use a *fixed maximum pressure* p_{02}. It may then be shown that the ratios of $(P_{out}/A_2 V_{a2})$ will be the same as those of ρC_p in table E.1.1, but the levels of p_{01} and ρ_1 will vary according to the ρ_1 factor (table E.1.3). The power output relative to the volumetric flow in the low pressure ducting is then given by the ratio $(P_{out}/A_1 V_{a1})$, which is favourable to the monatomic gas (table E.1.3).

In practice, the variable C_p and γ of a *real gas* and the losses in a *real engine cycle* reduce this advantage. Also, there is a limited range of practically usable gases, especially of polyatomic type, with which to make the comparison, but data for a few gases are shown in table E.1.4. Only a few figures are tabulated,

Table E.1 The use of Alternative Working Fluids

1 General Comparison, different Types of Gas

		Monatomic	Linear	Non-linear
Molecule type (section B.1):				
Nominal value of C_p/C_v	γ:	1.666	1.4	1.333
	$\gamma/(\gamma-1)$:	2.5	3.5	4.0
ISA conditions; value	ρC_p:	879.1	1230.7	1406.6
	ρC_p ratio:	0.714	1.0	1.143

2 Example with Fixed Pressure Ratio (Ideal)

At constant $p_{02}/p_{01} = 10$	T_{02}/T_{01}:	2.512	1.931	1.778
With $T_{01} = 288$ K	T_{02}:	723.5	566	512
With T_{03} limit of 1200 K	$T_{03}-T_{02}$:	476.5	634	688
From equation E.18, with	P_{out}:	252.3	376.4	423.7
$AV_a = 10^{-3}$	P_{out} ratio:	0.670	1.0	1.126

3 Optimum Ideal Case

At constant T_{02}/T_{01} = 2.041	p_{02}/p_{01}:	5.951	12.146	17.353
At constant max. p_{02}	ρ_1 factor:	2.041	1.0	0.700
ρ_1 factor x ρC_p ratio	(P_{out}/A_1V_{a1}) ratio:	1.457	1.0	0.800

4 Practical Optimum Cases

Molecule:	monat.	diatomic or linear structure				polyatomic	
Gas:	He	H_2	N_2	air	CO_2	NH_3	SO_2
Mol. wt:	4.00	2.016	28.02	28.964	44.00	17.03	64.06
p_{02}/p_{01}:	4.554	8.352	8.783	8.865	19.238	16.907	23.385
η_{th}:	0.257	0.269	0.264	0.265	0.245	0.242	0.256
$\dfrac{P_{out}}{A_1V_{a1}}$ ratio	1.222	1.013	0.996	1.0	0.676	0.785	0.593

5 Compressor Sizes for Same Power Output and Mach Number

c_1 ratio:	2.936	3.798	1.017	1.0	0.781	1.262	0.640
A_1/A_{air}:	0.279	0.260	0.987	1.0	1.894	1.009	2.635
D_1/D_{air}:	0.528	0.510	0.994	1.0	1.376	1.005	1.623

the fixed data being $T_{01} = 288$ K, $T_{02} = 588$ K, $T_{03} = 1200$ K, $p_{03}/p_{04} = 0.9 \times p_{02}/p_{01}$ and $\eta_{\infty c} = \eta_{\infty t} = 0.85$ (section C.3).

The comparison still requires a further qualification because the volumetric flow rate, pressure ratio and other parameters are dependent on the *sonic velocity* of the gas. Different gases can be compressed through the same ratio by a given number of stages provided that the flow Mach numbers are the same. As light gases have high sonic velocity (section B.4.2), the value of V_a must be raised proportionately and, by implication, so must the blade speed U. If this was possible without limitation, then the cross-sectional area of the low-pressure ducting would be reduced in proportion to maintain the required volumetric flow rate. This is shown in table E.1.5: the V_a ratio is proportional to the c_1 ratio and the area required relative to that for turbomachinery using air is given by

$$\frac{A_1}{A_{\text{air}}} = \frac{1}{(P_{\text{out}}/A_1 V_{a1}) \text{ ratio}} \times \frac{1}{c_1 \text{ ratio}}$$

The diametral ratio is then the square root of the area ratio.

This again favours helium, but the full benefit is unobtainable because raising V_a and U by the factor 2.93 increases the blade stress by the factor $2.93^2 = 8.585$ (equation E.16). It is fortunate that a lower pressure ratio is required in any case, but appropriate *blade design* helps to maximise the advantage. Reduced reaction in the compressor and increased reaction in the turbine (section E.2) would reduce the value of α_1 (figure E.1) and U while raising V_a further still to maintain the required M. In an extreme case, increasing V_a by 3.59 and U by 2.07 reduces the blade stress by 50 per cent at constant V_1 ($3.59^2 + 2.07^2 = 2 \times 2.93^2$). Tapering the blades from hub to tip in chord and thickness would further reduce the stress, as would the use of a slightly larger diameter to increase the hub/tip ratio (equation E.16). The relatively small diameter may permit the use of still higher internal pressure levels. Finally, the Mach number may be reduced a little, increasing the number of stages but still yielding smaller and cheaper turbo-machinery.

It is clear that the blades designed for one gas would be unsuitable for most other gases. This analysis clearly favours helium as do present manufacturing trends: the light gases also have greater thermal conductivity, which may reduce the heat exchanger sizes. Discussion in reference 72B presents a different method of assessment.

Appendix F

Compressor and Turbine Blade Vibrations

Compressor and turbine blades have a variety of natural modes of vibration, predominantly as blades alone but also in combination with flexing of the disc rim. These modes occur at characteristic frequencies, which are determined by the distribution of mass and stiffness (in bending or torsion), resulting from the variable thickness over the blade area. The nodal pattern pertinent to each mode can be determined statically by exciting the blade, held horizontally with its root in a vice: at resonant frequencies, sand sprinkled on the blade will move to lie along the characteristic nodal lines. High levels of cyclic stress may occur at the blade surface, at one or more of the anti-nodes, if the blade is excited at that frequency within the engine operating range. A part of the design and development process is ensuring that such stresses are not excessive, as indicated by testing with strain gauges fitted at the anti-nodes, by stress-coat paint or by deflection sensors. Reference 100 covers the subject in greater depth.

F.1 Compressor Vibration Modes

Typical blade vibration modes are shown in figure F.1, together with the nodal patterns and strain gauge position for maximum stress. The simplest mode and lowest in frequency is first flap or flexural (1F), in which the blade oscillates as a vibrating cantilever bending in its weakest direction. At higher frequencies the blade may vibrate in second flap, with one stationary node, or in third flap with two nodes, etc. In flap vibrations, the blade root remains relatively stationary in the disc rim, moving only within its clearance, but possibly causing damage by frettage (section F.5).

A related family is the pin-rolling series, which are akin to the flap modes but deliberately permit rocking of the blade root on a loose fitting pin. Such a fastening may be used to ease the blade stresses while operating in low-speed

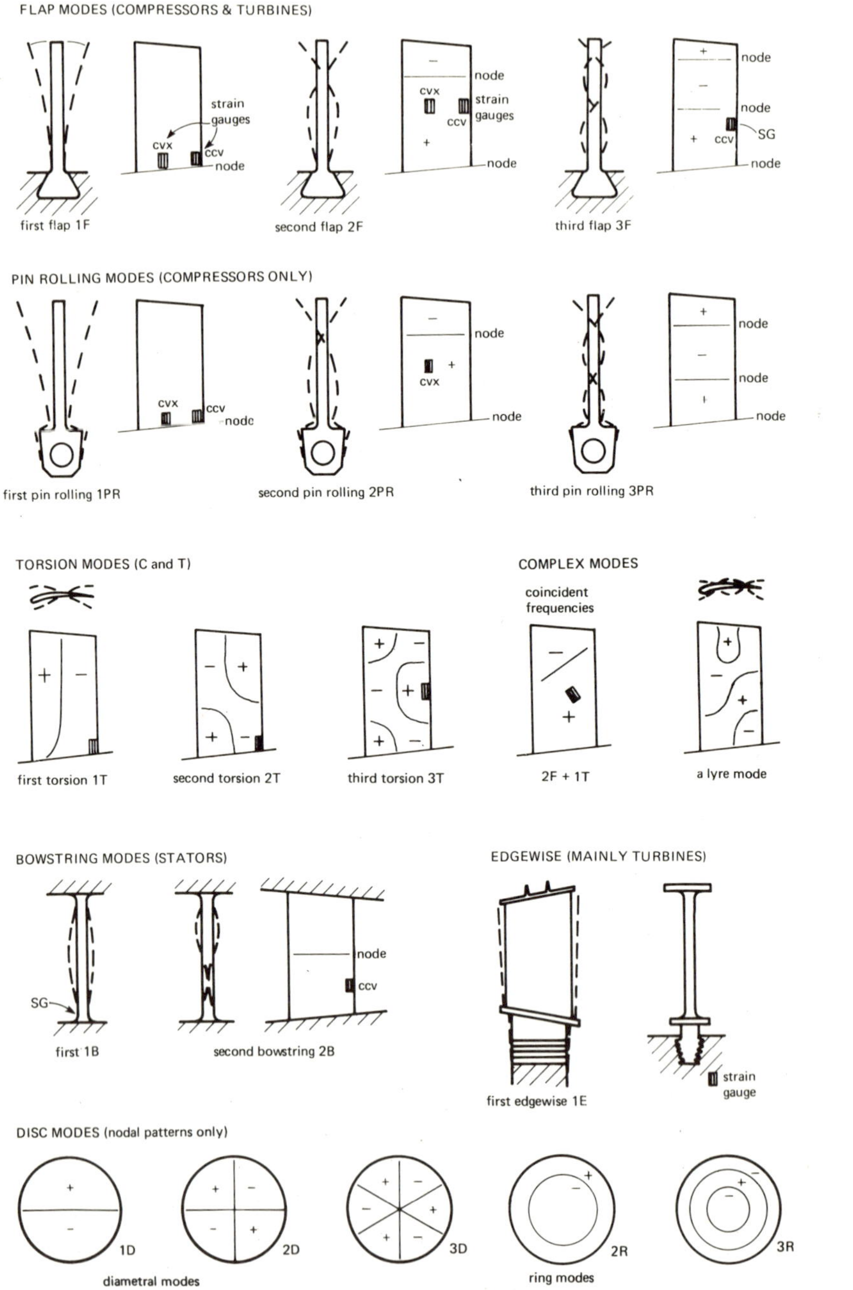

Figure F.1 *Compressor and turbine, blade and disc vibration modes showing the strain gauge position and orientation to measure maximum stress (usually at mid-chord on the convex (cvx) side or at the leading or trailing edge on the concave (ccv) side, to be as far as possible from the neutral axis in bending)*

stall (section 10.4); the clearance is adjusted deliberately to achieve the lowest stress levels.

The next main family is torsion, the simplest being first torsion, which is at quite a high frequency. More complex modes also occur at higher frequencies, involving flap and torsion.

It is quite common for discs to vibrate; singly, or with the rotor as an assembly. The modes depend strongly on the form of construction used and the presence or lack of inherent damping, such as by adjacent surfaces which rub together. The types of mode that may occur are also shown in figure F.1. If the blades are heavy in relation to the diaphragm under the disc rim, then a combined blade/disc rim vibration mode can occur, which in some cases may result in failure, depending on the structural design.

Stator vanes also vibrate, but are less prone to failure than rotor blades. The modes include flap, torsion and bowstring, the last being associated with shrouded assemblies in which the vane is located at both ends.

F.2 Turbine Vibration Modes

Turbine blade vibration modes are similar in nature to those in figure F.1, with two main exceptions. There are no pin-rolling modes in turbines, the root fixing being almost universally the fir-tree type (figure 8.2). There is, however, the significant first edgewise (1E) mode of vibration: the blade motion is similar in nature to, but normal to the direction of, a first flap vibration. It is associated with turbine disc rim fore-and-aft vibration as a ring mode or as a ripple with a number of diametral nodes.

Turbine vibrations are very susceptible to changes in design, because the blading may vary considerably (section E.7). Consecutive stages may also have quite different critical modes, because turbine geometry changes considerably between stages.

F.3 Compressor Blade Excitation

The conditions at which blades encounter vibration problems are shown in figure F.2. As the speed of rotation increases, the frequency of each mode of vibration is matched by successive engine order (EO) lines. The frequency of the vibration modes often changes with rotational speed, flap modes tending to rise, stiffening with centrifugal force, and torsion modes tending to fall as increasing temperature reduces the modulus of elasticity. These two influences are the predominant influences in their respective cases.

Blade passing orders and intake airflow distortion result in excitation at integral engine order frequencies (section G.5). Thus, a particular airflow distortion pattern associated with poor intake geometry may induce three

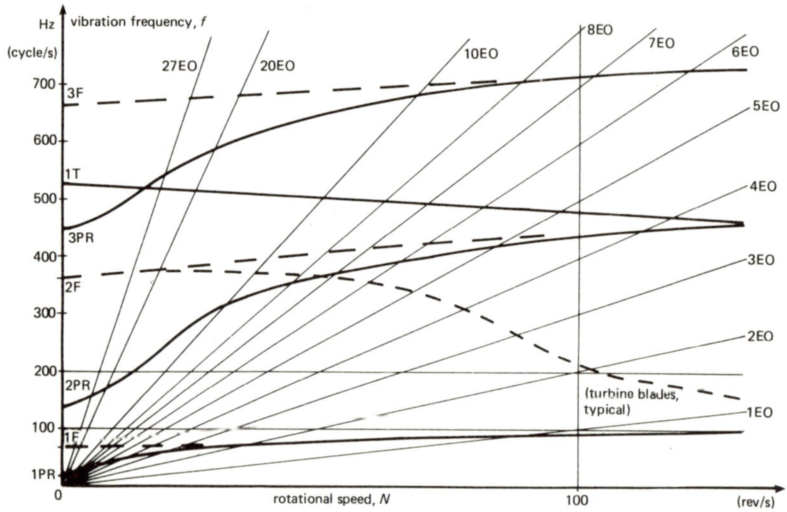

Figure F.2 *Potential sources of excitation for a typical pin-fixed compressor blade through the operating speed range. The loss of frequency with Young's modulus at higher temperatures for a typical turbine blade mode is also shown*

disturbances per revolution, at three times engine order (3EO). The presence of eight structural members across the annulus at inlet or outlet to a compressor would cause an 8EO excitation, or the wakes from twenty-seven rotor blades in one stage may cause a 27 EO excitation, resulting in a vibration mode in a nearby row of stators.

The influence of the *wakes* from the blades or stators of one stage can be felt at a considerable distance. It has been known for blades to break in fatigue as far as three stages away from the source of excitation, upstream as well as downstream. Thus, the influence is still strong after passing through six rows of blades; it is in fact an acoustical signal which also contributes to the over-all noise of a compressor (section 13.2).

These are all forcing excitations, which determine the frequency of vibration. Another source of excitation arises which may be of non-integral engine order but is still forcing. This is the rotating stall cells present in the front compressor stages at low speeds (section 10.4). Depending on the number of stall cells and their speed of rotation, the excitation could be at any intermediate frequency.

There is also a non-forcing excitation; the self-induced aerodynamic *flutter* which occurs near the low-pressure ratio limit of compressor operation (section 10.3). Flutter occurs with high-speed flows and negative incidence, and involves vibration in the first torsion mode or a combined second flap/first torsion mode: the latter mode has an angled nodal line, as shown in figure F.1. The blade motion cyclically varies the incidence angle of the flow (section E.4) and the lift force. The blade alternates in a loop between conditions of positive and negative lift, or different levels of positive lift: if the change of lift force emphasises the motion, then flutter ensues. At higher flow velocities, choke flutter occurs, with shock waves moving cyclically with the blade. The mechanism is discussed more fully in reference 32D, but a simple rule [100] for the avoidance of flutter is to keep the rotor blade torsional frequency parameter $2\pi fc/V$ above 1.6 (where c is the blade chord and V the air relative velocity).

The practical effect of flutter is to impose very high stresses leading to blade fatigue failures within a few seconds or minutes. The author has seen a monitored strain gauge output signal leap instantaneously from about 35 to 250 MPa at about 2.5 kHz, with the onset of flutter. This stress level tore the strain gauges off the titanium blades, which broke a few minutes later while running at about 4 per cent lower speed. It occurred in a blade three stages ahead of a peripheral air bleed outlet as a result of the bleed flow, and was overcome by thickening the blade, cutting off the leading and trailing edge tip corners and accepting only blades with nodal lines indicating more flap than torsional motion (section F.5).

F.4 Turbine Blade Excitation

Again, turbine blade excitation is similar in nature to that of compressors, but with a few major exceptions. The intake distortion effect is absent in engines with multiple combustors, but may occur in engines with a single combustion system located off the centre line. There are no problems equivalent to the stall bands and, because the flow chokes in the NGVs noise and excitation cannot pass to the next stage upstream. This limits the passing order excitation of a turbine rotor to the NGVs immediately ahead of and behind the rotor. Disc vibration with the blades is important, particularly at the later stages with longer blades. Finally, all the vibration modes fall in frequency as the speed rises, because the effect of temperature increase is so significant; the lower modes, 1F and 1E may drop considerably for this reason.

There are two new factors affecting turbine blades. The number of flame sources, if the combustion system is concentric with the shafts, causes circumferential temperature vibrations in the combustion outlet (section 7.5). The variations in density and velocity in turn provide cyclic variation of the gas forces acting on the blades. Secondly, the NGVs deteriorate during the overhaul life of the turbine until caution or performance loss dictates their replacement; their changing condition can enhance existing excitations or introduce new ones.

F.5 The Reduction of Blade Vibration Problems

The effect of any excitations present is that, through the speed range, stress peaks occur as the frequency of each vibration mode matches the excitation

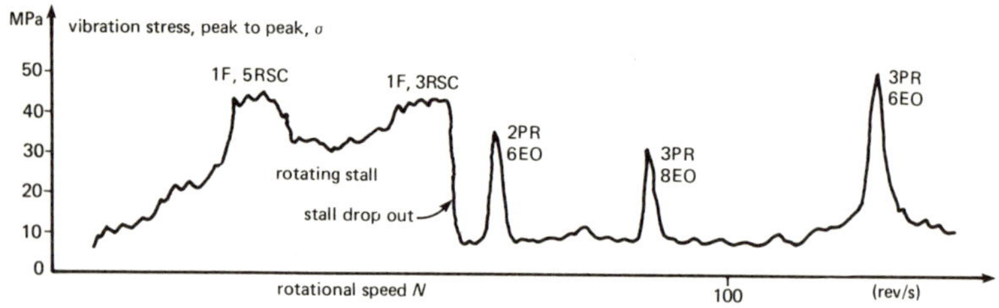

Figure F.3 *Typical vibration stress levels in the compressor blade of figure F.2, including the stresses in rotating stall, as measured at the characteristic position for each mode*

frequency. This is demonstrated for a compressor blade in figure F.3, which matches the example chosen for figure F.2.

The stresses are measured by sophisticated strain-gauging techniques throughout and, preferably, just beyond the speed range. The frequency of any stress peak identifies the vibration mode and the source of excitation, and the level of stress indicates the severity. If a blade or disc failure problem is involved there are several steps which may be taken to find a solution, in service or at the design stage

(1) Reduce the level of excitation: this can be done by duct changes or, in an extreme case, by changing the number of blades in a stage or increasing the spacing between blade rows. Use different, prime numbers of blades in each stage.

(2) Tune the vibration mode to a different frequency to avoid a frequent operating speed; this can be done by cutting off the leading and or trailing edge corners (figure F.4*a*); or by altering the thickness of the blade airfoil, all over or selectively, during manufacture.

(3) Introduce shrouds to provide a damping contact between adjacent blades, the angle of contact being chosen to counter torsional vibrations (figures

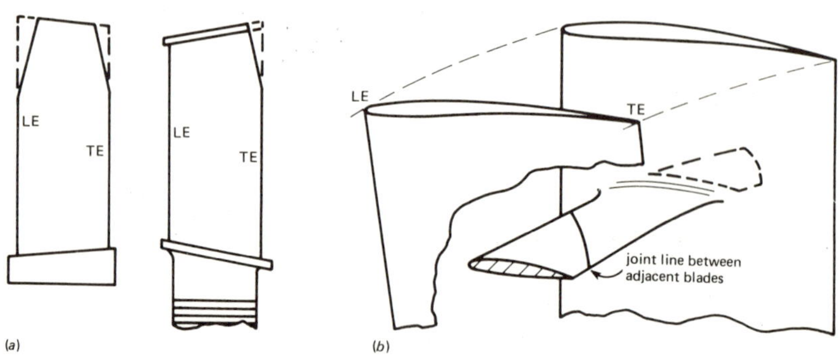

Figure F.4 *Two methods of reducing blade vibration amplitudes and stresses: (a) removal of leading and/or trailing edge tip corners to alter a mode shape or frequency, and; (b) the shroud ring of clappers on long compressor blades.*

F.4*b* and 8.2*c*). The interlocked turbine shroud is more effective at this than the straight shroud and both are more effective than platform damping alone. Wire lacing of turbine blades has the same effect.

(4) Design compressor blades with low aspect ratio, that is, short relative to the chord (section E.6) to reduce stress levels and avoid flutter modes.

(5) Reduce the clearance between turbine blade shrouds, to increase their interference when hot. This can convert flap modes into edgewise modes or bowstring modes.

(6) Weld adjacent turbine blades together, in groups of two or more.

(7) Restrict engine operation to avoid speeds at which vibration problems occur.

(8) Change the blade material (compressor blades only).

(9) Select appropriate material combinations or surface treatments for mating components, to avoid frettage.

The last two items need further explanation. The choice of *material* for the blading affects the balance between mass and stiffness, and considerably affects the frequency of each mode. The material for compressor blades, whether glass reinforced plastics (GRP) composite, aluminium, titanium, steel or nickel alloy is basically chosen for the temperature of its operating environment and the duty. Each has its own safe fatigue stress limit, perhaps 30 MPa for aluminium or 100 MPa for titanium or steel. Aluminium, being soft, tends to be easily damaged by ingested grit, especially at the trailing edge, concave side: its failures may often originate from such a mark. Steel and nickel alloys avoid such problems, but add to the weight. Titanium has an excellent strength-to-weight ratio but, in a fatigue situation, is very seriously weakened by the slightest notch or crack. Even the frettage marks, which develop on the blade root if suitable plating or other treatment is neglected, can initiate fatigue cracks which may lead to release of the blade from the disc rim. On top of fatigue stresses can be superimposed a creep problem affecting certain alloys; this combined *creep fatigue* condition has caused failure of titanium discs in aircraft engines. The material for turbine blades is chosen so much for its characteristics at high temperature that the differences between suitable nickel or cobalt alloys may not overcome a vibration problem.

The problem of *frettage* arises where minute rubbing motion continually wears through the protective oxide film of one or both components. This permits corrosion mechanisms and welding mechanisms to occur, depending on the materials. If micro-welds occur, they are torn apart by subsequent motion. The material surfaces are marked and damaged, making them capable of initiating early fatigue cracks. This may occur particularly at the contact surfaces between blade roots and discs, but also applies to all other surfaces in contact, whether in light or heavy contact. The solution may lie in the material combination, in coatings such as graphite or molybdenum disulphide, in silver plating, or in hard facing by metal sprays of various types.

Appendix G

Noise: Introduction and Blade Noise

This appendix contains material at two levels. The first four sections briefly cover the basics of noise and the terminology used, to help the reader unfamiliar with the subject to understand chapter 13. The later sections take the mechanism of blade noise generation beyond the level of chapter 13.

G.1 Noise and Hearing

Sound travels in the form of very slight pressure fluctuations in the air or medium. It involves a synchronised motion superimposed on the random molecular motion (section B.1). For instance, a blow on a drumskin moves the skin and all the adjacent molecules. As the skin bounces back, this layer of molecules is impelled into a second layer, which is impelled into a third, and so on. Successive layers bounce back and forth between their neighbouring layers, in a sequential fashion such that the effect of the motion of the first layer is transmitted to the subsequent layers.

The periodic closing and opening of the gaps between layers causes the air pressure to increase and decrease, in a wave which travels through the medium (figure G.1). If the forcing surface is now considered as a piston instead of a drumskin, and is driven backwards and forwards by a constant speed, rotating crank mechanism, the wave will be repeated at constant *frequency* f and the pressure will vary sinusoidally at any listener's position to peak levels above and below the mean level. The mean level in this case is the atmospheric pressure, and the magnitude of the acoustic signal is assessed by the root mean square (r.m.s.) *sound-pressure*, p. The value of p will vary according to the intensity of the noise (section G.2).

The layers of molecules may be flat (plane), as they usually are at a point remote from the source. Close to a point source, they will be spherical, and they

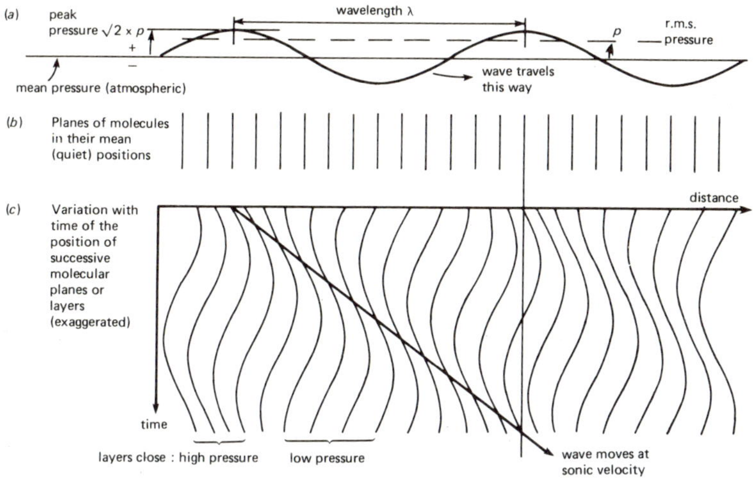

Figure G.1 *The manner in which phase changes between the motion of successive molecular planes cause sound waves to travel through an otherwise stationary gas*

would be cylindrical round a line source. Other shapes are also encountered (section G.5), and they are also the shape of the wave front, the contour linking the peak pressure of a wave generated at one instant. The direction of motion of the sound wave at any point is at right-angles to the wave front.

Sound waves travel at sonic velocity c, whose value depends on the nature and the temperature of the fluid medium (section B.4.2). The physical spacing between the pressure peaks of consecutive sound waves is the *wavelength* λ, where

$$c = f\lambda \tag{G.1}$$

The *receiver* of a noise signal is usually a surface, which can be moved back and forth by the acoustic pressure variations. The *ear drum* of humans and animals is a diaphragm whose motion approximately follows that of the piston or drumskin. This motion is transmitted by a linkage of small bones to the inner ear, in which a membrane vibrates to excite different nerve fibres at different frequencies. The preload on the linkage is adjusted automatically by muscles, to permit detection of the faint sound of an insect yet not to be damaged by the sound of a drum, or rocket. The ear-drum deflection amplitude [80D] is 10^{-11} m at 0 dB (section G.2) and 10^{-4} m at 140 dB.

A *microphone* also uses a diaphragm to sense sound-pressure signals. Its movement may be sensed by capacitive or electromagnetic inductive effects, and this signal is amplified electronically and recorded by tape or paper chart or displayed on a meter. This system has no automatic level compensation, so must be set in the right range to suit the noise.

Neither the ear nor the noise meter has perfect *frequency response*. The human ear in a healthy adolescent hears from about 20 Hz to 20 kHz, but hears increasingly poorly as the frequency departs above or below the middle range of about 1 to 4 kHz. The noise meter normally senses the true level quite accurately within prescribed frequency limits.

G.2 Quantifying Noise

The strength or *intensity* of a noise is measured in terms of the medium density ρ and p and c

$$I = p^2/\rho c$$

in W/m^2, where ρc is the characteristic impedance in rayls.

The audible range of sound pressure and intensity is too wide to be displayed on a simple linear scale. A logarithmic scale of *sound-pressure level* (SPL) is therefore used which relates the intensity of the incoming signal to a datum value. The datum chosen is the quietest sound normally audible at a frequency of 1 kHz, and corresponds to an r.m.s. sound pressure of $p_0 = 2 \times 10^{-5}$ N/m^2. The SPL is then defined as

$$\begin{aligned} \text{SPL} = L_p &= 10 \log_{10} (p^2/p_0{}^2) \text{ dB} \\ &= 20 \log_{10}(p/p_0) \text{ dB} \qquad \text{(G.3)} \end{aligned}$$

The value of L_p when $p = p_0$ is then 0 dB re 2×10^{-5} N/m^2. An r.m.s. sound pressure of 1 N/m^2 corresponds to a value of

$$L_p = 20 \log_{10}(10^5/2) = 20 \log_{10} 50\,000 = 20 \times 4.699 \approx 94 \text{ dB}$$

re 2×10^{-5} N/m^2. A signal of 2N/m^2 would correspond to L_p = 100 dB re 2×10^{-5} N/m^2. Thus, doubling the sound pressure adds 6 dB to the sound-pressure level anywhere throughout the scale, whereas doubling the sound intensity adds 3 dB to L_p. In the remainder of this chapter, it will be assumed that a value of L_p quoted in dB will always be 're 2×10^{-5} N/m^2', unless specified otherwise in each case.

These sound-pressure values, and even that at L_p = 140 dB where p = 200 N/m^2, are all negligible relative to the magnitude of atmospheric pressure at 10^5 N/m^2.

Whereas the sound-pressure level indicates the intensity of sound heard at any point, the *sound-power level* is used to indicate the acoustic energy emitted from the source. Assuming simple spherical radiation, the total sound power emitted, is given by the value of intensity times the spherical area at the radius of measurement

$$\begin{aligned} P &= 4\pi R^2 I \qquad \text{(G.4)} \\ &= 0.3\, R^2 p^2 \text{ W} \end{aligned}$$

at ISA ($\rho c = 418$). This power is related to a reference value, which is standardised as $P_0 = 10^{-12}$ W, by the sound-power level

$$\text{PWL} = L_w = 10 \log_{10}(P/P_0) \text{ dB} \tag{G.5}$$

It can be seen that, for a source of constant power P, and ignoring directivity and effects of the environment which may attenuate (reduce) the signal, the effect of changing the *distance* between the source and the ear may be expressed as

$$\frac{p_2^2}{p_0^2} = \frac{p_1^2}{p_0^2}\frac{R_1^2}{R_2^2}$$

where subscript 1 refers to the first listening position and 2 refers to the second. Taking logs and multiplying by 10

$$L_{p2} = L_{p1} + 10 \log_{10}(R_1^2/R_2^2)$$

or

$$L_{p2} = L_{p1} - 20 \log_{10}(R_2/R_1) \tag{G.6}$$

If R_2 is double R_1, then the measured sound-pressure level is reduced by: $20 \times 0.301 \approx 6$ dB and, if R_2 is ten times R_1, the reduction is 20 dB.

These values of L_p apply regardless of the frequency of the signal. However, the human ear does not respond uniformly to all frequencies (section G.1). The *frequency spectrum* is divided into a number of bands to assess the effects of signals in the different bands. The first division is into octave bands between frequencies of which the upper value is twice the lower, and which are identified by $\sqrt{2}$ and $1/\sqrt{2}$ times centre frequency. The next subdivision is often into one-third octave bands, and another common one is the 2 per cent band width where the spacing of the extremes is 2 per cent of the mean.

The measured noise is adjusted differently in the different frequency bands by *weighting scales*, which adjust the level in proportion to the way in which it affects the ear. The most common and practical is the A-weighting, which may be applied by electronics built into a meter so that its output may read directly in dBA. For meters without this facility, the dBA value in any frequency band may be calculated from the measured level by equation G.7 (within ±0.5 dB from 25 Hz to 10 kHz)

$$\text{dBA} = L_p - 11.1 \left| \log_{10}\left(\frac{2500}{f}\right) \right|^{2.1} + 1.3 \tag{G.7}$$

where f is the centre frequency of the band.

While the A-scale indicates the adjustments made to allow for hearing capability, it does not cater for the relative *annoyance* or apparent *loudness* of signals at different frequencies. There are many scales to make these subjective allowances [81], of which the one most used in aeronautics to assess annoyance is *perceived noise*. The adjustments made to the level in each frequency band are made according to over-all noise level as well as frequency, so the correction is not made simply as with the A-scale. The final reading is made in units of PNdB, and may be calculated as shown by reference 86.

A further allowance is made with moving aircraft because they are not heard at any one ground station for very long. The maximum value of PNdB is taken, and allowances are then made for pure tone content as well as duration. This calculation may take several hours for each measured point if done manually, but the result is the *effective perceived noise* level, in EPNdB.[86]

The over-all noise is the sum of the contributions in each frequency band, but the summation must be applied to the intensities (p^2), not to the levels. Adding two signals of equal level, say 88 dB, involves adding 3 dB to get 91 dB. This arises from

$$L_p = 10 \log_{10}(2 \times p^2/p_0{}^2) = 10 \log_{10} 2 + 10 \log_{10}(p^2/p_0{}^2)$$

where $10 \log_{10} 2$ is 3.01. For levels 3 dB apart, their sum is the higher level plus 2 dB. With greater differences, the contribution of the lower level is rapidly reduced.

G.3 Noise at Several Frequencies; Harmonics

Very few noises consist of pure tones at a single frequency. Often, noise from a source with nominally a single generating frequency contains multiples of that frequency. A drum-roll or other repetitive but interrupted sound is anything but sinusoidal, so its analysis shows the presence of a wide range of frequencies.

Fourier analysis permits any repetitive signal to be synthesised by combinations of pure tones at different frequencies, strengths and phase angles, where these frequencies are all *harmonics* or multiples of the *fundamental* frequency of repetition. For instance, the second harmonic is at twice the fundamental frequency. The signals with more rapid rates of pressure change or shorter pulse duration have more high-frequency content. An extreme case is the square wave, with instant pressure rise to a constant peak level, followed later by an instantaneous fall to a constant minimum, negative peak.

G.4 Noise Source Mechanisms

There are three basic mechanisms of acoustic signal generation. The *monopole* is a tiny spherical source which expands and contracts about a mean volume. Two touching monopoles, expanding and contracting out of phase, constitute a *dipole* source. Four monopoles arranged in a square, with diagonal opposites expanding while the others contract, form a *quadrupole*. These sources are in fact the mathematical models used to represent actual sources.

These mechanisms, perhaps millions at a time, apply to different physical sources. The monopole applies to combustion, and the turbulence past a blade at zero pitch. The dipole is asymmetrical and requires balance by surface contact, such as the interaction of flow turbulence with a loaded blade. The quadrupole is generated by fluid shearing, as in jet noise. While the monopole radiates equally in all directions, the others radiate directionally where the signal from each monopole is not cancelled by the effect of its neighbour(s). They all radiate at low efficiency.[80A]

G.5 The Propagation of Blade Interactive Noise

The predominant noise heard is at the blade passing order frequency, which implicitly contains harmonic tones (section G.3) because of the nature of the generating mechanism (section G.6). This noise is propagated along the inlet and outlet ducts by a resonant mode of the air in the duct, excited by the fan. While one duct mode involves plane wave travelling along the duct axis, there is nothing intermittent about the fan rotation to cause this except in a special case (section G.8). Nor are howl type modes (section 13.3) of any significance in this mechanism.

The family of duct modes which can carry blade noise is the spiral modes, such as the five-lobed mode ($m = 5$) shown in figure G.2. There is an intermittent feature able to generate waves of this type, being the small numbers of blades (B) and vanes (V) which pass each other. These spiral sound waves originate at the stator vanes (section 13.2) and also propagate rearwards.

G.6 The Generation of Blade Passing Order Tones

At a rotor speed of N rev/s, the frequency at which the blades pass the stators is BVN, assuming that B and V have no common factor. They are not heard at this

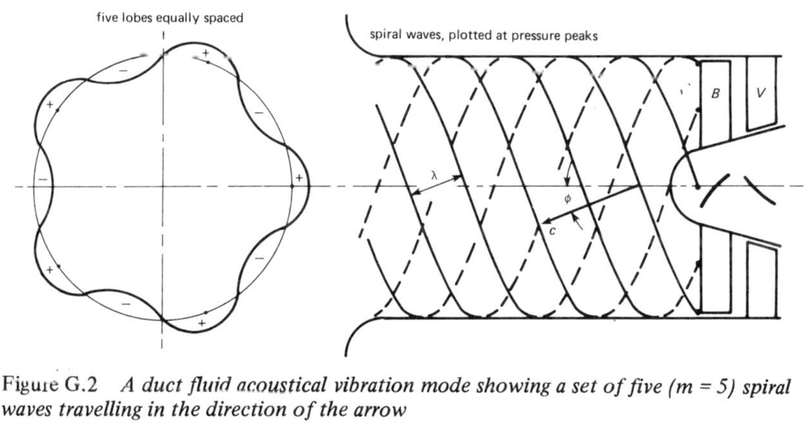

Figure G.2 *A duct fluid acoustical vibration mode showing a set of five ($m = 5$) spiral waves travelling in the direction of the arrow*

frequency, however, because the effect of one blade wake on one vane (section 13.2) is too slight. Figures G.3*a*, *b* and *c* show three conditions in which these small signals may interact. In the first, the small wave radiating from vane P (using the leading edge for convenience of drawing) and that from Q merge with those from previous, similarly spaced vanes to form the strong angled waves shown, originating at 0. The second diagram shows vane Q triggered before vane P, generating waves angled in the opposite sense. In the third diagram, relative to the first, the sound wave from vane P passed vane Q before it was triggered, so this case is unable to form a strong wave and is not heard. Finally, figure G.3*d* shows how the angles change when flow through the fan is considered: the wave radiates from a point moving along the line PP′, but the argument is unchanged.

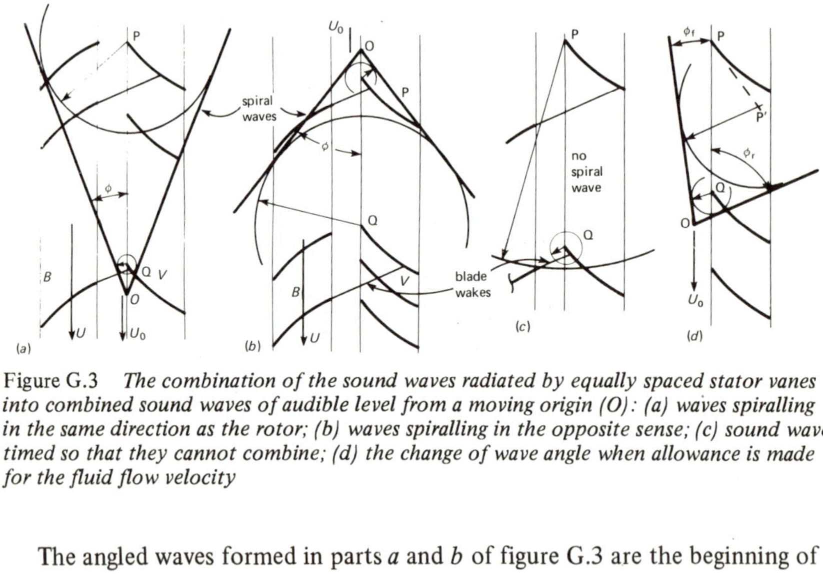

Figure G.3 *The combination of the sound waves radiated by equally spaced stator vanes into combined sound waves of audible level from a moving origin (O): (a) waves spiralling in the same direction as the rotor; (b) waves spiralling in the opposite sense; (c) sound waves timed so that they cannot combine; (d) the change of wave angle when allowance is made for the fluid flow velocity*

The angled waves formed in parts *a* and *b* of figure G.3 are the beginning of the spiral waves. They travel at sonic velocity c in the direction arrowed ϕ in figure G.2, where ϕ is measured at the blade tip radius R_t. Simple geometry requires the origin O to spin circumferentially at $U_0 = c/\sin\phi$, a supersonic velocity. These waves are akin in some ways to the shock waves from a supersonic body. In the non-flow case *a*, slowing the rotor permits the wave from P to approach closer to Q before Q is triggered. The angle ϕ increases until, at 90°, the wave rotates (at $U_0 = c$) round the duct and does not emerge. At slightly lower speed (figure G.3*c*), the wave does not even form and the mode is cut off, case *c*. Different relative blade spacing causes this at different speeds.

It was shown by Tyler and Sofrin [101] that the duct modes which may be excited by a given fan are limited to the m values

$$m = nB + kV \tag{G.8}$$

where n is the blade passing order harmonic number and k is any integer (positive, zero or negative). In practice, only $n = 1$ need be considered, as the other harmonics are dependent on and carried by the same signal as the fundamental. However, several modes (m values) could be propagating simultaneously at the same frequency.

The signal carried by these waves is at rotor blade passing order frequency, BN. The mode number m must be integral to create a continuing series of waves, and is negative if the waves spiral in the opposite sense from the rotor. The spacing of the waves (section G.1) is therefore, from figure G.2

$$\lambda = \frac{c}{BN} = \frac{2\pi R_t \sin \phi}{m}$$

from which

$$U_o = \frac{c}{\sin \phi} = \frac{2\pi R_t BN}{m} = \left(\frac{B}{m}\right) U_t = \left(\frac{B}{m}\right) M_t c \qquad \text{(G.9)}$$

where U_t and M_t are the blade tip tangential velocity and Mach number respectively.

G.7 The Cut-off Condition

Interactive tones propagate with intensity proportional to the sixth power of the fan tip velocity [80A], provided that the rotational speed is above the cut-off speed. Slowing down to cut-off, the frequency BN falls, λ and ϕ increase and the value of U_o falls towards c. Cut-off occurs therefore when the numerical value of U_o (with m positive or negative) falls below c; that is, when

$$\left| \left(\frac{B}{m}\right) M_t c \right| \leqslant c$$

from which

$$\left| \frac{m}{B} \right| \geqslant M_t$$

It has been shown [102] that, when making allowance for fluid flow with an approach Mach number M_a, the cut-off criterion is more accurately represented by

$$\left| \frac{m}{B} \right| \geqslant \frac{M_t}{\sqrt{(1 - M_a{}^2)}} \qquad \text{(G.10)}$$

The right-hand side of this equation has typical values between 0.5 and 1.0. Using 1.0 for the example of a high-speed fan, table G.1 shows the effect of different values of k and V/B, where m is determined from equation G.8.

Table G.1 Variation of Parameters Affecting Interactive Noise

k	V/B	m	No interaction noise if:	Comments
+ve	any	large, +ve	M_t may exceed 1.0	Supersonic noise (section 13.2)
0	any	+B	M_t below 1.0	Quiet
−1	<1	small, +ve	M_t very small	The worst case
−1	$1<\frac{V}{B}<2$	small, −ve	M_t very small	Better, m −ve (section G.9)
−1	>2	large, −ve	M_t below 1.0	Quiet
−2	<1	small, −ve	M_t very small	Some noise, m −ve
−2	>1	large, −ve	M_t below 1.0	Quiet

G.8 Rotor Order Tone Generation

It is shown in reference 85 that, if allowance is made for the production differences between the rotor blades, and their consequently different wakes and noise generation (as is allowed for in buzz-saw noise, section 13.2), equation G.8 may be rewritten

$$m = n + kV$$

Different values of n and k permit all the weak engine order tones to appear. The particular case of n = V and k = −1 gives m = 0, a plane wave along the duct axis at stator order frequency.

G.9 Discussion of Rotor–Stator Interactive Noise

The dipole noise radiated from the stator vanes is strong where it propagates nearly normal to the stator chord, and weak in line with the stator chord. This makes modes with negative values of m quieter in the inlet, and modes with positive values of m quieter in the outlet duct. If the ratio V/B cannot be chosen to obtain complete cut-off right up to maximum speed, it can at least be used to give more favourable values of m.

The value of ϕ rises with the value of m, as shown in equation G.9. At m = 0, the noise propagates axially, and larger values of m result in propagation from the fan duct inlet at increasing angles from the fan axis.

Compressor noise departs from the sixth power law (section 13.2) because of the numerous components in the noise signal; their different directivities; the different cut-off speeds of the different modes; and according to whether the fan as a source is well matched to the noise propagating capacity (a Bessel function) of the duct mode. Also of significance is the frequency of each main tone relative to the dBA or PNdB curves, which affects the noise subjectively.

All this discussion has assumed that the rotor blades and stator vanes are equally spaced. Either set could be variably spaced. The effect of variable spacing of the stator vanes is discussed in reference 103.

References

1 A. Meyer, The combustion gas turbine, its history and development, *Proc. Instn Mech. Engrs,* **141** (1939) 197–222

2 F. Whittle, The early history of the Whittle jet propulsion gas turbine, *Proc. Instn mech. Engrs,* **152** (1945) 419–35

3 M. L. Bramson, Report on the Whittle system of aircraft propulsion (theoretical stage) – October 1935, *Aeronaut. J.,* **74** (1970) 128–33

4 *Gas Turbines – Status and Prospects, I.Mech.E. Conference Publications C.P. 1976–1,* containing:

4A F. W. Armstrong, Gas turbine evolution, pp. 1–15

4B R. M. Denning and T. Jordan, 'The aircraft gas turbine – status and prospects, pp. 17–34

4C P. A. Yerbury, Gas turbine power for large hovercraft, pp. 117–24.

4D L. J. Spinks, Some observations on the selection of gas turbine generating plant, pp. 147–56

4E B. M. Huggett, The gas turbine in the gas transmission environment, pp. 125–35

5 J. Cox, Selection of compressors, *Chart. mech. Engr,* **24** (1977) 58–63

6 Technical Advances in Gas Turbine Design, *Proc. Instn mech. Engrs,* **183** (part 3N) (1969) containing:

6A R. N. Penny, Regenerators for high temperature gas turbine engines, pp. 110–20

6B T. B. Ferguson, Centrifugal compressors for gas turbines, pp. 1–10

6C G. Pilkington and D. R. Carlisle, Combustion aspects of industrially applied aero gas turbines, pp. 85–97

6D A. H. Lefebvre, The design of tubular gas turbine combustion chambers for optimum mixing performance, pp. 150–5

6E W. Deacon, A survey of the current state of the art in gas turbine combustion chamber design, pp. 20–7

6F J. F. Barnes and J. Dunham, Aerodynamic and thermal considerations in designing axial flow turbine blades, pp. 11–19

6G R. J. E. Glenny, Advanced materials for gas turbines, pp. 71–84

6H N. Stephensons, Cast high temperature materials, pp. 98–109

6I M. C. S. Barnard and R. S. Benson, Radial gas turbines, pp. 59–70

6J E. H. Warne, Gas turbine fuel and control systems, pp. 121–27

6K M. J. McArthur and P. W. Charles, Recent developments in the control of industrial gas turbines, pp. 128–37

6L G. A. Abraham, The advancement of controls with modern gas turbines, pp. 138–49

6M J. P. Gostelow, J. H. Horlock and H. Marsh, Recent developments in the aerodynamic design of axial flow compressors, pp. 50–8

7 E. M. Eltis and G. L. Wilde, The Rolls-Royce RB211 turbofan engine, *Proc. Instn mech. Engrs,* **188** (1974) 549–75

8 H. C. Hottel and J. B. Howard, *New Energy Technology – Some Facts and Assessments* (M.I.T. Press, Cambridge, Mass., 1971)

9 *The Jet Engine* (Rolls-Royce Ltd, Derby, 1969)

10 *Diesel and Gas Turbine Worldwide Catalogue* (Diesel and Gas Turbine Progress, Oshkosh, Wis., annual)

11 K. Fulton, International turbine engine directory, *Flight International* (7 January 1978) 29–70

12 D. C. Whittley, The augmentor wing: powered lift STOL a proven concept, *Interavia,* 2 (1974) 143–45

13 J. Hartley, Will we ever drive gas turbine cars, *Autocar* (28 January 1978) 52–5

14 G. Clark, Turbine engine powers army's new battle tank, *Diesel and Gas Turbine Progress Worldwide,* **9** (2 March 1977) 12–13

15 H. Czermak and K. Goebel, Operating experience with a combined cycle gas turbine and steam turbine plant at the Hohe Wand power station of the Niederösterreichische Electrizitätswerke Aktiengesellschaft, *Modern Steam Plant Practice* (Institution of Mechanical Engineers, London, 1971) 107–13

16 R. Coats, The efficient use of gas turbines for production, *Chart. mech. Engr,* **25** (1978) 51–8

17 H. J. Perkins and W. J. R. Stocks, Gas turbine total energy – the key to high thermal utilization, *Proceedings of the Conference on Energy Recovery in Process Plants*, CP2–1975 (Institution of Mechanical Engineers, London, 1975) 81–8

18 A. Meyer, The Velox steam generator, *Trans. Am. Soc. mech. Engrs,* **57** (1935) 469–78

19 R. A. Harmon, C. A. Kinney and W. M. Crim, Jr., Cogeneration in Europe and the combined cycle gas turbine, *Turbomachinery International,* **20** (March 1979) 29–34

20 K. Bammert, *A General Review of Closed Cycle Gas Turbines using Fossil, Nuclear and Solar Energy* (Verlag Karl Thiemig, Munich, 1975)

21 W. L. Nelson, *Petroleum Refinery Engineering*, 2nd ed. (McGraw-Hill, New York, 1941)

22 Nitric Acid (SABAR) process, *Hydrocarbon Processing*, **54** (November 1975) 164

23 R. T. C. Harman and A. G. Williamson, Gas turbine topping for increased energy recovery in sulphuric acid manufacture, *Applied Energy*, **3** (January 1977) 23–40

24 A. R. Valdes, Use expander cycles for LPG recovery, *Hydrocarbon Processing*, **53** (December 1974) 89–91

25 M. Bergman (Ed.), Storage in Excavated Rock Caverns, *Proceedings of the First International Symposium in Stockholm, Rockstore 77* (Pergamon, Oxford 1977)

26 *Turborefrigerating Machine*, TXM-300 (Techmashexport, Moscow)

27 W. F. Rush, S. A. Weil, R. J. Dufour and M. J. Lassota, Brayton refrigeration systems, in *An Assessment of Selected Heat Pump Systems*, Project HC-4-20 (American Gas Association 67–137

28 J. R. Powell, F. J. Solzano, W. S. Yu and J. Milan, *A Closed Brayton Cycle using Hydrogen as a Work Fluid* (Brookhaven National Laboratory, Upton, New York)

29 R. W. Haywood, *Analysis of Engineering Cycles* (Pergamon, Oxford, 1967)

30 W. R. Thomson, *Preliminary Design of Gas Turbines* (Emmott, London, 1963)

31 H. Cohen, G. F. C. Rogers and H. I. H. Saravanamuttoo, *Gas Turbine Theory*, 2nd ed. (Longmans, London, 1972)

32 W. R. Hawthorne and W. T. Olson (Eds), *Design and Performance of Gas Turbine Power Plants*, Vol. 11 of *High Speed Aerodynamics and Jet Propulsion* (Oxford University Press, London, 1960) containing:

32A D. H. Mallinson and S. J. Moyes, Performance, pp. 463–552

32B L. C. Gibbons, Fuels for aircraft gas turbine engines, pp. 245–88

32C W. T. Olson, Combustion chamber development, pp. 289–350

32D J. R. Schnittger, Flutter problems in gas turbines, pp. 425–59

33 G. L. Wilde, Future large civil turbofans and power plants, 17th Short Bros. Commemorative Lecture, R. Ae. Soc paper No. 587, *Aeronaut. J.* (July 1978) 281–99

34 R. E. Neitzel, Future subsonic transport engine technology improvements and resultant propulsion alternatives, *J. Energy* (American Institution of Aeronautics and Astronautics) **3** (May/June 1977) 145–50

35 Gas Turbine Acceptance Tests, International Standard ISO 2314 (BS 3135)

36 I. E. Treager, *Aircraft Gas Turbine Engine Technology* (McGraw-Hill, New York, 1970)

37 J. Belson, Disc failures: a cause for concern, *Flight International*, (31 July 1975) 163–66

38 J. Chauvin, Ed., *Supersonic Turbo-Jet Propulsion Systems and*

Components, AGARDograph 120 (Technivision Services, Slough, 1969) containing:

38A J. Chauvin, F. Breugelmans and A. Janigro, Supersonic compressors, pp. 59–178

38B G. A. Halls, Air cooling of turbine blades and vanes, pp. 241–85

39 H. S. Fowler, An investigation into the flow processes in a centrifugal compressor impeller, Mechanical Engineering Report ME-220 (National Research Council of Canada, 1966)

40 A. Whitfield, Slip factor of a centrifugal compressor and its variation with flow rate, *Proc. Instn mech. Engrs*, **188** (1974) 415–21

41 F. G. Groh, G. M. Wood, R. S. Kulp and D. P. Kenny, Evaluation of a high hub/tip ratio centrifugal compressor, *J. Bas. Engng*, **92** (1970) 419–29

42 H. S. Fowler, The distribution and stability of flow in a rotating channel, *Jl. Engng Pwr*, **90** (1968) 229–36

43 L. J. Cheshire, The design and development of centrifugal compressors for aircraft gas turbines, *Proc. Instn mech. Engrs*, **153** (1945) 426–40

44 W. Jansen, Rotating stall in a vaneless radial diffuser, *J. bas. Engng*, **86** (1964) 750–58

45 D. P. Kenny, A novel low cost diffuser for high performance centrifugal compressors, *J. Engng Pwr*, **91** (1969) 37–47

46 R. C. Voysey, Some vibration problems in gas turbine engines, *Proc. Instn mech, Engrs*, **153** (1945) 483–95

47 A. R. Howell and R. P. Bonham, Overall and stage characteristics of axial flow compressors, *Proc. Instn mech. Engrs*, **163** (1950)

48 R. A. Wallis, Dust erosion of fan materials, *Mech. chem. Engng, Trans. Instn Engrs Aust.*, **MCII** (1975) 33–9

49 H. Constant, The early history of the axial type of gas turbine engine, *Proc. Instn mech. Engrs*, **153** (1945) 411–26

50 D. E. Glenny, Ingestion of debris into intakes by vortex action, C. P. 1114 Aeronautical Research Council (H.M.S.O., December 1968)

51 G. C. Oates (Ed.), *A Comprehensive Study of Aircraft Gas Turbine Engines* (U.S. Aero-Propulsion Laboratory, Ohio, 1977) including:

51A W. S. Blazowski, Fundamentals of combustion, Chapter 15

51B R. M. Henderson and W. S. Blazowski, Turbopropulsion combustion technology, Chapter 20

52 Gas turbines – Procurement, International Standard ISO 3977–1978, (E) (BS 3863)

53 *Joint Conference on Combustion*, Proc. Am. Soc. mech. Engrs (Boston, June 1955) and Instn Mech. Engrs (London, Oct. 1955), containing:

53A B. Karlowitz, Applications of electronic probes to measurements in turbulent flames, pp. 327–29

53B R. P. Probert, Application of research to gas turbine combustion problems, pp. 400–6

53C J. P. Longwell and M. A. Weiss, Heat release rates in hydrocarbon combustion, pp. 334–9

53D H. R. Hazard, Coal firing for the open cycle gas turbine: a comparison of methods, pp. 340–6

54 F. W. Lipfert, Correlations of gas turbine emission data, A.S.M.E. Paper 72–GT–60, presented at the 17th International Gas Turbine

55 J. J. Faitani, Smoke reduction in jet engines through burner design, SAE paper 680348 (1968)

Conference and Products Show, San Francisco, March 1972

56 A. H. Lefebvre, Theoretical Aspects of Gas Turbine Combustion Performance, CoA Note Aero. No. 163 (College of Aeronautics, Cranfield, August 1966)

57 N. Syred and J. M. Beér, Combustion in swirling flows: a review, *Combustion and Flame*, **23** (1974) 143–201

58 R. C. Adkins, A short diffuser with low pressure loss, *J. Fluids Engng,* **97** (September 1975) 297–302

59 A. A. Riskalla and A. H. Lefebvre, The influence of air and liquid properties on airblast atomisation, *J. Fluids Engng,* **97** (September 1975) 316–20

60 D. G. Ainley and G. C. R. Mathieson, A method of performance estimation for axial flow turbines, R and M 2974 (Aeronautical Research Council, H.M.S.O. 1957)

61 H. R. M. Craig and H. J. A. Cox, Performance estimation of axial flow turbines, *Proc. Instn mech. Engrs,* **185** (1970–71) 407–24

62 J. H. Horlock, *Axial Flow Turbines* (Butterworths, London, 1966)

63 W. R. Hawthorne (Ed.), *Aerodynamics of Turbines and Compressors*, Vol. 10 of *High Speed Aerodynamics and Jet Propulsion* (Oxford University Press, 1964) containing:

63A A. R. Howell, Flow in cascades, pp. 270–312

63B J. R. Erwin, The supersonic compressor, pp. 368–432

63C E. Duncombe, Aerodynamic design of axial flow turbines, pp. 433–523

64 H. Daneshyar, K. J. Edwards, J. H. Horlock, M. S. Janota, R. D. Pearson and R. Shaw, A comparison of the performance of three model axial flow turbines tested under both steady and pulse flow conditions, *Proc. Instn mech. Engrs,* **184** (1969–70) 1119–31

65 P. R. Sahm and M. O. Speidel (Eds), *High Temperature Materials in Gas Turbines* Proceedings of a symposium at Baden, 1973 (Elsevier, Amsterdam, 1974) containing:

65A D. J. Lines, Ceramic materials for gas turbine components, pp. 155–77

66 F. W. Armstrong, Tomorrow's power, *Flight International* (15 January 1977) 113–16 and 134

67 T. Alexander, Hot prospects for the new ceramics, *Fortune* (April 1976) 152–62

68 Hot corrosion problems associated with gas turbines, ASTM STP 421, Proceedings of a Symposium held at Atlantic City, 1966 (American Society for Testing and Materials, 1967.)

69 D. G. Ainley, The high temperature turbo-jet engine, *J. R. Aeronaut. Soc.* **60** (September 1956) 563–89

70 G. F. Hiett and I. H. Johnston, Experiments concerning the aerodynamic performance of inward flow radial turbines, *Proc. Instn mech. Engrs,* **178**, 3I (1963–64) 28–42

71 R. S. Benson, An analysis of the losses in a radial gas turbine, *Proc. Instn mech. Engrs*, **180**, 3J (1965–66) 41–53

72 H. R. Cox (Ed.), Gas Turbine Principles and Practice (George Newnes, London, 1955) containing:

72A A. W. H. Jamieson, The radial turbine, Chapter 9

72B E. Glaister and A. G. Smith, Thermodynamics, Chapter 3

73 I. J. Day, E. M. Greitzer and N. A. Cumpsty, Prediction of compressor performance in rotating stall, *J. Engng Pwr,* **100** (January 1978) 1–14

74 I. J. Day and N. A. Cumpsty, The measurement and interpretation of flow within rotating stall cells in axial compressors, *J. mech. Engng Sci.,* **20** (No 2) (1978) 101–14

75 E. M. Greitzer, Surge and rotating stall in axial flow compressors, *J. Engng Pwr,* **98** (April 1976)190 –217

76 M. L. Monaghan, Boosting for a purpose, in *Turbocharging and Turbochargers*, Conference Publication 1978–2, pp. 59–69 (Institution of Mechanical Engineers, London, 1978)

77 A. J. Fawke and H. I. H. Saravanamuttoo, Digital computer simulation of the dynamic response of a twin-spool turbofan with mixed exhausts, *J. R. aeronaut. Soc.* (No. 753) (September 1973) 471–8

78 O. N. Lawrence and R. D. Powell, The application of servo-mechanism analysis to fuel control problems, *Proc. Instn mech. Engrs,* **172** (1958) 439–69

79 K. A. Hatchett and B. E. Sparkes, The new Lucas control system for the Rolls-Royce Spey engine, Automotive Engineering Congress, Detroit, Mich. 10–14 January 1966, SAE paper No. 660049

80 C. M. Harris (Ed.), *Handbook of Noise Control* (McGraw-Hill, New York, 1957) containing:

80A H. E. von Gierke, Aircraft noise sources, Chapter 33

80B R. K. Cooke and P. Chrzanowski, Transmission of noise through walls and floors, Chapter 20

80C H. E. von Gierke, Aircraft noise control, Chapter 34

80D H. Davis, The hearing mechanism, Chapter 4

81 L. L. Beranek (Ed.), *Noise and Vibration Control* (McGraw-Hill, New York, 1971)

82 *Aerodynamic Noise*, AGARD Lecture Series No. 80 containing:

82A B. W. Lowrie, Fan Noise, paper No. 5

82B K. W. Bushell, Gas turbine engine exhaust noise, paper No. 4

83 B. W. Lowrie, Simulation of flight effects on aero engine fan noise, American Institution of Aeronautics and Astronautics, paper No. 75–463, 1975

84 J. S. B. Mather, Noise generative processes in high and low speed rotating machinery, Institution of Chemical Engineers Symposium No. 45

85 J. S. B. Mather, J. Savidge and M. J. Fisher, New observations on tone generation in fans, *J. Sound Vibr.*, **16** (March 1971) 407–18

86 *Aircraft Noise*, Annex 16 to the Convention on International Civil Aviation (ICAO, Montreal, 1971)

87 M. J. T. Smith, Quietening a quiet engine – the RB 211 demonstrator programme, Society of Automotive Engineers paper No. 760897, 1976

88 P. A. E. Stewart, Radiography of gas turbines in motion, *Chart. mech. Engr*, **19** (May 1972) 65–7

89 J. H. Bompas-Smith, *Mechanical Survival: the Use of Reliability Data* (McGraw-Hill, London, 1973)

90 G. L. B. Hall, Aero gas-turbine engines: integrity of the high energy rotating components as governed by low-cycle fatigue, Institute of Metals and Metallurgy, Monograph and Report series No. 32, 1967

91 F. W. Morley, A philosophy of aircraft engine manufacture, 1965 Lord Sempill Paper presented to Institution of Production Engineers, London, 10 March 1965

92 A. B. Shearer, On-site generation: turbine generators, in series *Power Supply Basics, International Power Generation*, 1 (no. 2) (December/January 1977–78) 23–6

93 W. P. Auer, Operating experience of installed gas turbines, in Operating Experience of High Duty Prime Movers, *Proc. Instn mech. Engrs*, **178** (Part 3K) (1963–64) 100–15

94 T. C. Heard and R. P. Lang, The concept of availability for gas turbine application evaluation, *J. Engng Pwr*, **100** (July 1978) 452–6

95 J. H. Horlock, *Axial Flow Compressors* (Butterworth, London, 1958)

96 D. G. Ainley, Performance of axial flow turbines, *Proc. Instn mech. Engrs*, **159** (W.E.P. No. 41) (1948)

97 G. L. Mellar and G. M. Wood, An axial compressor end-wall boundary layer theory, *J. Bas. Engng*, **93** (June 1971) 300–16

98 D. G. Ainley and G. C. R. Mathieson, An examination of the flow and pressure losses in blade rows of axial flow turbines, R and M 2891 (Aeronautical Research Council, H.M.S.O., 1955)

99 J. Dunham and P. M. Came, Improvements to the Ainley-Mathieson method of turbine performance prediction, *J. Engng Pwr*, **92** (July 1970) 252–6

100 E. K. Armstrong and M. A. Stevenson, Some practical aspects of

compressor blade vibration, *J. R. aeronaut. Soc.*, **64** (59) (March 1960) 117–30

101 J. M. Tyler and T. G. Sofrin, Axial compressor noise studies, *S.A.E. Trans.*, **70** (1962) 309–32

102 T. G. Sofrin and J. C. McCann, Pratt and Whitney experience in compressor noise reduction, Paper presented at 72nd meeting of the Acoustical Society of America, 1966

103 P. E. Duncan and B. Dawson, Reduction of interaction tones from axial flow fans by non-uniform distribution of the stator vanes, *J. Sound Vibr.*, **38** (No. 3) (1975) 357–71

Index